COVID-19: Tackling Global Pandemics through Scientific and Social Tools

COVID-19: Tackling Global Pandemics through Scientific and Social Tools

Edited by

SAPTARSHI CHATTERJEE, PHD
Associate Professor of Microbiology,
Adamas University, Kolkota,
India

ELSEVIER

ACADEMIC PRESS
An imprint of Elsevier

Academic Press is an imprint of Elsevier
125 London Wall, London EC2Y 5AS, United Kingdom
525 B Street, Suite 1650, San Diego, CA 92101, United States
50 Hampshire Street, 5th Floor, Cambridge, MA 02139, United States
The Boulevard, Langford Lane, Kidlington, Oxford OX5 1GB, United Kingdom

Library of Congress Cataloging-in-Publication Data
A catalog record for this book is available from the Library of Congress

British Library Cataloguing-in-Publication Data
A catalogue record for this book is available from the British Library

ISBN: 978-0-323-85844-1

For information on all Academic Press publications visit our website at https://www.elsevier.com/books-and-journals

Publisher: Andre Gerhard Wolff
Acquisitions Editor: Linda Versteeg-buschman
Editorial Project Manager: Sam W. Young
Production Project Manager: Kiruthika Govindaraju
Cover Designer: Christian J. Bilbow

Typeset by TNQ Technologies

Contributors

Balogun Muhsin Adekunle
Adeniran Ogunsanya College of Education
Otto/Ijanikin, Lagos, Nigeria

Varun Agiwal
Indian Institute of Public Health
Hyderabad, Telangana, India

Anisuzzaman
Department of Parasitology
Faculty of Veterinary Science
Bangladesh Agricultural University
Mymensingh, Bangladesh

Ali-Erin Balikel
Faculty in management
London School of Commerce
Istanbul, Turkey

Georges Bellos
Graduate Student
Lebanese International University
Beirut

Swarnav Bhakta
School of Life Science and Biotechnology
Adamas University
Kolkata, West Bengal, India

Arijit Bhattacharya
School of Life Science and Biotechnology
Adamas University
Kolkata, West Bengal, India

Shatarupa Biswas
Department of Microbiology
School of Life Science and Biotechnology
Adamas University
Kolkata, West Bengal, India

Debleena Biswas
Research Scholar, Department of Sociology
Ravenshaw University, Cuttack
Odisha, India

Folami Ahmadu Bolanle
Osun State College of Education, Ilesa
Lagos Annex, Nigeria

Dibyajnan Chakraborty
Dr. KPC Bioinnovations and Diagnostics
Kolkata, West Bengal, India

Pyali Chatterjee
MLS, MATS University
Raipur, Chhattisgarh, India

Saptarshi Chatterjee
Associate Professor of Microbiology
Adamas University
Kolkata, West Bengal, India

Suvendu Choudhury
Capgemini India
Kolkata, West Bengal, India

Megha Dutta
Department of Microbiology
School of Life Science and Biotechnology
Adamas University
Kolkata, West Bengal, India

Sanmitra Ghosh
Department of Microbiology
School of Life Science and Biotechnology
Adamas University
Kolkata, West Bengal, India

Zobayda Farzana Haque
Department of Microbiology and Hygiene
Faculty of Veterinary Science
Bangladesh Agricultural University
Mymensingh, Bangladesh

Viana Hassan
Institute for Tourism
Travel and Culture
University of Malta
Msida, Malta
Lebanese University
Faculty of Tourism
Borj El Brajneh, Lebanon

Muhammad Tofazzal Hossain
Department of Microbiology and Hygiene
Faculty of Veterinary Science
Bangladesh Agricultural University
Mymensingh, Bangladesh

Hadeya Jeha
Lebanese International University
Bekaa, Lebanon

Subhendu Karmakar
Royal Society-Newton International Fellow
School of Chemistry, University of Birmingham
Birmingham, United Kingdom
PBC Postdoctoral Fellow
School of Pharmacy
The Hebrew University of Jerusalem
Jerusalem, Israel

Mohamad Knio
Assistant Professor In Economics at CUCA
City University College of Ajman, Ajman
United Arab Emirates
Saloumi, Lebanon

Ashok Kumar
SHKM Government Medical College
Nalher, Haryana, India

Jitendra Kumar
Department of Statistics
Central University of Rajasthan
Ajmer, Rajasthan, India

Saurabh Kumar
Department of Management
Invertis University
Bareilly, Uttar Pradesh, India

Tanmay Majumdar
National Institute of Immunology
New Delhi, Delhi, India

Budhaditya Mukherjee
School of Medical Science and Technology
Indian Institute of Technology
Kharagpur, West Bengal, India

Musolihu Majeed Olayori
Federal College of Education
Nigeria

Amulya K. Panda
National Institute of Immunology
New Delhi, Delhi, India

Joydeep Paul
School of Life Science and Biotechnology
Adamas University
Kolkata, West Bengal, India

Asha Shelly
National Institute of Immunology
New Delhi, Delhi, India

Snehlata
School of Medical Science and Technology
Indian Institute of Technology
Kharagpur, West Bengal, India

Kora Bhanu Teja
School of Medical Science and Technology
Indian Institute of Technology
Kharagpur, West Bengal, India

Prabhat Trivedi
Department of Statistics
Central University of Rajasthan
Ajmer, Rajasthan, India

Contents

CHAPTER 1

Application of CRISPR-Based Diagnostic Tools in Detecting SARS-CoV-2 Infection

SNEHLATA • KORA BHANU TEJA • BUDHADITYA MUKHERJEE

1 INTRODUCTION

In early 2020, numerous cases of pneumonia were reported in Wuhan, China. These assumed cases were later found to be the evolution of the novel Coronavirus (CoV), SARS-CoV-2 (severe acute respiratory syndrome coronavirus 2), and the disease is referred to as COVID-19 (coronavirus disease 2019) [1]. CoVs cause respiratory infections leading to a range of symptoms from the common cold (human coronavirus [HCoV] OC43, HCoV NL63, and HCoV 229E) to acute respiratory distress syndrome (SARS-CoV-2, SARS-CoV, and Middle East respiratory syndrome [MERS]). Although this is the first worldwide pandemic caused by CoVs, other members of the same genera were reported to cause severe epidemics in the recent past, including SARS, 2002–03, and MERS, 2012–13 [2]. CoVs contain mainly four structural proteins, namely, spike (S) protein, envelope (E) protein, membrane (M) protein, and nucleocapsid (N) protein (Fig. 1.1), of which the S protein plays a major role in attachment and virus-cell membrane fusion during virus infection [3]. This outbreak has drastically increased with over 30.87 million reported cases and 0.95 million deaths globally as of 19th September 2020 [4], signifying its rapid transmission through varying means. Similarly, person-to-person transmissions have been noticed in most of the reported cases from infected individuals who show least or no symptoms, increasing the risk of disease transmission. These asymptomatic cases have created a state of fear factor and panic, globally, among healthy individuals [5]. As of now, the WHO suggests the use of quantitative reverse transcription-polymerase chain reaction (qRT-PCR) to diagnose COVID 19 and some antigen-based serologic tests for the purpose of seroprevalence. However, certain limitations of both these methods have driven researchers to turn to other innovations such as RT-LAMP (reverse transcription-loop-mediated isothermal amplification) or its variants and the CRISPR (clustered regularly interspaced short palindromic repeats)-Cas system [6]. CRISPR was discovered in *Escherichia coli* in 1987, which were later observed in other bacterial species as well. The title role of these repeat sequences was unclear until, in 2005, numerous researchers detailed the alikeness of the sequences' DNAs, directing to the hypothesis that the sequences are part of the adaptive immune system in bacteria [7], which terminates viral DNA by using Cas endonuclease in a sequence-specified manner [8]. By redirecting this immune response to specific, chosen parts of genetic material, scientists were able to make meticulous genetic alterations, which formed the groundwork of the CRISPR-based therapeutic and diagnostic tool in different models of infectious pathogens. In this chapter, we will first sequentially look into several diagnostic tools that are currently used for detecting SARS-CoV-2 infections, with their advantages and major drawbacks. Also, we will majorly focus on the rapidly developing CRISPR technology and its applications in detecting several infectious pathogens, with a particular focus on SARS-CoV-2 infection.

2 CURRENT STATUS OF WIDELY APPLIED DIAGNOSTIC METHODS FOR DETECTING SARS-COV-2 INFECTION

2.1 Quantitative Reverse Transcription-Polymerase Chain Reaction-Based Detection

qRT-PCR is the current gold standard method for the diagnosis of an active SARS-CoV-2 infection. Quantitative detection is carried out using specialized instruments that detect fluorescent probes that track the size

COVID-19: Tackling Global Pandemics through Scientific and Social Tools. https://doi.org/10.1016/B978-0-323-85844-1.00002-7

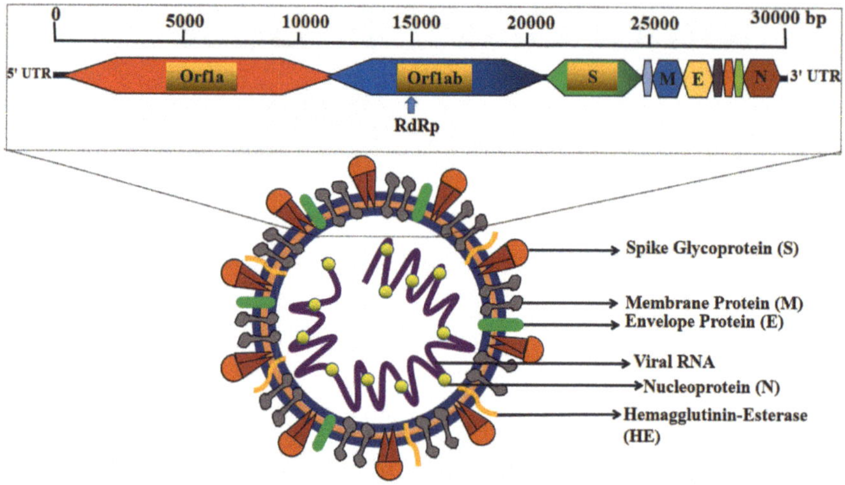

FIG. 1.1 Schematic representation of the severe acute respiratory syndrome coronavirus 2 (SARS-CoV-2) structure, along with its genome architecture, representing two accessory genes ORF1a and ORF1b as well as four specific gene fragments encoding for four structural proteins, namely, S (spike), M (membrane), E (envelope), and N (nucleocapsid). *UTR*, untranslated region.

of the target sequence amplification in real-time. RNAs isolated and purified from respiratory tract specimens are reverse-transcribed, which converts viral RNA to complementary DNA (cDNA). Then this cDNA is amplified using a thermal cycler and detected in real-time with the use of fluorescence probes. Detection probes (e.g., TaqMan probes) annealed to the complementary region of amplicons, get cleaved by the 5′-3′ exonuclease activity of DNA polymerase resulting in the separation of quencher and fluorophore, thus producing fluorescence. With each cycle of PCR, more dye molecules are released with increasing fluorescence intensity, which is directly proportional to the number of amplicons synthesized. Additionally, dsDNA (double-stranded DNA)-binding dyes such as SYBR Green are used for the generation of fluorescence but are less specific than the TaqMan probe [9].

In SARS-CoV-2, the specific target genes for qRT-PCR are N, E, and S protein-encoding genes; the Orf (open reading frame) 1ab; and RNA-dependent RNA polymerase genomic regions residing within Orf1ab (Fig. 1.1) [10,11]. The N and E genes are used for first-line screening, as both are highly conserved among the beta-CoVs and may show cross-reactivity with other CoVs; therefore the RdRp and S genes are used in confirmatory tests for differentiating SARS- CoV-2 from other CoV infections, as they are highly divergent [12].

The main advantage of this method is high sensitivity. This method can be used for the early detection of trace amounts of viral proteins present in noninvasively collected samples such as saliva [13]. Although the sensitivity is high, there are several limitations of qRT-PCR that limit its practicality. qRT-PCR requires expensive and sophisticated instruments that are typically confined to specialized laboratory facilities in terms of equipment, biosafety, and biosecurity, a challenge in limited resource availability [14]. Furthermore, sample handling and data processing require specialized staff. Several qRT-PCR platforms such as open system qRT-PCR machines, TrueNat, and CBNAAT (cartridge-based nucleic acid amplification test) are currently being used for diagnosing COVID-19 in India and all these systems require huge laboratory facilities, as the testing samples are contagious. The minimum time required from sample collection to test is 3–5 h, which is a big disadvantage for qRT-PCR-based diagnosis. To address these issues, some attempts have been made, including optimization of detection protocols and primer sets as well as the development of low-cost qRT-PCR machines. However, most of these developments are still in their infancy and will need a lot of standardization to be ready for use in a clinical setup.

Apart from these technical difficulties, the leading drawback of qRT-PCR is the false-negative as well as the false-positive results. False-negative results could arise due to sampling errors, variations in viral load of specimens, improper handling of clinical samples, operational procedure, the limit of detection differences between several qRT-PCR kits, and the presence of PCR inhibitors in inappropriately treated samples [15,16].

On the other hand, the qRT-PCR system is highly sensitive to minute levels of RNAs, and because of this reason, minimal contamination of the sample may cause false-positive results indistinguishable from true-positive results. Data handling error is another probable cause of false-positive results in qRT-PCR [17]. Previously also, for Zika, Ebola, SARS-CoV-1, and MERS-CoV, this false-positive qRT-PCR-based diagnosis had created lots of trouble for the medical healthcare system [17,18]. A comparison between these diagnostic methods and other widely applied and currently used diagnostic tools is represented in Table 1.1.

2.2 LAMP/RT-LAMP-Based Detection

In the view of point-of-care (POC) diagnostic applications, in recent years a novel isothermal nucleic acid amplification-based technique called loop-mediated isothermal amplification (LAMP) has evolved for the detection of virus-specific genes. This alternative exponential amplification technique does not require specialized equipment, well-trained personnel for interpretation of results, or highly purified samples and yet provides high sensitivity and specificity as compared to RT-PCR assays [18–20]. For amplification of target nucleic acid, LAMP operates between 60°C and 65°C without the need for temperature cycling. It utilizes a total of six primers that include two of both inner and outer primers (this can recognize six different regions of target) and two loop primers [21,22]. The target nucleic acid (RNA/DNA) binds to both the inner and the outer primers and their extension leads to the formation of stem-loops at both ends. Then the loop and inner primers bind with the stem-loop region for initiation of DNA amplification leading to the formation of dsDNA. The strand-displacing polymerase (such as *Bst* polymerase) allows the separation of dsDNA for several cycles of amplification [23].

Furthermore, combing the ultrasensitive LAMP with RT-PCR resulted in the development of several RT-LAMP assays for the direct detection of SARS-CoV-2 genomic regions with the fluorescence or colorimetric readouts [24,25]. For this purpose, calcein, a fluorescein complex that quenched when bound to manganese, was used by Yan et al. [26] in RT-LAMP assays for the visual detection of the SARS-Cov-2 viral Orf1ab and S genes. During the amplification of DNA, manganese is sequestered by the released pyrophosphates, thereby releasing the free calcein. Calcein can then bind to Mg^{2+} ions to generate an increasing amount of fluorescence. Yin et al. developed the iLACO (an isothermal LAMP-based method for COVID-19) platform, based on the RT-LAMP, for rapid colorimetric detection of SARS-CoV-2 also using calcein. This technique involves six primers for the amplification of the ORF1ab gene fragment. The reaction was performed at 65°C in a water bath for 20 min to see the visible color change from pink to yellow due to the presence of a colorimetric pH indicator. The detection capability of iLACO was also confirmed by the inclusion of SYBR Green dye in the reaction mix to observe the color change. The system was also modified by including GeneFinder (a nucleic acid dye with improved fluorescent signal and sensitivity), which resulted in the generation of green fluorescence after exposure to blue light, with the negative control remaining pink [25]. In this regard, phenol red was used by Baek et al. [27] as a pH indicator for visual readout, which changed color from pink to yellow showing the generation of amplicons at 65°C, with detection capabilities of 200 copies of COVID-19 RNA in 30 min.

However, designing as well as optimization of primers and reaction conditions are some of the drawbacks of LAMP. Moreover, RT-LAMP also suffers from false-positive readouts like qRT-PCR because of not only its high sensitivity but also the use of the fluorescent dye calcein or pH indicators, which can also act as a source of contamination. The false results can also be due to various nonspecific amplifications such as the formation of primer-dimers or nontarget sequences resulting in detectable false signals [9].

2.3 Serologic Immune Assay-Based Detection

Although PCR-based methods serve well for detecting active infections, they are not practical for repeated testing of a large population for seroprevalence purposes. From that perspective, antigen-based low-cost rapid test kits are ideal for frequent testing at the community level. Indirect diagnosis of SARS-CoV-2 infection is done by detecting viral proteins and specific antibodies such as IgM and/IgG in serum or plasma samples. Various immunoassays that are being commonly used include ELISA (enzyme-linked immunosorbent assay), CLIA (automated chemiluminescence immunoassay), and LFIA (rapid lateral flow immunoassay) [28]. After 7–8 days (second week) of viral infection, IgM and IgG antibody production start, and from 10 to 30 days of the initial infection, IgM, as well as IgG, can be well detected [29]. ELISA is used for detecting viral antigens as well as for indirect antibody-based detection. Antigen detection is the direct method, in which antibodies against SARS-CoV-2 N protein are precoated on a microtiter plate and

TABLE 1.1
Comparison of Different Diagnostic Methods in the Application for SARS-CoV-2 Detection, Along With Their Advantages and Limitations.

Diagnostic Method	Principle	Sensitivity	Time	Advantages	Drawback
qRT-PCR	Amplification of genetic material	High [34],	3–5 h/test	• Used for early detection of trace amounts of viral proteins • High specificity [34]	• Presence of inhibitors leads to false-negative results [34] • Data handling error is another probable cause of false-positive results • The need for real-time thermocyclers, data tools, and trained data interpretation personnel
LAMP/RT-LAMP	Isothermal amplification	High compared to RT-PCR	20–40 min	• Does not require well-equipped instruments • Amplification at constant temperature • Disbarring of thermocyclers • Inexpensive • Test results are obtained quickly	• Designing primers, optimization of primers and reaction conditions, • False results arise due to the formation of primer-dimer
ELISA	Antigen or antibody detection	Low sensitivity	1–5 h	• More accurate • Specific • Easy to perform • 90 Samples can be run together • Less expensive • Needs BSL 2 and specific equipment	• Time consuming • Costly • Its use is imposed due to the diagnosis of acute infections around the time of symptom onset when the risk of viral shedding and transmission is very high
Lateral flow immunoassay	Antigen or antibody detection	Less sensitivity	Less than 20 min	• This test requires very little training to perform and does not rely on clinical laboratory personnel or scientists to analyze • Fast, low cost, and low sample volume is required	• False-negative results obtained in early disease • Cross-reactivity • Accuracy of this test still being evaluated • Low throughput
CLIA	Quantitative detection of antibody titers by measuring the amount of emitted signal	High sensitivity	1–2 h	• High throughput of samples can be analyzed in less amount of time • High signal intensity • Stable results	• Needs sophisticated instruments

CLIA, automated chemiluminescence immunoassay; *ELISA*, enzyme-linked immunosorbent assay; *qRT-PCR*, quantitative reverse transcription-polymerase chain reaction; *RT-LAMP*, reverse transcription-loop-mediated isothermal amplification.

secondary antibodies labeled with horseradish peroxidase (HRP) are used against the same N protein. ELISA tests are mostly used for the identification of antibodies generated against the N and S proteins by detecting the anti-SARS-CoV-2 IgM and IgG responses. These show similar sensitivity and specificity as compared to other POC tests but are limited in terms of diagnosis of acute infections around the time of symptom onset when the risk of viral shedding and transmission is maximum [28]. Alternatively, modifications of this technique have been developed, such as CLIA that makes use of emitted electromagnetic radiation generated by a chemical reaction in which the substrate reacts with the avidin HRP resulting in the production of light. Quantitative detection of IgM and IgG antibody titers is done by measuring the amount of emitted luminous signals. CLIA is faster than ELISA because an automated chemiluminescence analyzer is used, thus high throughput of sample analysis is possible in less time [28,29]]. Clinical tests for detecting other biomarkers, such as C-reactive protein, can also be performed in patients with COVID-19 [25].

Multiple rapid POC diagnostics based on LIFAs is another well-established alternative that allows the detection of IgM and IgG in SARS-CoV-2-infected persons in a period of about 10–20 min [30]. Some of these rapid tests require relatively low sample volumes; for example, in the case of finger-prick method, 20 μL of blood or even approximately 10 μL of serum/plasma is sufficient. These types of tests generally contain SARS-CoV-2 recombinant protein conjugated with gold nanoparticle, mouse antihuman IgG in IgG test line, and mouse antihuman IgM in IgM test line, with the corresponding antibody in the control area [14]. LIFA can be performed anywhere and at any time for rapid testing of COVID-19 with limited healthcare resources, as it does not require specialized infrastructure or any trained personnel support. However, rigorous evaluation is needed for clinical precision of these rapid test methods, as a major disadvantage of this detection method is the high percentage of false-negative results, particularly in the early 6–7 days after the onset of symptoms of disease detection [31]. Hence, these tests are invariably constrained by poor predictive capabilities, commonly yielding false outcomes. Two recent studies comparing the RT-PCR and antigen-based testing to evaluate the efficacy of antigen tests reported massive false-negative result outcomes with an overall sensitivity of 30.2% and 45% [32,33], respectively. Furthermore, false-positive results from cross-reactivity due to an existing coinfection and the antigenic drift of SARS-CoV-2 have made this method less relevant to screen a large population size.

2.4 Imaging-Based Detection and Confirmation

Radiologic imaging plays an important role in the COVID-19 detection and control. Mostly, radiologic methods are used by trained medical professionals in a hospital setup in confirmed clinical cases. Hence, they this can be broadly regarded as a confirmatory test that is hugely advanced from early diagnostic tools. In this method, radiobiologists examine the images to look for any pathologic characteristics that can help in the diagnosis of SARS-CoV-2 infection. Depending on the stage of infection following the beginning of symptoms, imaging features will bear distinct characteristics (patchy alveolar consolidations, peripherally distributed opacities in different lung regions, etc.) of infection progression. In patients with an adequate medical history, clinical symptoms, and distinctive lung anomalies, COVID-19 can be diagnosed with radiologic images. For this purpose, chest computed tomography (CT) and X-ray imaging are the critical approaches that are highly relevant for early detection of the positive cases and to proceed toward treatment even without the RT-PCR results [34].

Chest CT imaging is a conventional and noninvasive imaging method that requires the processing of cross-sectional images by taking multiple X-ray images at various angles around the chest of a patient [35,36]. This can detect some characteristic manifestations in the lungs, bilateral as well as peripheral ground-glass opacities, and vascular dilation in the lesion associated with COVID-19, which is more prominent after 0–10 days of symptom onset [37,38]. With the subsequent progress of infection, irregular paved stone patterns develop, which are further followed by the growing consolidation of the lungs [38]. Along with high accuracy and speed, it is reported that detection sensitivity for SARS-CoV-2 is higher by chest CT than by qRT-PCR, with reduced false-negative results [39,40]. In this regard, Feng et al. have reported that patients who were recently exposed to SARS-CoV-2 and show symptoms of dry cough, fatigue, and fever should be imaged with CT regardless of negative qRT-PCR test results [41]. Despite of its high accuracy, chest CT has a particular limitation in COVID-19 diagnosis; that is, in some cases, it is very difficult to differentiate between the imaging features of COVID-19 and other types of viral pneumonia, thus causing an overlap leading to lower specificity [42]. Furthermore, CT imaging is costly and requires trained technicians and a huge amount of infrastructural support.

3 CRISPR-CAS SYSTEM AS A DIAGNOSTIC TOOL IN INFECTIOUS DISEASES

3.1 CRISPR-Cas Systems

CRISPR is an acronym that stands for clustered regularly interspaced short palindromic repeats. CRISPR-Cas system frameworks have revolutionized cutting-edge technology in modern molecular biology with their huge application in clinical research. In general, the essential components of the CRISPR-Cas system include a CRISPR RNA (crRNA), a protospacer adjacent motif (PAM), and the Cas protein, which performs the endonuclease function of the CRISPR-Cas complex. There are different types of CRISPR-Cas systems, most of which are DNA targeting. However, with the advancement of the SARS-CoV-2 pandemic, major attention has been shifted to RNA-targeting CRISPR. Based on the diversity of the interference machinery of the CRISPR-Cas systems, they are divided into two main classes [43], which are further subdivided into six types and several subtypes. The Class I CRISPR-Cas system (I, III, and IV) consists of a multiprotein complex, while the Class II CRISPR-Cas system (II, V, and VI) uses only one effector protein. Most work on DNA-editing technology and most diagnostic tools have included the Class II, Cas9 system, with the Cas12 and Cas13 systems being recently used.

3.1.1 Type II: CRISPR-Cas9 system

Cas9 is a DNA endonuclease that can cut DNA like a pair of molecular scissors. Cas9 recognizes a specific PAM in the proximal region of the target DNA in the noncomplementary strand. When the crRNA and tracer-RNA (trRNA) complex binds to the complementary strand near the PAM, it helps guide the Cas9 for introducing a dsDNA break. Cas9 protein consists of six domains REC I, REC II, bridge helix, PAM interacting, HNH, and RuvC. REC I is responsible for binding to gRNA (guide RNA; crRNA:trRNA duplex), the role of REC II is not well defined, the bridge helix is rich in arginine and is crucial for initiating cleavage activity upon binding of target DNA, the PAM-interacting domain confers PAM specificity, and HNH and RuvC perform the nuclease activity.

3.1.2 Type V: CRISPR-Cas12 system

The type V system is composed of four subtypes: V-A, -B, -C, and -D. Cas12a (also known as Cpf1) and Cas12b are the effector proteins encoded by the V-A and V-B subtypes, respectively [44]. Cas12a is an RNA-guided endonuclease that does not require tracrRNA (trans-activating crRNA) additionally for the dsDNA cleavage. These effectors lack the HNH domain but have a RuvC-like endonuclease domain [45]. After recognition of crRNA, Cas12a binds to the T-rich (5′-TTN-3′) sequence of the PAM in the target dsDNA, and then a conformational change occurs resulting in the formation of R-loop [29]. This leads to activation of the effector protein active site, i.e., RuvC-like domain cleaves the target forming a dsDNA staggered end with 5′ overhang.

3.1.3 Type VI: CRISPR-Cas13 system

The Class II CRISPR-Cas13 system is composed of four subtypes: VI-A (Cas13a) to VI-D (Cas13d [46]. The Cas13 effector protein in this system targets ssRNA (single-stranded RNA) rather than dsDNA, which makes it an exceptional case in the CRISPR world. It has some unique aspects that distinguish it from other Cas systems (Cas9/Cas12), which include the presence of two HEPN (higher eukaryotes and prokaryotes nucleotide-binding) domains as a catalytic site replacing the DNase domain [45]. When Cas13 recognizes the protospacer flanking site in the target RNA, conformational changes occur in the effector that bring HPEN1 and HPEN2 closer to each other resulting in the activation of the two HPEN catalytic domains [46]. This further leads to indiscriminate cleavage of the target ssRNA and also nonspecifically cleaves all other RNAs present in trans.

3.2 Application of the CRISPR-Cas System in Infectious Diseases Before the Emergence of SARS-CoV-2

Before its application as a diagnostic tool for detecting the SARS-CoV-2 infection, CRISPR has been already adapted as a potential diagnostic tool for detecting other infectious pathogens (Table 1.2). Pardee et al. developed a portable, low-cost diagnostic platform for the detection of Zika virus by combining nucleic acid sequence-based amplification (NASBA) with Cas9-based cleavage with toehold switch-based detection. The toehold switch is an RNA that encodes the gene for the protein of interest; however, translation of the gene is initially turned off. A hairpin structure in the toehold switch hides the ribosome-binding site and the starting sequence for the gene, thus repressing the production of the protein. The toehold switch is designed to detect a target RNA. The toehold switch contains a sequence that is complementary to the target RNA. When the target RNA is present the toehold switch recognizes and binds with the target RNA based on

TABLE 1.2
Overview of Different Existing CRISPR-Cas System-Based Diagnostic Tools for Several Infectious Diseases.

Detection Protein	Amplification/ Readout	Pathogens	Limit of Detection	Target	Catalytic Activity
Cas9 [47]	NASBA/calorimetric	Zika and dengue viruses	1–3 fm of RNA amplicons	DNA	RuvC domain and the C-terminal HNH domain
dcas9 [48]	PCR/fluorescence	*Staphylococcus aureus*	10 CFU/mL of bacteria	DNA	Mutation of RuvC domain and the C-terminal HNH domain No catalytic activity
Cas9 (ctPCR) [49]	PCR/qPCR	HPV16/18	5 ng of DNA amplicons	DNA	RuvC domain and the C-terminal HNH domain
Cas9 (CARP) [72]	PCR/electrophoresis/ qPCR	HPV16/18	2 pg of DNA amplicons	DNA	RuvC domain and the C-terminal HNH domain
Cas12a (HOLMES) [51]	PCR/RT-PCR/ fluorescence	JEV and PRV	1–10 aM of DNA/cDNA	RNA and DNA	Ruv like
Cas12a (DETECTR) [52]	RT-RPA/fluorescence	HPV16/18	8 zm	RNA	Ruv like
Cas13 [44]	RPA/fluorescence/ lateral flow assay	Viruses, bacteria, mutations	Less than 2 aM of RNA	RNA	HEPN 1 and HEPN 2

CARP, CRISPR-associated reverse PCR; *cDNA*, complementary DNA; *CFU*, colony forming units; *CRISPR*, clustered regularly interspaced short palindromic repeats; *ctPCR*, CRISPR typing PCR; *DETECTR*, DNA endonuclease-targeted CRISPR trans reporter; *HOLMES*, a one-hour low-cost highly efficient multipurpose system; *HPV*, human papillomavirus; *JEV*, Japanese encephalitis virus; *NASBA*, nucleic acid sequence-based amplification; *PCR*, polymerase chain reaction; *PRV*, pseudorabies virus; *qPCR*, quantitative PCR; *RT-PCR*, reverse transcription-polymerase chain reaction; *RT-RPA*, reverse transcriptase-recombinase polymerase amplification.

complementarity, resulting in color change. This method is capable of differentiating between the American and African strains of Zika with potential for detecting 1–3 fm of RNA amplicons [47].

Guk et al. discovered a new method to detect methicillin-resistant *Staphylococcus aureus* (MRSA), a group of gram-positive bacteria, by combining dead Cas9 (dCas9) and fluorescence in situ hybridization. dcas9 is a mutant form of Cas9 in which both the cleavage domains of the Cas9 are inactivated without affecting its DNA-binding ability. In this method, dCas9 and gRNA target the *mecA* gene (responsible for antibiotic resistance, such as methicillin). In this method, the clinical isolates of MRSA are incubated with dCas9 and single-guide RNA (sgRNA) for 15 min, which recognizes the target DNA. Then the magnetic bead based pull down is performed by using magnetic beads to isolate the dCas9/sgRNA complex-bound DNA, which is then detected using fluorescence spectroscopic intensity with SYBR Green as the probe [48].

Modified Cas9-based detection is also available for human papillomavirus (HPV) 16 and 18 detections by using CRISPR typing PCR (ctPCR), which detects target DNA target mainly in three steps. Initially, the target DNA is subjected to amplification through PCR with universal primers (PCR 1), which is used to find DNA samples containing the target sequence. This step is followed by digestion of PCR 1 products by CAT, which includes Cas9 cutting, adenine (A) tailed, and T-adapter ligand. Finally, the treated PCR 1 products are subjected to the second round of amplification by using specific primers that can differentiate which genotype (such as virus subtype) is present in the DNA sample. This method can detect both the L1 and E6/E7 genes, two high-risk HPVs (HPV16 and HPV18) in human cervical carcinoma cells, with a limit of detection of 5 ng of DNA [49].

Similar to ctPCR, CRISPR-associated reverse PCR (CARP) also utilizes three main steps including Cas9-mediated cleavage of target DNA, guided by a pair of sgRNAs. This is preceded by ligation of the cleaved DNA with DNA ligase and specific amplification of the target DNA with PCR. Amplification is done using a pair of reverse primers; this technique also detects the L1 and E6/E7 genes of two high-risk HPVs, HPV16 and HPV18 [50].

Apart from Cas9, the trans-cleavage activity of Cas12a is used to establish a diagnostic tool called HOLMES (a one-hour low-cost highly efficient multipurpose system). In HOLMES, the presence of target DNA or RNA is confirmed when Cas12a/crRNA binary complex forms a ternary complex by recognizing the target nucleotide, leading to Cas12-based cleavage activity. Cas12 (HOLMES) is used to identify DNA viruses such as the pseudorabies virus and RNA viruses such as the Japanese encephalitis virus. Once activated, Cas12 cleaves nontargeted reporter single-stranded DNA (ssDNA) molecules present as a probe in the system, which results in fluorescence [51].

Chen et al. also created another new method termed DNA endonuclease-targeted CRISPR trans reporter (DETECTR) by Cas12a that cleaves dsDNA and attains promiscuous nuclease activity. In this method, ssDNA is used as the nucleic acid detector, which requires initial isothermal amplification by recombinase polymerase amplification (RPA) and then the gRNA paves a way for Cas12 to bind to the target DNA. A fluorescent signal is generated when Cas12 cleaves the ssDNA reporters. The DETECTR system was implanted successfully to diagnose HPV strains 16/18 [52].

Apart from these, recently, a Cas13-based diagnostic tool called SHERLOCK has revolutionized CRISPR-based detection. SHERLOCK is a Cas13b-based detection system, which stands for specific high-sensitivity enzymatic reporter unlocking. SHERLOCK-based nucleic acid diagnostics have a lot of applications including viral detection, bacterial discrimination with respect to antibiotic-resistant genes, detection of cancer-related mutations, etc. In addition, the most remarkable utilization of SHERLOCK is in its ability to discriminate between targets with a single base pair difference, to effectively differentiate between the African and American strains of Zika virus, and to recognize distinct serotypes of dengue viruses. In this regard, it should be mentioned that an updated method based on SHERLOCK, SHERLOCK v2, can identify the dengue and Zika viruses circulating in the blood through the lateral flow assay system. This system, developed by Myhrvold et al. recognizes the viral nucleic acid directly from body fluids through SHERLOCK-HUDSON (heating unextracted diagnostic samples to obliterate nucleases) [53], which with its high sensitivity identifies dengue virus in the saliva, entire blood, and serum of patients.

3.3 Application of the CRISPR-Cas System in SARS-CoV-2 Detection

CRISPR-Cas system has become one of the promising methods for the rapid and portable detection of nucleic acids [43,44]. Owing to the rapid spread of the COVID-19 pandemic, there has been a need for the advanced diagnosis of SARS-CoV-2. In this regard, CRISPR has been recently adapted for detecting nucleic acid in vitro, making it an effective and reliable molecular diagnostic method for diagnosing different viral infections [52]. Lucia et al. [54] described the use of a CRISPR-Cas12-based system for the generation of fast, precise, and portable diagnostic tools to detect SARS-CoV-2 RNA sequences. Cas12 is an RNA-guided DNase that belongs to the Class II type V-A system that causes unspecific collateral ssDNA cleavage following recognition of the target. ssDNA, as a probe, was conjugated with reporter molecules and therefore upon cleavage will emit a fluorescent signal as well as it can be detected by lateral flow-based paper strip assay [53]. The assay was performed by the use of synthetic SARS-CoV-2 viral RNA fragments for various genes such as RdRp, ORF1b, and ORF1ab [54].

In another study, the AIOD-CRISPR (all-in-one dual CRISPR-Cas12a) assay method was reported for the detection of SARS-CoV-2 and HIV as a quick, simple, ultrasensitive, one-point, and visual detection of nucleic acid [55]. This assay involved Cas12a-crRNA complexes produced by two individual crRNAs, which when introduced binds to the target sequence at two sites close to primers. CRISPR detection and nucleic acid amplification components such as two RPA primers, DNA polymerase with strand displacement activity, SSB (ssDNA-binding protein), target sequences, fluorophore quencher (FQ)-labeled ssDNA reporters, and recombinase are incubated in a single reaction system at a constant temperature of 37°C. Cas12a-crRNA binding sites get exposed after RPA and bind to target sites resulting in Cas12a activation. The ssDNA-FQ reporters present nearby are cleaved by activated Cas12a endonuclease resulting in fluorescent signal. Furthermore, CRISPR-Cas12a system-based collateral cleavage activity continuously occurs by the RPA products, which too function as the substrates.

Broughton and colleagues developed one of the pioneering diagnostic platforms, the CRISPR-based SARS-CoV-2 DNA Endonuclease-Targeted CRISPR Trans Reporter (DETECTR) assay, for detecting SARS-CoV-2 RNA in respiratory swab samples [56]. RNAs extracted from samples were subjected to concurrent RT-LAMP, accompanied by predefined CoV sequence detection by Cas12 in the presence of reporter molecules. Detection of viral RNA results in the activation of Cas12 endonuclease leading to the cleavage of reporter molecules in trans, thus allowing visual detection of the virus. The assay is assembled in a lateral flow rapid assay system that can be completed in 30–40 min, and results are visualized on the lateral flow strips in a colorimetry-based naked-eye detection. The E and N genes of SARS-CoV-2 are targeted for primer designing, and the FAM-biotin reporter molecule is used for visualizing the Cas12 detection reaction. Visual readouts can be observed on the lateral strip as reporter molecules that are not cleaved get captured on the control line, and nonspecific cleavage by Cas12 produces a signal at the test line. Furthermore, the DETECTR assay has similar precision as the qRT-PCR and has some advantages such as not requiring thermocycling as the isothermal signal amplification occurs, providing single-nucleotide target specificity, fast turnover time, and simple readouts.

Wang et al. [57] also developed a similar but slightly advanced detection method for the detection of SARS-CoV-2 infection, which is a CRISPR/Cas12a-based naked-eye readout assay, called CRISPR/Cas12a-NER.

For SARS-CoV-2 detection, the CRISPR-Cas13-based strategy has also been developed as an alternative approach. Clinical validation of Cas13-based assay for SARS-CoV-2 RNA detection has driven us one more step toward the POC detection method [58]. A previously developed CRISPR-based diagnostic platform, termed the SHERLOCK assay, for the rapid detection of nucleic acid is now validated for diagnosing the COVID-19 infection [59]. This assay involves Cas13a enzyme/protein from *Leptotrichia wadei* and integrates recombinase-mediated polymerase preamplification with CRISPR-Cas enzymology for detection of the desired SARS-CoV-2 RNA sequences. Viral gene fragments are amplified isothermally by reverse transcriptase-recombinase polymerase amplification (RT-RPA), and then the amplified gene sequences are detected by CRISPR-Cas13a-mediated collateral cleavage of reporter molecules as fluorescence or colorimetric lateral flow readouts in less than 1 h. The RT-RPA reaction at 42°C for 25 min generates dsDNA products, which are transcribed into RNA using T7 RNA polymerase, and then Cas13 detects RNA targets, leading to collateral cleavage [9]. For clinical validation, Patchsung and colleagues [58] designed and tested four pairs of RPA-primer and the corresponding crRNAs against the SARS-CoV-2 genes of S, N, and Orf1ab (replicase polyprotein 1ab). This two-step SHERLOCK assay system was reported to be capable of detecting SARS-CoV-2 viral RNA with very high sensitivity and specificity as well as favors multiplexed and portable detection of viral RNA from clinical samples [58]. Furthermore, Joung et al. developed an advanced version of the SHERLOCK system called STOP (SHERLOCK Testing in One Pot) that can be performed at a single temperature (60°C) in less than 1 h in a test tube, thus negating the requirement of expensive equipment and trained personnel [60]. In STOP, a magnetic bead purification method is used to ease out the RNA extraction step and also to increase sensitivity. SARS-CoV-2 RNA genomes are enriched by the magnetic beads from the respiratory swabs into one particular STOPCovid reaction mixture. Moreover, this assay was again simplified (STOPCovid, version 2) by bringing together the lysis and magnetic bead-binding steps and removing both the washing and the elution steps, thereby minimizing the time frame of sample extraction to 15 min [60].

In a recent advancement, Maiti et al. (2020) [61] reported FnCas9 Editor Linked Uniform Detection Assay (FELUDA), which provided a faster and affordable alternative to sequence-based platforms and can be utilized in several pathologic conditions such as infections and genetic disorders. The assay uses a Cas9 ortholog from *Francisella novicida*, which exhibits high sensitivity because of its insignificant binding affinity to substrates that contain mismatches and this feature makes it different from other engineered Cas proteins. For the detection of nucleotide sequences, FELUDA utilizes a direct Cas9-based enzymatic readout, which means it is a direct and affinity-based method of detection. It also identifies nucleotide bases without the trans-cleavage of reporter molecules, thus providing a non-cleavage-based diagnosis of single nucleotide variations with 100% accuracy. With respect to the ongoing COVID-19, FELUDA was adapted on the paper strip to give lateral flow readout within 1 h with high sensitivity (100%) and specificity (97%) comparable to the gold standard method of qRT-PCR and is further optimized to decrease the detection time.

Some key advantages of this CRISPR-based approach over RT-PCR include the following: isothermal signal amplification to get rid of the need for thermocycler, requires less time, cheap, reliable, single nucleotide target specificity, integration with accessible and easy-to-use reporting formats such as lateral flow strips, and no requirement for complex laboratory infrastructure.

Apart from Cas12, Cas13, and Cas9, Mashimo and his group [62] have developed an in vitro nucleic acid detection platform termed CONAN (Cas3-operated Nucleic Acid Detection) to show the potential of Cas3-based assay as a rapid, cheap, sophisticated, and instrument-free diagnostic tool for SARS-CoV-2. It was shown that this Class I CRISPR-Cas3 system can also mediate collateral trans-cleavage activity to detect viral RNA. However, CONAN requires numerous Cas proteins such as Cas5, 6, 7, 8, and 11 along with Cas3. It provides rapid (within 40 min) and sensitive detection as it also uses a lateral flow strip for one-pot assay with RPA, which makes it a portable method for POC diagnostics. The strip has an intact FQ-ssDNA reporter at the control line (negative) with an anti-FITC-gold nanoparticle conjugate antibody, whereas the cleaved reporter produces a fluorescent signal on the test line (positive). The CONAN-based detection assay is comparable with real-time RT-PCR and Cas12-based assays in its speed and sensitivity, but unlike the DETECTR assay, the CONAN assay provides higher specificity for single-base-pair discrimination within PAM sequences. A schematic representation of the general workflow for the four CRISPR-Cas (3, 9, 12, and 13)-based detection of SARS-CoV infection is represented in Fig. 1.2.

4 CONCLUDING REMARKS AND FUTURE PERSPECTIVES

Infectious diseases have a major impact on millions of people around the world, resulting in large economic losses [63]. Therefore the development of rapid, POC molecular testing for the diagnosis and management of diseases has become a matter of utmost importance/concern. Conventional PCR-based methods for the diagnosis of DNA/RNA of pathogens have several limitations, such as the need for sophisticated infrastructure, equipment, reagents, and trained personnel for performing the assays. Ideally, POC diagnostic assays should provide on-site, fast, sensitive, specific, affordable, and portable detection of pathogens. Lately, CRISPR-Cas system-based methods for the diagnosis of infections have positively exploded with the capacity to drive pathogen detection to another advanced level. CRISPR-Cas-based methods are simple, rapid, more specific, and sensitive for the detection of several infectious pathogens such as the Zika virus, Ebola virus, HPV, SARS-CoV-2, etc. The SHERLOCK and DETECTR diagnostic methods have reached the sensitivity of attomolar (aM) levels and HOLMES reached up to zeptomole (zm) levels [64]. Moreover, by using CRISPR-Cas methods, results of the diagnosis can be visualized on the lateral flow strip or by fluorescence with the naked eye, thus providing equipment-free signal readout.

Moreover, it should also be mentioned that apart from its application in diagnostics, CRISPR has also found utilization in terms of therapy. Cas13 serves as both a companion diagnostic and therapy itself, as Cas13 has been developed for a variety of applications in mammalian cells, including RNA knockdown, detecting, and editing. CRISPR-Cas13-based strategy, known as PAC-MAN (prophylactic antiviral CRISPR in human cells) was proposed recently [29]. It is an antiviral approach that inhibits viruses by first recognizing the viral genome and the viral mRNAs produced in the host and degrading them to stop their expression. A similar approach was adapted for human respiratory epithelial cells, Class II type VI-D CRISPR-Cas13d system from *Ruminococcus flavefaciens* [65,66]. In this system, crRNAs have 22-nucleotide spacer sequences that guide the Cas13d to the target RNA for degradation [67]. This method is also being tested to adopt for SARS-CoV-2 diagnosis by developing a bioinformatic pipeline to characterize extremely conserved regions across several SARS-CoV-2 gene sequences, which can be targeted using the CRISPR-Cas13d system. Some groups have also reported both *E. coli* and *Thermobifida fusca* and isolated Class 1 type I-E CRISPR-Cas and a Cas3 endonuclease that are capable of large-region deletions in the cells of humans by inducing targeted DNA cleavage [68,69]. They both have an antiviral cascade system based on the crRNA-Cas complex that can be utilized for therapy.

CRISPR-Cas-based systems also made it possible to diagnose non-infectious diseases. For example, the CRISPR-Cas9 system has been considered as an important tool in the field of cancer for identifying various mediators of cancer as well as oncogenes [70]. The SHERLOCK diagnostic assay that utilizes Cas13 effector protein is also used for the identification of new therapeutic biomarkers involved in cancer development [70,71]. Furthermore, CRISPR-Cas diagnostic tests can be linked with artificial intelligence to create more quick, precise, and smarter early alarm systems for the detection of infectious pathogens [72].

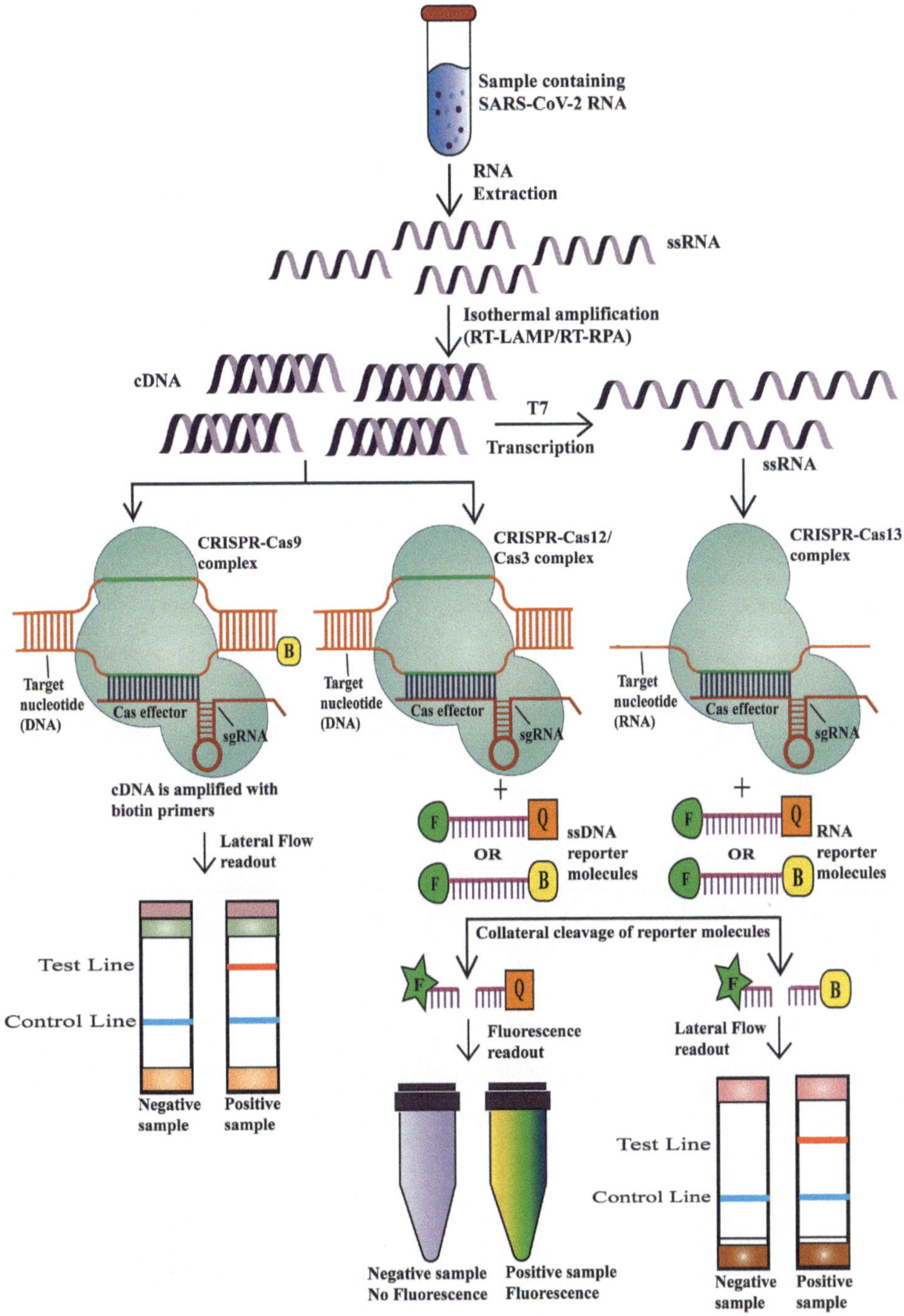

FIG. 1.2 Assembled generic schematic workflow for severe acute respiratory syndrome coronavirus 2 (SARS-CoV-2) viral RNA diagnosis by a different CRISPR (clustered regularly interspaced short palindromic repeats)-Cas system-based detection procedure. Different steps include isolation of viral RNA, RNA amplification by either RT-LAMP (reverse transcription-loop-mediated isothermal amplification) or RT-RPA (reverse transcriptase-recombinase polymerase amplification), and visualization by calorimetric or fluorescence readout. A schematic comparison of different CRISPR-based detection methods involving Cas9, Cas12, or Cas13 is presented. *B*, biotin; *cDNA*, complementary DNA; *F*, fluorophore; *Q*, quencher; *sgRNA*, single-guide RNA; *ssRNA*, single-stranded RNA.

REFERENCES

[1] Zhu N, et al. A novel coronavirus from patients with pneumonia in China, 2019. N Engl J Med 2020;382(8):727–33.
[2] Jiang S, et al. A distinct name is needed for the new coronavirus. Lancet 2020;395(10228):949.
[3] Fehr AR, Perlman S. Coronaviruses: an overview of their replication and pathogenesis. Methods Mol Biol 2015; 1282:1–23.
[4] WHO global corona virus pandemic update.
[5] Rothe C, et al. Transmission of 2019-nCoV infection from an asymptomatic contact in Germany. N Engl J Med 2020;382(10):970–1.
[6] Lamb LE, et al. Rapid detection of novel coronavirus (COVID-19) by reverse transcription-loop-mediated isothermal amplification. medRxiv; 2020. 2020.02.19.20025155.
[7] Barrangou R. The roles of CRISPR-Cas systems in adaptive immunity and beyond. Curr Opin Immunol 2015;32:36–41.
[8] Makarova KS, et al. Evolution and classification of the CRISPR-Cas systems. Nat Rev Microbiol 2011;9(6):467–77.
[9] Feng W, et al. Molecular diagnosis of COVID-19: challenges and research needs. Anal Chem 2020;92(15):10196–209.
[10] Corman VM, et al. Detection of 2019 novel coronavirus (2019-nCoV) by real-time RT-PCR. Euro Surveill 2020; 25(3).
[11] Chu DKW, et al. Molecular diagnosis of a novel coronavirus (2019-nCoV) causing an outbreak of pneumonia. Clin Chem 2020;66(4):549–55.
[12] Zhou P, et al. A pneumonia outbreak associated with a new coronavirus of probable bat origin. Nature 2020; 579(7798):270–3.
[13] Al-Tawfiq JA, Memish ZA. Diagnosis of SARS-CoV-2 infection based on CT scan vs RT-PCR: reflecting on experience from MERS-CoV. J Hosp Infect 2020;105(2):154–5.
[14] Russo A, et al. Current status of laboratory diagnosis for COVID-19: a narrative review. Infect Drug Resist 2020; 13:2657–65.
[15] Lin C, et al. Comparison of throat swabs and sputum specimens for viral nucleic acid detection in 52 cases of novel coronavirus (SARS-Cov-2)-infected pneumonia (COVID-19). Clin Chem Lab Med 2020;58(7):1089–94.
[16] Oh MD, et al. Viral load kinetics of MERS coronavirus infection. N Engl J Med 2016;375(13):1303–5.
[17] Lau KA, et al. Ebola preparedness: diagnosis improvement using rapid approaches for proficiency testing. J Clin Microbiol 2017;55(3):783–90.
[18] Cohen AN, Kessel B, Milgroom MG. Diagnosing COVID-19 infection: the danger of over-reliance on positive test results. medRxiv; 2020. 2020.04.26.20080911.
[19] Zhao Y, et al. Isothermal amplification of nucleic acids. Chem Rev 2015;115(22):12491–545.
[20] Ouyang W, et al. Alert to potential contagiousness: a case of lung cancer with asymptomatic severe acute respiratory syndrome coronavirus 2 infection. J Thorac Oncol 2020; 15(6):e82–3.
[21] Notomi T, et al. Loop-mediated isothermal amplification of DNA. Nucleic Acids Res 2000;28(12):E63.
[22] Notomi T, et al. Loop-mediated isothermal amplification (LAMP): principle, features, and future prospects. J Microbiol 2015;53(1):1–5.
[23] Kashir J, Yaqinuddin A. Loop mediated isothermal amplification (LAMP) assays as a rapid diagnostic for COVID-19. Med Hypotheses 2020;141:109786.
[24] Dao Thi VL, et al. A colorimetric RT-LAMP assay and LAMP-sequencing for detecting SARS-CoV-2 RNA in clinical samples. Sci Transl Med 2020;12(556).
[25] Yu L, et al. Rapid detection of COVID-19 coronavirus using a reverse transcriptional loop-mediated isothermal amplification (RT-LAMP) diagnostic platform. Clin Chem 2020;66(7):975–7.
[26] Yan C, et al. Rapid and visual detection of 2019 novel coronavirus (SARS-CoV-2) by a reverse transcription loop-mediated isothermal amplification assay. Clin Microbiol Infect 2020;26(6):773–9.
[27] Baek YH, et al. Development of a reverse transcription-loop-mediated isothermal amplification as a rapid early-detection method for novel SARS-CoV-2. Emerg Microbes Infect 2020;9(1):998–1007.
[28] Vashist SK. In vitro diagnostic assays for COVID-19: recent advances and emerging trends. Diagnostics 2020;10(4).
[29] Abbott TR, et al. Development of CRISPR as an antiviral strategy to combat SARS-CoV-2 and influenza. Cell 2020;181(4):865–876 e12.
[30] Li Z, et al. Development and clinical application of a rapid IgM-IgG combined antibody test for SARS-CoV-2 infection diagnosis. J Med Virol 2020;92.
[31] Cheng MP, et al. Diagnostic testing for severe acute respiratory syndrome-related coronavirus 2: a narrative review. Ann Intern Med 2020;172(11):726–34.
[32] Scohy A, et al. Low performance of rapid antigen detection test as frontline testing for COVID-19 diagnosis. J Clin Virol 2020;129:104455.
[33] Mak GC, et al. Evaluation of rapid antigen test for detection of SARS-CoV-2 virus. J Clin Virol 2020;129:104500.
[34] Zu ZY, et al. Coronavirus disease 2019 (COVID-19): a perspective from China. Radiology 2020;296(2):E15–25.
[35] Whiting P, Singatullina N, Rosser J. Computed tomography of the chest: I. Basic principles. BJA Edu 2015;15(6): 299–304.
[36] Lee EYP, Ng MY, Khong PL. COVID-19 pneumonia: what has CT taught us? Lancet Infect Dis 2020;20(4):384–5.
[37] Bernheim A, et al. Chest CT findings in coronavirus disease-19 (COVID-19): relationship to duration of infection. Radiology 2020;295(3):200463.
[38] Pan F, et al. Time course of lung changes at chest CT during recovery from coronavirus disease 2019 (COVID-19). Radiology 2020;295(3):715–21.
[39] Ai T, et al. Correlation of chest CT and RT-PCR testing for coronavirus disease 2019 (COVID-19) in China: a report of 1014 cases. Radiology 2020;296(2):E32–40.
[40] Yang W, Yan F. Patients with RT-PCR-confirmed COVID-19 and normal chest CT. Radiology 2020;295(2):E3.

[41] Feng H, et al. A case report of COVID-19 with false negative RT-PCR test: necessity of chest CT. Jpn J Radiol 2020; 38(5):409–10.
[42] Li L, et al. Using artificial intelligence to detect COVID-19 and community-acquired pneumonia based on pulmonary CT: evaluation of the diagnostic accuracy. Radiology 2020;296(2):E65–71.
[43] Wright AV, Nunez JK, Doudna JA. Biology and applications of CRISPR systems: harnessing nature's toolbox for genome engineering. Cell 2016;164(1–2):29–44.
[44] Gootenberg JS, et al. Nucleic acid detection with CRISPR-Cas13a/C2c2. Science 2017;356(6336):438–42.
[45] Yan F, Wang W, Zhang J. CRISPR-Cas12 and Cas13: the lesser known siblings of CRISPR-Cas9. Cell Biol Toxicol 2019;35(6):489–92.
[46] Koonin EV, Makarova KS, Zhang F. Diversity, classification and evolution of CRISPR-Cas systems. Curr Opin Microbiol 2017;37:67–78.
[47] Pardee K, et al. Rapid, low-cost detection of Zika virus using programmable biomolecular components. Cell 2016; 165(5):1255–66.
[48] Guk K, et al. A facile, rapid and sensitive detection of MRSA using a CRISPR-mediated DNA FISH method, antibody-like dCas9/sgRNA complex. Biosens Bioelectron 2017;95:67–71.
[49] Wang Q, et al. CRISPR-typing PCR (ctPCR), a new Cas9-based DNA detection method. Sci Rep 2018;8(1):14126.
[50] Zhang B, et al. Detection of target DNA with a novel Cas9/sgRNAs-associated reverse PCR (CARP) technique. Anal Bioanal Chem 2018;410(12):2889–900.
[51] Li SY, et al. CRISPR-Cas12a-assisted nucleic acid detection. Cell Discov 2018;4:20.
[52] Chen JS, et al. CRISPR-Cas12a target binding unleashes indiscriminate single-stranded DNase activity. Science 2018;360(6387):436–9.
[53] Myhrvold C, et al. Field-deployable viral diagnostics using CRISPR-Cas13. Science 2018;360(6387):444–8.
[54] Lucia C, Federico P-B, Alejandra GC. An ultrasensitive, rapid, and portable coronavirus SARS-CoV-2 sequence detection method based on CRISPR-Cas12. bioRxiv; 2020. 2020.02.29.971127.
[55] Ding X, et al. All-in-One dual CRISPR-Cas12a (AIOD-CRISPR) assay: a case for rapid, ultrasensitive and visual detection of novel coronavirus SARS-CoV-2 and HIV virus. bioRxiv; 2020.
[56] Broughton JP, et al. CRISPR-Cas12-based detection of SARS-CoV-2. Nat Biotechnol 2020;38(7):870–4.
[57] Wang X, et al. Rapid and sensitive detection of COVID-19 using CRISPR/Cas12a-based detection with naked eye readout, CRISPR/Cas12a-NER. Sci Bull (Beijing) 2020; 65(17):1436–9.
[58] Patchsung M, et al. Clinical validation of a Cas13-based assay for the detection of SARS-CoV-2 RNA. Nat Biomed Eng; 2020.
[59] Kellner MJ, et al. SHERLOCK: nucleic acid detection with CRISPR nucleases. Nat Protoc 2019;14(10):2986–3012.
[60] Joung J, et al. Point-of-care testing for COVID-19 using SHERLOCK diagnostics. medRxiv; 2020.
[61] Roping in 'Feluda' to detect Covid-19. 2020.
[62] Yoshimi K, et al. Rapid and accurate detection of novel coronavirus SARS-CoV-2 using CRISPR-Cas3. medRxiv; 2020. 2020.06.02.20119875.
[63] Lozano R, et al. Global and regional mortality from 235 causes of death for 20 age groups in 1990 and 2010: a systematic analysis for the Global Burden of Disease Study 2010. Lancet 2012;380(9859):2095–128.
[64] Li Y, et al. CRISPR/Cas systems towards next-generation biosensing. Trends Biotechnol 2019;37(7):730–43.
[65] Konermann S, et al. Transcriptome engineering with RNA-targeting type VI-D CRISPR effectors. Cell 2018; 173(3):665–676 e14.
[66] Yan WX, et al. Cas13d is a compact RNA-targeting type VI CRISPR effector positively modulated by a WYL-domain-containing accessory protein. Mol Cell 2018;70(2): 327–339 e5.
[67] Gootenberg JS, et al. Multiplexed and portable nucleic acid detection platform with Cas13, Cas12a, and Csm6. Science 2018;360(6387):439–44.
[68] Dolan AE, et al. Introducing a spectrum of long-range genomic deletions in human embryonic stem cells using type I CRISPR-Cas. Mol Cell 2019;74(5):936–950 e5.
[69] Morisaka H, et al. CRISPR-Cas3 induces broad and unidirectional genome editing in human cells. Nat Commun 2019;10(1):5302.
[70] Tian X, et al. CRISPR/Cas9 - an evolving biological tool kit for cancer biology and oncology. NPJ Precis Oncol 2019;3:8.
[71] Khambhati K, Bhattacharjee G, Singh V. Current progress in CRISPR-based diagnostic platforms. J Cell Biochem 2019;120(3):2721–5.
[72] Wang X, Shang X, Huang X. Next-generation pathogen diagnosis with CRISPR/Cas-based detection methods. Emerg Microbes Infect 2020;9(1):1682–91.

CHAPTER 2

COVID-19 Pandemic: Animal Cross Talk and Comparison Between nSARS-CoV-2 and Animal Coronaviruses

ANISUZZAMAN[a] • ZOBAYDA FARZANA HAQUE[a] •
MUHAMMAD TOFAZZAL HOSSAIN

1 INTRODUCTION

In this modern era, the world faces many complex problems, an emerging infection is one of them and a single discipline, institution, or country cannot battle alone against them. Most of the emerging and reemerging diseases of humans in the past couple of decades are animal-linked with noted cross-species transmissions [1]. Transmission of the disease-causing agents from one species to another involves several steps and intermediate hosts. Animals causing the linkage in transmission may act as vector or reservoir hosts or as both. It is not mandatory that they are always mammals or birds. In fact, rodents, arthropods, gastropods, and even parasites are capable of transmitting human and animal diseases whether they are viral, bacterial, or parasitic diseases [2–5]. This phenomenon can cause devastating threats to the present and future biodiversity, including human beings. More interdisciplinary approaches are required rather than anthropocentric views to address these threats. A combined research endeavor of biologists, ecologists, microbiologists, parasitologists, veterinarians, epidemiologists, and human health professionals can help find possible solutions to the problems resulting from these threats [6].

One Health is a holistic approach and deals with ways to attain an optimum status of human health, which is closely related to animal health and their environment. It explains that integrated health largely depends on the interaction among humans, animals, and their shared environment [7–9]. It implies the interdependence of humans and animals with an interface to the soundness of ecosystems, termed as eco-health, in which they exist [10]. This concept reveals the context of infectious disease ecology, where the relevant and significant relationship of both animals and the environment is responsible for the occurrence of the newly emerging or reemerging infectious diseases infecting humans and animals [11–22]. So adequate knowledge regarding the triangular interrelationship among the host, pathogen, and environment, along with their ecology, is crucial to counter infectious agents.

In general, zoonotic diseases are good examples of this mark. The threat of emerging and reemerging zoonoses has now increased globally because of several factors such as bloom in trade and travel, climate change, rapidly evolving pathogens, population explosion, changing habits and lifestyle of humans, and intensive integrated animal farming. Zoonotic diseases are caused by a wide range of infectious entities such as viruses, bacteria, parasites, fungi, and even prions [11,18–45], and emerging infectious diseases are, generally but not strictly, associated with environmental and animal factors. Zoonotic pathogens commonly undergo mutations, and after host jumping, pathogens struggle for adaptation to the hostile environmental conditions before spillover to humans [7,46]. The interaction between jump hosts and humans is critical for understating what happens during the outbreak of an emerging infectious disease such as severe acute respiratory syndrome (SARS), the Middle East respiratory syndrome (MERS) and presently the coronavirus disease 2019 (COVID-19) pandemic, caused by the novel SARS coronavirus 2 (nSARS-CoV-2) [7,47,48].

The entire world is anxiously watching the spread of COVID-19 from country to country covering the entire world (215 countries), following modern travel routes.

[a]Contribute equally to the article and names are in alphabetic order.

COVID-19: Tackling Global Pandemics through Scientific and Social Tools. https://doi.org/10.1016/B978-0-323-85844-1.00009-X

It was first reported to the WHO Country Office in China on December 31, 2019 [49]. After that, it inevitably crossed the boundaries of the country and has become a pandemic [50,51].

COVID-19 pandemic is the most recent infectious disease outbreak that emerged at the human-animal-environment interface. Although there is a controversy, it is generally believed and has been scientifically proved that the virus has jumped to human hosts from bats [52–54]. In this century, several other pandemics have emerged from close contact between humans and wildlife. For example, the first known severe illness caused by a coronavirus was the SARS pandemic in 2002–03, and it was introduced to humans due to interaction with civets and civets got infected by bats [55–59]. It spread to 26 countries, with more than 8000 infections and 774 deaths and a case fatality rate of 9.5% [60–63]. A second outbreak of severe illness began in 2012 in Saudi Arabia caused by the MERS virus and humans got the virus after direct exposure to camels [52,64]. The spread of the MERS virus was reported in 27 countries with 2494 infections and 858 deaths and a case fatality rate of 34.4% [64]. Another deadliest bat-borne pathogen was the Nipah virus (NiV) that was discovered in 1998 during the first reported outbreak in Sungai Nipah, a village in Malaysia, where humans contracted NiV from pigs, the intermediate hosts of the virus [65,66]. Later in 2001, it caused a severe outbreak in Bangladesh and India [67,68].

The Centers for Disease Control and Prevention (CDC) estimates that 60% of known infectious diseases and up to 75% of new or emerging infectious diseases are zoonotic in origin [69–72]. Of them, more than 60 zoonotic viruses are hosted by bats [73]. Among these 60 bat-associated viral species, 59 are representing the currently evolving and emerging and reemerging viral infectious diseases of humans, containing an RNA genome [74,75]. The most recent bat-adaptive zoonotic transmission by the nSARS-CoV-2 is happening with rapid spread. It is a combined effect of human activities causing increased exposure to bats [76] and the diversified regional temperature and humidity due to global warming [77]. So at this point, COVID-19 is not only a global pandemic occurring due to the nSARS-CoV-2 infection but also a perilous pivot point of modern times, reflecting the consequences of a few episodes practiced in recorded history.

However, similar to the other hypervirulent human coronaviruses (hCoVs), nSARS-CoV-2 has a presumed animal origin and seems to be condescended from a related bat coronavirus (CoV) that spilled over to humans. The spillover or host jump of the nSARS-CoV-2 postulated either directly from bat or after adaptation in another animal species, such as the Malayan pangolin [78–81]. So far it has been scientifically proved that genetically nSARS-CoV-2 is highly related (96% nt) to a SARS-like bat CoV [82,83]. Despite this proved data, further investigation is needed to define the roles of bats as the origin and natural reservoir and the role of pangolins as the intermediate host. Other than these two animals, there might be more evolutional origin of this outbreak. Although several pieces of research have been done and many are ongoing, the exact animal source of this disastrous outbreak remains a question until now. Therefore it is imperative to understand which animals may be susceptible to this virus, so as to take precautionary measures to minimize the spread of COVID-19 [84,85].

As veterinary researchers, we have been following the development of COVID-19 and we will briefly discuss the COVID-19/nSARS-CoV-2 infection with a particular focus on the role of animals and the associated zoonotic links. We will also describe the structure of CoVs and compare nSARS-CoV-2 with other CoVs affecting humans, animals, and birds.

2 CORONAVIRUSES: TAXONOMY, CLASSIFICATION, AND STRUCTURES

CoVs belong to a large family (Coronaviridae) of viruses causing acute, mild upper respiratory infection (common cold) in humans, animals, and birds. The name "coronavirus" is derived from Latin *corona*, meaning "crown" or "wreath" [86]. This name refers to the characteristic appearance of virions (the infective form of the virus) under electron microscopy, which has a fringe of large, bulbous surface projections creating an image reminiscent of the solar corona or halo [87]. This morphology is due to the viral spike peplomers, which are proteins on the surface of the virus [88]. CoV usually causes mild to moderate diseases, but some CoVs are lethal in animals, birds, and humans. Additionally, several strains of CoVs are zoonotic (transmitted between animals and humans), but many strains are not.

Taxonomically, CoVs are classified under the subfamily Orthocoronavirinae, family *Coronaviridae*, order Nidovirales, and realm *Riboviria* [89,90] (Fig. 2.1). As the number increased, in 2009, based on the rooted and unrooted genetic trees and partial nucleotide sequence of RNA-dependent RNA polymerase, the subfamily Coronavirinae has been recognized and divided into four genera: *Alphacoronavirus* (α-CoV) and

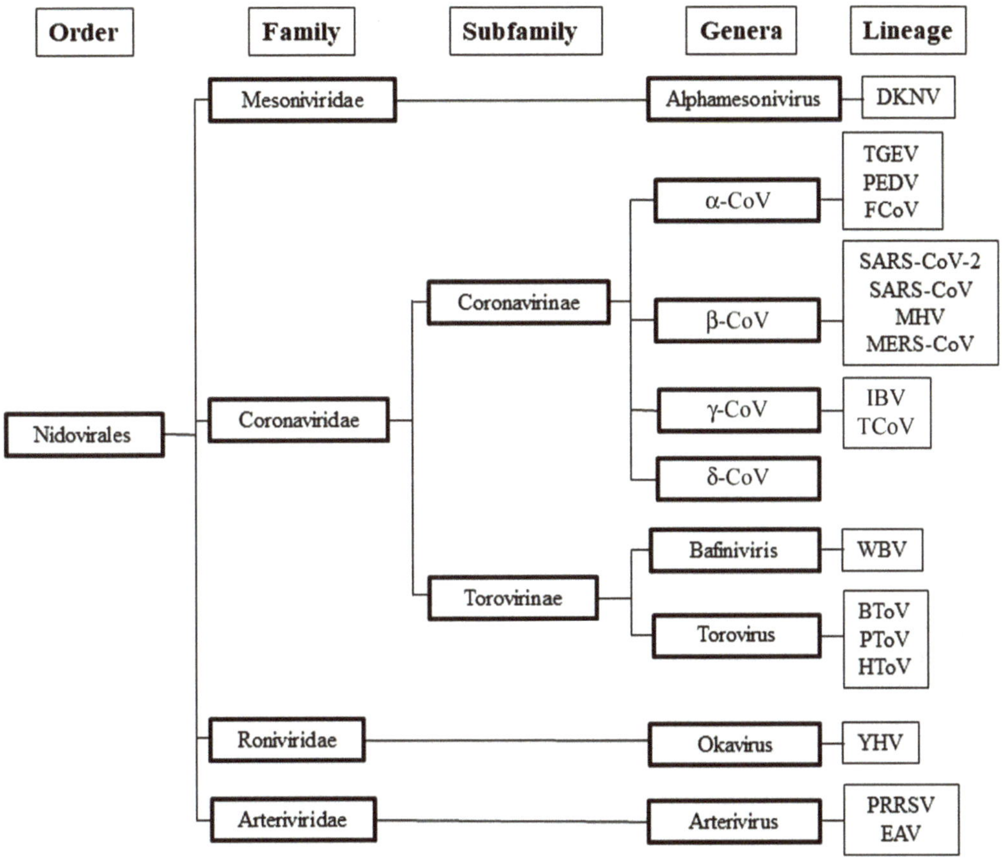

FIG. 2.1 Classification of the coronaviruses (CoVs). *BToV*, bovine torovirus; *DKNV*, dak nong virus; *EAV*, equine arteritis virus; *FCoV*, feline CoV; *HToV*, horse torovirus; *IBV*, infectious bronchitis virus; *MERS-CoV*, Middle East respiratory syndrome CoV; *MHV*, mouse hepatitis virus; *PEDV*, porcine epidemic diarrhea virus; *PRRSV*, porcine reproductive and respiratory syndrome virus; *PToV*, porcine torovirus; SARS-CoV-2, severe acute respiratory syndrome coronavirus 2; *TCoV*, turkey CoV; *TGEV*, transmissible gastroenteritis virus of swine; *WBV*, white bream virus; *YHV*, yellow head virus.

Betacoronavirus (β-CoV) that infect only warm-blooded animals and have a probable bat origin and the *Gammacoronavirus* (γ-CoV) and *Deltacoronavirus* (δ-CoV) that infect not only birds but also fish, and some mammals, and are assumed to have swine origin [91,92].

The common name coronavirus is used to refer to any member of the subfamily Orthocoronavirinae [90]. As of 2020, 45 species are officially recognized [93]. CoVs are relatively large viruses coated with an envelope and commonly termed as enveloped viruses. The average diameter of the virus particles is around 125 nm. The diameter of the envelope is 85 nm and the spikes are 20 nm long [94,95], and its total molecular weight is on an average 40,000 kDa [96]. The envelope of the virus in electron micrographs appears as a distinct pair of electron-dense shells. The viral envelope consists of a lipid bilayer, in which the membrane (M) protein, envelope (E) protein, and spike (S) structural proteins are anchored.

On an average, a CoV particle has 74 surface spikes [97]. A subset of CoVs (specifically the members of β-CoV subgroup A) also have a shorter spikelike surface protein called hemagglutinin esterase [98].

Inside the envelope, there is a nucleocapsid, which is formed from multiple copies of the helical symmetry nucleocapsid (N) protein that are bound to the positive-sense single-stranded RNA genome. The genome size of CoVs ranges from 26.4 to 31.7 kb, which is one of the largest among RNA viruses. The genome has a 5′ methylated cap and a 3′ polyadenylated tail.

The lipid bilayer envelope, membrane proteins, and nucleocapsid protect the virus when it is outside the host cell [94,99,100].

Virulence or pathogenicity of the CoVs varies significantly and largely depends on risk factors. Some can kill more than 30% of the infected individuals, e.g., MERS-CoV [101,102], and some are relatively harmless, like the common cold virus [100]. CoV-related common colds are clinically characterized by fever, runny nose, and a sore throat from swollen adenoids, which may lead to the development of bronchitis (either viral bronchitis or secondary bacterial bronchitis) and pneumonia (either direct viral pneumonia or secondary bacterial pneumonia) [103,104]. The hCoVs, e.g., SARS-CoV that causes SARS, have a unique pathogenesis because they cause both upper and lower respiratory tract infections [104].

So far, six species of hCoVs have been discovered and one species has been subdivided into two different strains, making seven strains of hCoVs altogether. Of them, hCoV OC43 (HCoV-OC43) and hCoV HKU1 (HCoV-HKU1) are β-CoVs and hCoV 229E (HCoV-229E) and hCoV NL63 (HCoV-NL63) are α-CoVs, which generally produce mild symptoms. In contrast, three hCoVs, namely, MERS-CoV, SARS-CoV, and nSARS-CoV-2 are β-CoVs and produce potentially severe symptoms [105,106].

This classification has been confirmed by unique features of the chemistry and replication of these viruses. Most of the hCoVs belong to either 229E-like or OC43-like group, which differ both antigenically and in ex vivo culture requirements. Of them, 229E-like CoVs can usually be isolated from human embryonic fibroblast cultures, while OC43-like viruses can be isolated from, or adapted for growth in, suckling mouse brain. There is little antigenic cross-reaction between these two types; however, they cause independent epidemics of indistinguishable diseases [106,107].

3 TRANSMISSION OF CORONAVIRUSES

Most of the animal CoVs and hCoVs are transmitted through the fecal-oral route, and shedding of CoVs into the environment occurs through respiratory secretion and feces. These viruses spread from infected hosts to susceptible host through contact or contaminated materials [108,109]. Different species of bats are considered as the natural host for most of the CoVs affecting humans, and different species of mammals act as intermediate hosts to disseminate viruses from their natural reservoir hosts to humans depending on their ability to access the host cell [110]. Bats transmitted SARS and MERS to the intermediate host, civet cat and camels, respectively, and then the viruses were finally introduced to humans (Fig. 2.3).

Similarly, nSARS-CoV-2 is thought to be transmitted from bats to an unidentified intermediate host to humans [108]. Thus wildlife and livestock might play an important role in interspecies transmission in the emergence of these devastating CoVs in humans. It has been proved that the main transmission route of nSARS-CoV-2 is from person to person [111]. When infected people sneeze, cough, talk, laugh, or breathe heavily the virus particles are transmitted to other persons in their close contact like less than 1 m apart through respiratory droplets or aerosols. Research shows that the nSARS-CoV-2 virus particles can spread through the air current and survive in the air for up to 3 h. It can get into the lungs of healthy people while they breathe out and breathe in the contaminated air. The virus can also be shed in the environment while the coughing and sneezing particles of infected people or their touch contaminates the surface of objects [112]. A few studies reported that virus particles can be found in an infected person's poop, albeit experts are not sure whether the infection can spread through contact with that [113,114]. But many research reports strongly support that nSARS-CoV-2 transmission from humans to animals and from animals to animals occurs by airway as well as direct contact [115–121]. The virus most often spreads through infected people showing symptoms. But it is possible to pass it on without presenting any signs, that is, asymptomatic spread. People can also pass it on before they notice any signs of infection, that is, presymptomatic spread. Thus community spread occurs when someone gets the infection without any known contact with a sick person. A few pet animals have tested positive for the new coronavirus with mild symptoms. The pets may have caught the virus from close contact with their infected owner. However, the transmission of nSARS-CoV-2 from affected animal to humans has not yet been reported [112].

4 ANIMAL CORONAVIRUSES

Bats and birds are major reservoirs for CoVs, which infect a diverse group of domestic and wild animals. Infections usually remain subclinical in most cases, but a few of them are responsible for severe infection and have been recognized to cause pathologic conditions in veterinary medicine since the 1930s [122–126]. The animal CoVs cause enteritis and/or upper respiratory tract disease in ruminants, companion animals, and poultry [47,127]. The CoVs infect a wide range of animals including cattle, horses, swine, dogs, cats, camels, rabbits, rodents, birds, ferrets, mink, bats, snake (such as Chinese cobra and krait), frogs, marmots, hedgehogs (*Erinaceus europaeus*), Malayan or Javan or Sunda pangolin (*Manis javanica*), and many other wild animals (Table 2.1) [128,129]. Most of the

TABLE 2.1
Coronaviruses and Their Different Natural Hosts.

Group	Coronaviruses	Abbreviation	Natural Host
Human	Human coronavirus 229E	HCoV-229E	Humans
	Human coronavirus NL63	HCoV-NL63	Humans
	Human coronavirus OC43	HCoV-OC43	Humans
	Human coronavirus HKU1	HCoV-HKU1	Humans
	Severe acute respiratory syndrome-related coronavirus	SARS-CoV, SARS-CoV-1, SARS-CoV-2	Humans
	Middle East respiratory syndrome-related coronavirus	MERS-CoV	Humans
	Human enteric coronavirus	HECoV	Humans
Domestic and domesticated avian species	Infectious bronchitis virus	IBV	Chicken
	Turkey coronavirus	TCoV	Turkey
	Pheasant coronavirus	PhCoV	Pheasant
	Quail coronavirus	QCoV	Quail
	Guinea fowl coronavirus	GfCoV	Guinea fowl
	Puffinosis coronavirus	PCoV	Shearwater
Domestic and domesticated ruminants	Bovine coronavirus	BCoV	Cattle
	BCoV-like coronavirus	–	Sheep
	BCoV-like coronavirus	–	Goat
	Bubaline coronavirus	BuCoV	Water buffalo
	BCoV-like coronavirus	–	*Llama*
	Alpaca (beta) coronavirus	ACoV	Alpaca
	Alpaca (alpha) coronavirus	ACoV	Alpaca
	Dromedary camel coronavirus UAE-HKU-23	DcCoV UAE-HKU23	Dromedary camel
	Dromedary camel MERS-CoV	–	Dromedary camel
	Dromedary camel alphacoronavirus	–	Dromedary camel
Domestic and domesticated carnivores	Canine coronavirus I	CCoV-I	Dog
	Canine coronavirus II	CCoV-II	Dog
	Canine respiratory coronavirus	CRCoV	Dog
	Feline coronavirus I	FCoV-I	Cat
	Feline coronavirus II	FCoV-II	Cat
	Mink coronavirus	MCoV	Mink
	Ferret coronavirus	FCoV	Ferret
Domestic swine	Transmissible gastroenteritis virus	TGEV	Pig
	Porcine respiratory coronavirus	PRCoV	Pig
	Porcine epidemic diarrhea virus	PEDV	Pig
	Severe acute diarrhea syndrome-coronavirus	SADS-CoV	Pig
	Swine enteric alphacoronavirus	SeACoV	Pig
	Porcine hemagglutinating encephalomyelitis virus	PHEV	Pig
	Porcine deltacoronavirus	PDCoV	Pig
Laboratory animals	Murine hepatitis virus	MHV	Mouse
	Rat coronavirus	RtCoV	Rat
	Sialodacryoadenitis virus	SDAV	Rat
	Rabbit enteric coronavirus	RECoV	Rabbit

animals infected with CoVs get infections through the fecal-oral route and the gastrointestinal tracts are affected [130]. Usually each CoV is host specific and infects a single animal species, and interspecies transmission occurs rarely in a restricted way. Although some CoVs have a zoonotic impact, several CoVs are known to cause diseases only in animals and have not yet been proved to infect humans [131]. Significant research efforts have been focused on elucidating the viral pathogenesis and the epidemiology of the animal CoVs, especially by multidisciplinary approaches (One Health) and veterinarians are playing pivotal roles.

4.1 Coronaviruses in Ruminants

β-CoVs affect ruminants, which include cattle, water buffalo, pig, goat, sheep, *Llama*, alpaca, and dromedary camel, and cause respiratory diseases and gastroenteritis. Alpaca and dromedary camels are also affected by α-CoV that causes mild respiratory diseases [108,132]. Bovine CoV (BCoV) is endemic in cattle worldwide that infect the upper and lower respiratory tracts and the intestine and is shed through feces and upper respiratory tract secretions. It causes three distinct clinical syndromes in cattle: (1) calf diarrhea, which is an enteric disease with high mortality rates in neonatal calves of 1–3 weeks of age in both dairy and beef cattle herds; (2) winter dysentery in adult dairy, beef cattle and in captive wild ruminants is an acute disease characterized by hemorrhagic diarrhea, frequent respiratory signs, anorexia, and decreased milk production in dairy cows with significant economic losses; and (3) respiratory infections in cattle of various ages, with a higher prevalence in 2- to 3-month-old calves, including the bovine respiratory disease complex or shipping fever of feedlot cattle [108]. Although there are distinct disease syndromes, all BCoV isolates from both enteric and respiratory infections are antigenically similar, comprising a single serotype with two to three subtypes [108].

Among six CoVs of swine, four are α-CoVs such as transmissible gastroenteritis virus of swine (TGEV), porcine respiratory coronavirus (PRCoV), porcine epidemic diarrhea virus (PEDV), and severe acute diarrhea syndrome-coronavirus (SADS-CoV); one is β-CoV, e.g., porcine hemagglutinating encephalomyelitis virus (PHEV); and one is δ-CoV, e.g., porcine deltacoronavirus (PDCoV). PRCoV causes a mild respiratory disease, PHEV is the causative agent of neurologic and/or digestive disease in pigs, and TGEV, PEDV, SADS-CoV, and PDCoV are responsible for acute gastroenteritis in swine, with fatal infections in piglets, and greatly threaten the swine industries worldwide [7,133–136]. PRCoV and TGEV share some epitopes, so the extensive circulation of PRCoV in swine herds produces neutralizing antibodies and thus resulted in a drastic reduction of transmissible gastroenteritis outbreaks worldwide [132]. In the 1970s, PEDV was first detected in the pig population as a consequence of a spillover event from bats. On the basis of the S gene, the PEDV genotypes are (1) G1a PEDV, classical European and Asian strains with moderate virulence; (2) G2 PEDV, highly virulent strains originated in Asia; (3) G1b PEDV, associated with mild clinical outbreaks; and (4) S1 N-terminal domain-deletion (NTD-del) strains associated to mild clinical forms [137]. SADV-CoV, at present known as swine enteric α-CoV, is another virulent swine enteric virus that originated from bats and produces severe enteritis in suckling piglets, causing significant mortality. On the other hand, PHEV, which causes encephalomyelitis, has not been derived from bat CoVs, instead they descend from a rodent CoV [82,138]. PDCoV infection seems to be a host-switching event between avian and mammal CoVs [132].

Bovinelike CoVs are responsible for enteritis and neonatal mortality in domestic sheep, neonatal diarrhea in kids, severe diarrhea with neonatal mortality in water buffalo calves, severe gastroenteritis in dromedary camel calf, and severe diarrhea in llamas and alpacas throughout the world [139].

4.2 Coronaviruses in Companion Animals

In cats, two α-CoV-1 genotypes known as feline CoV type I (FCoV-I) and type II (FCoV-II) have an affinity to the respiratory tract, central nervous system, abdominal cavity, and gastrointestinal tract to produce enteritis and infectious peritonitis with high mortality [128]. Among three CoVs of dogs, two are α-CoVs, namely, canine coronavirus I (CCoV-I) and II (CCoV-II), and one is β-CoVs, namely, canine respiratory coronavirus (CRCoV). CCoVs (species α-CoV-1) are commonly responsible for mild, self-limiting enteritis in pups, and CRCoV, together with other viral and bacterial agents, causes canine infectious respiratory disease with mild respiratory signs [140]. There are two types of CoVs that infect ferrets. Ferret enteric coronavirus causes a gastrointestinal syndrome known as epizootic catarrhal enteritis, and a more lethal systemic strain of the virus (like feline infectious peritonitis [FIP] in cats) known as ferret systemic coronavirus causes anorexia, weight loss, diarrhea, and a palpable intra-abdominal mass, less frequently hind limb paresis and central nervous system dysfunction [141–143].

Only one CoV of horse known as equine CoV, a BCoV-descendant β-CoV, was first isolated from the feces of a diarrheic foal and was believed to affect only foals. But now the virus has been recognized as a new, clinically important enteric virus of adult horses [144].

4.3 Coronaviruses in Laboratory and Wild Animals

Murine coronavirus of the genus β-CoV has been reported to cause infection in rodents. The prototype virus, mouse hepatitis virus, is a natural pathogen of mice, normally infecting the liver, gastrointestinal tract, and central nervous system, causing a wide range of diseases, including hepatitis, gastroenteritis, and acute and chronic encephalomyelitis. Two prototype strains of rat coronavirus (RCoV) are sialodacryoadenitis virus (SDAV) and Parker's RCoV (RCoV-P), both infect the lower respiratory tract causing interstitial pneumonia in young rats. SDAV can also infect the eye, salivary gland, and lacrimal gland of rats [132]. Guinea pig coronavirus and rabbit coronaviruses cause hepatitis, enteritis, and respiratory infections in laboratory animals [122]. Mink coronavirus 1 is the causative agent of mink epizootic catarrhal gastroenteritis. CoV-infected rhesus macaques are also asymptomatic but exhibit diffuse interstitial pneumonia or show mild symptoms, such as weight loss and rapid respiration associated with virus replication both in the upper and lower respiratory tracts. A novel CoV named SW1 has been identified by using panviral microarray technology in the liver tissue of captive beluga whales (*Delphinapterus leucas*) [122,145].

Six species of wild ruminants under the Cervidae family, namely, caribou/reindeer (*Rangifer tarandus caribou*), elk/wapiti (*Cervus elaphus*), sambar deer (*Cervus unicolor*), white-tailed deer (*Odocoileus virginianus*), sika deer (*Cervus nippon yesoensis*), and water deer (*Hydropotes inermis*), were found positive for BCoV-like CoVs [140]. Giraffe (*Giraffa camelopardalis*), several species of antelopes, wisent (*Bison bonasus*), Himalayan tahr (*Hemitragus jemlahicus*), and dromedary camels (*Camelus dromedarius*) [149] were also found positive for bovinelike CoV.

Wild carnivores such as wolves (*Canis lupus*), red foxes (*Vulpes vulpes*), Eurasian otters (*Lutra lutra*), and common genets (*Genetta genetta*) [150,151] were positive for CCoVs. CCoV-like viruses were also found in spotted hyenas (*Crocuta crocuta*) and silver-backed jackals (*Canis mesomelas*) [152]. FCoVs with FIP cases were reported in servals (*Felis serval*), cheetah (*Acinonyx jubatus*), mountain lion (*Puma concolor*), and European wild cat (*Felis silvestris*) [153–156]. Divergent α-CoV-1 viruses were detected in Chinese ferret badger (*Melogale moschata*) and raccoon dog (*Nyctereutes procyonoides*) [157].

More than 200 novel CoVs have been identified in bats and only 10 α-CoVs and 7 β-CoVs have been officially recognized by the International Committee on Taxonomy of Viruses (ICTV). MERS-related CoV and SARS-related CoV (SARS-CoV-1 and SARS-CoV-2) are mostly related to bat CoVs [132].

4.4 Coronaviruses of Poultry and Other Birds

All CoVs of the genus γ-CoV have been reported to cause infections only in birds. Birds of the order Galliformes, i.e., domestic fowl (*Gallus gallus*), turkeys (genus *Meleagris*), and pheasants (Phasianidae), are infected with three CoVs, namely, infectious bronchitis virus (IBV), turkey coronavirus (TCoV), and pheasant coronavirus (PhCoV), respectively. In the poultry industry, IBV causes extensive economic loss by producing respiratory illness, urinary tract infection, and reproductive problems [47,127], while TCoV causes enteritis and PhCoV causes respiratory and kidney disease [132]. Quail and guinea fowl also suffer from enteric diseases produced by quail coronavirus and guinea fowl coronavirus, respectively. IBV-like CoVs are detected in penguins, pigeons, peafowl, parrots, waterfowl, partridge, teal, quail, duck, and whooper swan [132,158–160]. Duck coronavirus in ducks, pigeon coronavirus in pigeons, or goose coronavirus in geese are the CoVs distinct from avian coronaviruses [132,161,162].

5 NSARS-COV-2 AND OTHER HUMAN CORONAVIRUSES

nSARS-CoV-2 belongs to the genus β-CoV and subgenus Sarbecovirus. Although there are currently seven known hCoVs, namely, HCoV-NL63, HCoV-229E, HCoV-OC43, HCoV-HKU1, SARS-CoV, MERS-CoV, and nSARS-CoV-2, nSARS-CoV-2 (Figs. 2.2 and 2.4) shares the highest overall genome sequence similarity to the bat CoV RaTG13. However, the receptor binding domain (RBD) shares more sequence similarity with the pangolin SARS-CoV [163], which both bind to angiotensin-converting enzyme 2 (ACE2) suggesting evidence of a recombination event in the evolution of nSARS-CoV-2 due to animal mixing or close exposures between bats and pangolins before evolving to infect humans. SARS-CoV and MERS-CoV belong to the genus β-CoV, but nSARS-CoV-2 is different from these two at the genetic level. nSARS-CoV-2 has been found to have 88%–89% identity to two bat origin SARS-CoVs (bat-SL-CoVZC45 and bat-SL-CoVZXC21, also named as ZC45 and ZXC21, respectively), while it has 82% identity to human SARS-CoV Tor2 and human SARS-CoV

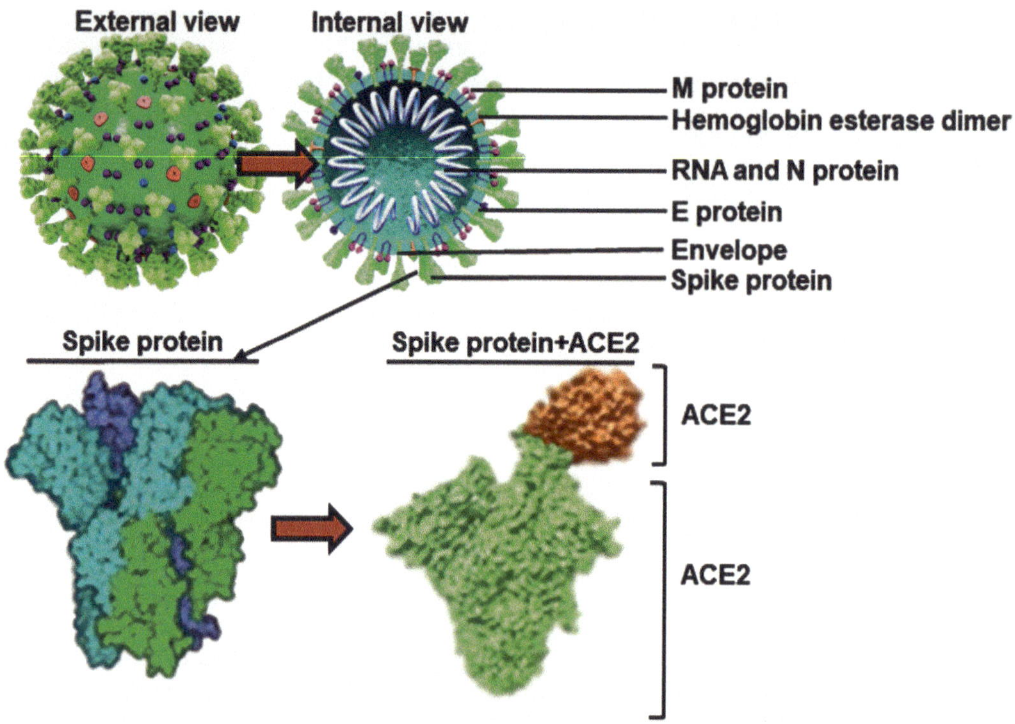

FIG. 2.2 Structures and receptors of the coronaviruses. *ACE2*, angiotensin-converting enzyme 2.

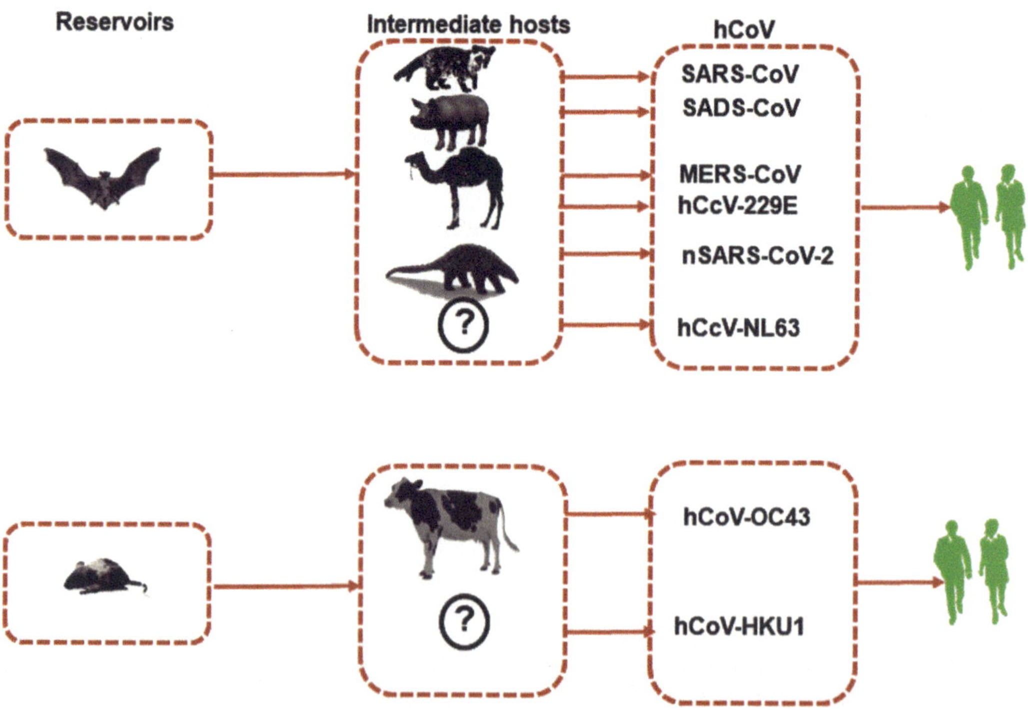

FIG. 2.3 Transmission of human coronaviruses (hCoVs) from reservoirs to humans via intermediate hosts. *MERS-CoV*, Middle East respiratory syndrome CoV; *nSARS-CoV-2*, novel SARS coronavirus 2; *SADS-CoV*, severe acute diarrhea syndrome-coronavirus; *SARS-CoV-2*, severe acute respiratory syndrome coronavirus 2.

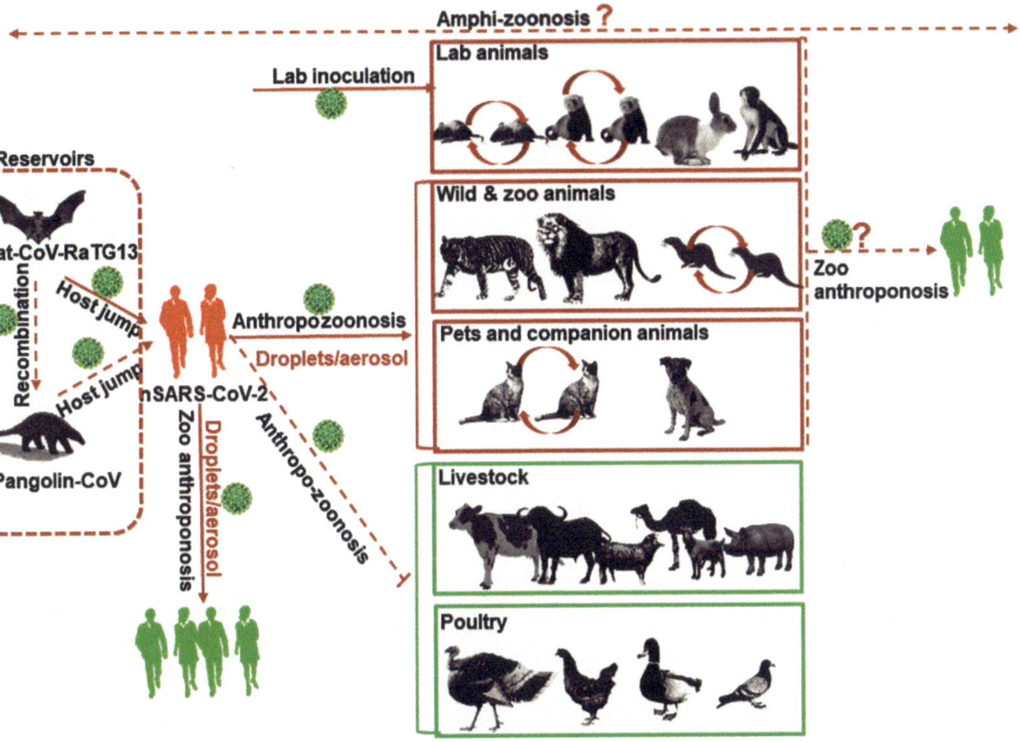

FIG. 2.4 Origin and various animal interactions of novel severe acute respiratory syndrome coronavirus 2 (nSARS-CoV-2). *Anthropozoonosis*, diseases transmitted from humans to animals. *Zoo anthroponosis*, diseases transmitted from animals to humans.

BJ01 2003 at the nucleotide level [164–168]. In contrast, only 50%–51.8% identity was observed between nSARS-CoV-2 and MERS-CoV and 79% between nSARS-CoV-2 and SARS-CoV. Further phylogenetic analyses reveal that nSARS-CoV-2 is more close to bat-origin SARS-CoV [164,169–171]. Advanced and in-depth genome analysis identified the presence of 380 amino acid substitutions among the sequences of nSARS-CoV-2 (HB01) when compared to the corresponding consensus sequences of SARS-CoV and SARS-CoV-like viruses. This amino acid substitution might have contributed to the functional as well as pathogenic divergence of this novel virus [172].

The genome sequence analysis of nSARS-CoV-2 reveals that it shares a sequence identity of 79.6% to SARS-CoV and 96% to a novel bat CoV RatG13, indicating its bat origin. The receptor binding spike protein of nSARS-CoV-2 had less than 75% sequence homology with the previously described all SARS-related CoVs but the virus uses ACE2 receptor for cellular entry like other SARS-CoVs (Fig. 2.2) and it did not show any affinity to other cellular binding domains such as dipeptidyl peptidase 4 (DPP4) and aminopeptidase N. Some isolates of nSARS-CoV-2 genes show less than 80% sequence identity with SARS-CoV. Also, there are differences in the spike protein gene between nSARS-CoV-2 and SARS-CoV in four out of five key residues of receptor binding motif and three short insertions in N-terminal domain of S gene [80,108,122,173].

All CoVs contain 26,000–32,000 bp genomes with a variable 6–11 open reading frames, which encode nonstructural proteins as well as structural proteins. nSARS-CoV-2 is related to SARS-CoV, the virus that causes SARS, in addition to SARS-like bat CoVs (RaTG13/CoVZC45/CoVZCX21), and all these viruses bind to the ACE2 receptor due to their similar S domain genomic sequences. Furthermore, a large proportion of SARS-CoV and nSARS-CoV-2 genomes is identical; however, notable differences occur in key locations, such as the absence of 8a protein and longer 8b protein in nSARS-CoV-2. Another analysis shows 380 amino acid substitutions between nSARS-CoV-2 and SARS/SARS-like viruses, specifically located in nsp2 and nsp3. Six substitutions were found in the RBD at amino acids located at 357–528, with a further four substitutions in the C-terminus region of the S1 domain. However, no substitutions were found in nsp7, nps13, E, M, and accessory proteins p6 and 8b, suggesting substitutions of RBD probably contributed to the increased pathogenicity and transmission potential of nSARS-CoV-2. nSARS-CoV-2 possesses an intact open reading frame 8 without the 29-nt deletion that is found in a majority of human SARS-CoV. Furthermore, SARS-CoV-2 contains a distinct proteolytically sensitive activation loop (polybasic furin cleavage site) at the S1-S2 junction that is not typically found in other human lineage b β-CoVs. However, this is a feature of several other animal and human CoVs, such as HKU1 (lineage a). Compared to SARS-CoV, the addition of such a cleavage site is thought to enhance cell-cell fusion without affecting viral entry, although the exact role of this site is still to be fully determined. MERS-CoV diverges from SARS-CoV and nSARS-CoV-2 at the RBD because MERS-CoV binds to DPP4. Furthermore, encoded proteins pp1ab, pp1a, E, M, accessory protein 7a, and N genes vary considerably between SARS-CoV and CoV-2 genomes. Other major respiratory viruses include influenza viruses, which belong to the Orthomyxoviridae family of the Orthornavirae kingdom. Influenza A and B types infect humans, mammals, and birds, whereas types C and D primarily affect animals. All major flu pandemics have been caused by type A influenza viruses. Influenza viruses are negative-sense RNA viruses compared to the positive-sense CoVs. Seasonal flu vaccines are designed to incorporate recent seasonal influenza type A strains and, as such, offer little to zero protection against CoVs. Aside from the vast genomic differences between lineage b β-CoVs and type A influenza viruses, the presence of an efficient cleavage site in nSARS-CoV-2, and hemagglutinin on influenza strains, may be the causative factor for rapid viral replication and transmission and increased pathogenicity of both viruses. As this is not typically a feature of both bat and pangolin CoVs (although the knowledge about these strains is limited), it may be that nSARS-CoV-2 acquired this from the first human-human transmission, or through a yet unidentified bat or pangolin SARS-like CoV [174,175].

In summary, nSARS-CoV-2 shares a high degree of genomic similarity with other β-CoVs including human SARS-CoV and bat SARS-like-CoVs/RaTG13. While their genomes are largely identical, key genomic differences are suggesting a different evolution for nSARS-CoV-2, most likely involving bat SARS-like-CoVs. Understanding the genomic differences will help scientists develop highly effective vaccines against nSARS-CoV-2, as well as learn how viruses evolve [176].

6 RELATIONSHIP BETWEEN SARS-COV-2 AND ANIMAL CORONAVIRUSES

In humans and birds, clinical signs represented by CoVs are respiratory tract infections that can range from mild to lethal. Mild illnesses in humans include common cold (which is also caused by other viruses, predominantly rhinoviruses), while more lethal varieties can cause SARS, MERS, and COVID-19. On the other hand, in cows and pigs they cause diarrhea, while in mice they cause hepatitis and encephalomyelitis. Animal CoVs and hCoVs resemble each other in morphology and chemical structure, and in some cases, for example, the CoVs of humans (causal agents of common cold) and cattle are antigenically related. There is no evidence, however, that hCoVs can be transmitted by animals. But CoVs have great ability to mutate, which may be transmitted from animals to humans, resulting in mild respiratory infections, such as the common cold, but can lead to serious illnesses, such as pneumonia.

In animals, various CoVs invade many different tissues and cause a variety of diseases, but in humans, except a few lethal pandemics, they are only proved to cause mild upper respiratory tract infections, i.e., common colds. On rare occasions, gastrointestinal CoV infection has been associated with outbreaks of diarrhea in children, but these enteric viruses are not well characterized [100]. Birds also carry CoVs, except nSARS-CoV-2. To our knowledge, until today, a single case of COVID-19 has been reported from neither live poultry (such as chickens, ducks, and other table birds) nor livestock (e.g., cattle, sheep, goats, buffaloes, camels, and pigs), suggesting rearing, marketing, and consumption of meat and other products derived from livestock and poultry do not pose any threat of COVID-19 to humans [71]. Nevertheless, the Chinese Disease Control Center reported nSARS-CoV-2 in a frozen wing sample of chicken imported from Brazil to Shenzhen, which might be because of contamination.

7 ORIGIN, ANIMAL LINKS OF NSARS-COV-2, AND CONTROVERSY REGARDING ZOONOSES

Initially, clusters of pneumonia with unknown cause were detected in Wuhan, China, in December 2019 [177], and shortly the outbreak was traced to be caused by a novel strain of CoV [178], which was given the interim name 2019-nCoV by the World Health Organization (WHO) on January 10, 2020 [179], and later renamed as nSARS-CoV-2 by the ICTV. The virus spread rapidly to other parts of China and many other countries, thus declared a Public Health International

Emergency by the WHO on January 31, 2020. Consequently, it was categorized as a high risk on February 29, 2020, and gained the pandemic status on March 11, 2020 [180], despite tremendous efforts to be contained by the WHO, and was further referred to as the coronavirus disease 2019 (COVID-19). COVID-19 has presently spread to more than 215 countries/territories while causing 4,251,676 confirmed deaths out of cumulative confirmed infected asymptomatic or symptomatic cases accounting to more than 199,866,646 as of August 4, 2021 [181].

The first patient (P_0, patient zero) with unexplained pneumonia, identified in December 8, 2019, came from Wuhan South China Seafood Market. Initially, other hospitalized patients in Wuhan were linked to the same seafood and live animal market where poultry, bats, snakes, frogs, rabbits, marmots, and hedgehogs are sold, suggesting an animal origin for the initial spread to humans. Subsequent investigations revealed that the crowded seafood market only boosted circulation of the novel CoV and spread it to the whole city in early December 2019, whereas based on the genome data the virus likely began spreading from person to person in early December or even as early as late November. The first documented human case has been dated back to November 17, 2019.

The literature published so far suggest that nonhuman primates showed no clinical signs of respiratory illness when infected with SARS-CoV; however, when infected with nSARS-CoV-2, initial clinical signs such as loss of body weight, hunched posture (leaning forward with shoulders raised and head down), and loss of appetite are evident. Experimental infections have been established in monkeys, and viral replication in monkeys is also evident in respiratory organs rendering histopathologic changes in lungs but extrapulmonary replication is more evident and pronounced in nSARS-CoV-2 infection as compared with SARS-CoV, indicating more virulence and pathogenicity [182]. Clinical symptoms and pathologic manifestations were observed in common marmosets (*Callithrix jacchus*) in SARS-CoV infection, but data on nSARS-CoV-2 infection in marmosets is currently not reported.

Higher virulence and fatality in juvenile cats infected with SARS-CoV-2 have been reported [133]. Cats are one of the most popular companion animals of humans and are highly vulnerable in COVID-19 pandemic, and cat-to-cat intraspecies transmission has also been reported [91,111]. There is also a possibility of reverse transmission of virus from infected cats to humans, although sufficient data is not available to establish the fact. Ferrets infected with both the viruses show similarities of clinical illness, but infection in ferrets by nSARS-CoV-2 is restricted to the upper respiratory tract while viral RNA could also be detected in lungs of ferrets in SARS-CoV infection and intraspecies transmission has been demonstrated [111]. Hamsters challenged with the SARS-CoV strain showed no clinical signs of illness including weight loss, but infection with nSARS-CoV-2 results in clinical signs of disease. Pigs and chickens are asymptomatic in infections with both SARS-CoV and nSARS-CoV-2. Collectively, although it is widely believed and scientifically proved that nSARS-CoV-2 has been introduced to humans through a rare evolutionary process called host jump, usual transmission from humans to animals or vice versa is extremely rare, and infection in food animals and birds has not been proved (Fig. 2.4). Based on the origin of the COVID-19 pandemic, it is obviously a zoonotic disease. However, the transmission of the pathogen from neither naturally infected animals (other than bats and pangolins) nor experimentally infected laboratory animals to humans (zoo anthroponosis: diseases transmitted from animal to human) has been reported, suggesting livestock, pets, zoo and wild animals, and poultry do not act as intermediate hosts or reservoirs of nSARS-CoV-2.

8 CONCLUSION

Research anchored in the recognition that the health of our planet hinges on symbiotic relationships among humans, animals, and the environment that we share and in the understanding that we are interconnected by default must gather evidence for a framework within which to interpret and apply this evidence toward preventing further pandemics or transboundary diseases. From our present level of understanding, it is quite clear that although COVID-19 originated in and was introduced to humans from animals such as bats/or pangolins, and several sporadic cases were reported to spread from infected humans to animals, the diseases are not largely transmitted from humans to animals or vice versa. Livestock, pets, poultry or birds, and zoo or wild animals have not yet been clearly proved to play major roles in the rapid spread of the COVID-19 pandemic. Human-to-human transmission through aerosol or droplet is the major means of spreading of nSARS-CoV-2 and has given the disease a pandemic status.

REFERENCES

[1] El Zowalaty ME, Järhult JD. From SARS to COVID-19: a previously unknown SARS- related coronavirus (SARS-CoV-2) of pandemic potential infecting humans - call for a One Health approach. One Health 2020;9:100124.

[2] Rahman MA, Alim MA, Hossain MM, Yeasin MG, Alam MZ, Hatta T, Anisuzzamn Tsuji N, Mondal MMH. Snail-borne trematodiasis in cattle in Barind Tract of Bangladesh, an overlooked problem. Jpn J Vet Parasitol 2019;18.

[3] Yasin MG, Alim MA, Anisuzzaman, Ahasan SA, Munsi MN, Chowdhury EH, Hatta T, Tsuji N, Mondal MMH. Trematode infections in farm animals and their vector snails in Saint Martin's Island, the southeastern offshore area of Bangladesh in the Bay of Bengal. J Vet Med Sci 2018;80(4):684–8.

[4] Anisuzzaman, Islam MK, Alim MA, Miyoshi T, Hatta T, Yamaji K, Matsumoto Y, Fujisaki K, Tsuji N. Longistatin, a plasminogen activator, is key to the availability of blood-meals for ixodid ticks. PLoS Pathog 2011;7: e1001312.

[5] Yamaji K, Tsuji N, Miyoshi T, Hatta T, Alim MA, Anisuzzaman, Kushibiki S, Fujisaki K. Hlcyst-1 and Hlcyst-2 are potential inhibitors of HlCPL-A in the midgut of the ixodid tick *Haemaphysalis longicornis*. J Vet Med Sci 2010;72:599–604.

[6] Bell D, Roberton S, Hunter PR. Animal origins of SARS coronavirus possible links with the international trade in small carnivores. Phil. Trans. R. Soc. B 2004;359: 1107–14.

[7] Bonilla-Aldana DK, Dhama K, Rodriguez-Morales AJ. Revisiting the one health approach in the context of COVID-19: a look into the ecology of this emerging disease. Adv Anim Vet Sci 2020;8(3):234–6.

[8] Shrestha K, Acharya KP, Shrestha S. One health: the interface between veterinary and human health. Int. J. One Health 2018;4:8–14.

[9] Maeda H, Hatta T, Tsubokawa D, Mikami F, Nishimaki T, Nakamura T, Anisuzzaman, Matsubayashi M, Ogawa M, da Costa CP, Tsuji N. Positive phototropism is accelerated in *Biomphalaria glabrata* snails by infection with Schistosoma mansoni. Parasitol Int 2018;67(5):609–11.

[10] OIE. One health. 2020. https://www.oie.int/en/for-the-media/onehealth/.

[11] Labony SS, Alim MA, Hasan MM, Hossain MS, Islam A, Alam MZ, Tsuji N, Anisuzzaman. Fish-borne trematode infections in wild fishes in Bangladesh. Pathog Glob Health 2020;14:1–8.

[12] Bonilla-Aldana DK, Suarez JA, Franco-Paredes C, Vilcarromero S, Mattar S, Gomez-Marin JE, Rodriguez-Morales AJ. Brazil burning! What is the potential impact of the Amazon wildfires on vector-borne and zoonotic emerging diseases? – a statement from an international experts meeting. Trav Med Infect Dis 2019; 31:101474.

[13] Frahm S, Anisuzzaman A, Prodjinotho UF, Vejzagić N, Verschoor A, Prazeres da Costa C. A novel cell-free method to culture Schistosoma mansoni from cercariae to juvenile worm stages for in vitro drug testing. PLoS Neglected Trop Dis 2019;13(1):e0006590.

[14] Dutta A, Kabir SML, Hossain MT. Occurrence of *Salmonella* and *Vibrio* species in fresh fishes collected from different markets of Mymensingh, Gazipur and Sherpur districts of Bangladesh and their characterization. Asian Aust J Food Saf Secur 2018;3(1):45–51.

[15] Sharifuzzaman, Shahiduzzaman M, Anisuzzaman, Parvez MS, Barragan A. Isolation and identification of *Cryptosporidium* from water sample in Bangladesh. Int J Nat Soc Sci 2015;2:118–24.

[16] Anisuzzaman, Hatta T, Miyoshi T, Matsubayashi M, Islam MK, Alim MA, Anas MA, Hasan MM, Matsumoto Y, Yamamoto Y, Yamamoto H, Fujisaki K, Tsuji N. Longistatin in tick saliva blocks advanced glycation end-product receptor activation. J Clin Invest 2014; 124:4429–44.

[17] Dhama K, Chakraborty S, Kapoor S, Tiwari R, Kumar A, Deb R, Rajagunalan S, Singh R, Vora K, Natesan S. One world, one health - veterinary perspectives. Adv Anim Vet Sci 2013;1(1):5–13.

[18] Alam MK, Siddique MP, Hossain MT, Rahman MA, Khan MSR. Microbiological quality of market fruits and their handling implication in relation to public health. Microb Health 2012;1(2):76–80.

[19] Anisuzzaman, Islam MK, Miyoshi T, Alim MA, Hatta T, Yamaji K, Matsumoto Y, Fujisaki K, Tsuji N. Longistatin, a novel EF-hand protein from the ixodid tick *Haemaphysalis longicornis*, is required for acquisition of host blood-meals. Int J Parasitol 2010;40:721–9.

[20] Sultana N, Hossain MT, Siddique MP, Uddin MI, Dina MA, Farhana Z. Microbial quality of dried fish of different areas of Chittagong and Mymensingh districts of Bangladesh. Int J BioRes 2010;2(8):01–5.

[21] Zinnah MA, Haque MH, Islam MT, Hossain MT, Bari MR, Babu SAM, Rahman MT, Islam MA. Drug sensitivity pattern of *Escherichia coli* isolated from samples of different biological and environmental sources. Bangladesh J Vet Med 2008;6(1):13–8.

[22] Zinnah MA, Bari MR, Islam MT, Hossain MT, Rahman MT, Haque MH, Babu SAM, Ruma RP, Islam MA. Characterization of *Escherichia coli* isolated from samples of different biological and environmental sources. Bangladesh J Vet Med 2007;5(1–2):25–32.

[23] Anisuzzaman, Tsuji N. Schistosomiasis and hookworm infection in humans: disease burden, pathobiology and anthelmintic vaccines. Parasitol Int 2020;75: 102051.

[24] Pillai VS, Krishna G, Veettil MV. Nipah virus: past outbreaks and future containment. Viruses 2020;12:465.

[25] Paul TR, Begum N, Shahiduzzaman M, Hossain MS, Labony SS, Anisuzzaman, Dey AR. Balantidiasis, a zoonotic protozoan infection, in cattle and domestic pigs. Bangladesh J Vet Med 2019;17(1):31–7.

[26] Neepa SF, Haque ZF, Sabuj AA, Islam MA, Saha S. Serological detection of avian reovirus in different poultry

flocks of Gazipur and Mymensingh districts of Bangladesh. Vet World 2019;12(7):1126–31.

[27] Sobur A, Haque ZF, Sabuj AAM, Ievy S, Rahman AAMT, El Zowalaty ME, Rahman T. Molecular detection of multidrug and colistin-resistant *Escherichia coli* isolated from house flies in various environmental settings. Future Microbiol 2019;14(10):847–58.

[28] Pondit A, Haque ZF, Sabuj AAM, Khan MSR, Saha S. Characterization of *Staphylococcus aureus* isolated from chicken and quail eggshell. J Adv Vet Anim Res 2018; 5(4):466–71.

[29] Islam R, Anisuzzaman, Hossain MS, Alam Z, Islam A, Khan AHNA, Kabir ME, Hatta T, Alim A, Tsuji N. Linguatula serrata, a food-borne zoonotic parasite, in livestock in Bangladesh: some pathologic and epidemiologic aspects. Vet Parasitol Reg Stud Rep 2018;13:135–40.

[30] Seyedmousavi S, Bosco S, de Hoog S, Ebel F, Elad D, Gomes RR, Jacobsen ID, Jensen HE, Martel A, Mignon B, Pasmans F, Piecková E, Rodrigues AM, Singh K, Vicente VA, Wibbelt G, Wiederhold NP, Guillot J. Fungal infections in animals: a patchwork of different situations. Med Mycol 2018;56:165–87.

[31] Ahmed R, Rafiquzaman SM, Hossain MT, Lee JM, Kong IS. Species-specific detection of *Vibrio alginolyticus* in shellfish and shrimp by real-time PCR using the *groEL* gene. Aquacult Int 2016;24:157–70.

[32] Seyedmousavi S, Guillot J, Tolooe A, Verweij PE, Hoog GS. Neglected fungal zoonoses: hidden threats to man and animals. Clin Microbiol Infect 2015;21: 416–25.

[33] Hossain MT, Kim YR, Kong IS. PCR–restriction fragment length polymorphism analysis using groEL gene to differentiate pathogenic *Vibrio* species. Diagn Microbiol Infect Dis 2014;78:9–11.

[34] Hossain MT, Kim YO, Kong IS. Multiplex PCR for the detection and differentiation of *Vibrio parahaemolyticus* strains using the *groEL, tdh* and *trh* genes. Mol Cell Probes 2013;27:171–5.

[35] Hossain MT, Kim EY, Kim YR, Kim DG, Kong IS. Development of a groEL gene-based species-specific multiplex polymerase chain reaction assay for simultaneous detection of *Vibrio cholerae, Vibrio parahaemolyticus* and *Vibrio vulnificus*. J Appl Microbiol 2013;114:448–56.

[36] Hossain MT, Kim YR, Kim EY, Lee JM, Kong IS. Detection of *Vibrio cholerae* and *Vibrio vulnificus* by duplex PCR using *groEL* gene. Fish Sci 2013;79:335–40.

[37] Kim YR, Kim EY, Kim DG, Kim YO, Hossain MT, Kong IS. DNA array with the *groESL* intergenic sequence to detect *Vibrio parahaemolyticus* and *Vibrio vulnificus*. Anal Biochem 2012;424:32–4.

[38] Hossain MT, Kim EY, Kim YR, Kim DG, Kong IS. Application of *groEL* gene for the species-specific detection of *Vibrio parahaemolyticus* by PCR. Lett Appl Microbiol 2012;54:67–72.

[39] Heo WS, Kim EY, Kim YR, Hossain MT, Kong IS. Salt effect of nisin Z isolated from a marine fish on the growth inhibition of *Streptococcus iniae*, a pathogen of streptococcosis. Biotechnol Lett 2012;34:315–20.

[40] Sarker MAH, Jahan M, Parvin MN, Malek MA, Hossain MT. Identification of bacterial flora isolated from apparently healthy water birds of Dhaka Zoo of Bangladesh. Bangladesh J Vet Med 2012;10(1–2):21–6.

[41] Islam M, Rahman MM, Khan MSR, Akhter J, Hossain MT, Siddique MP. Identification and characterization of bacteria isolated from respiratory tract of man. Int J BioRes 2010;1(6):27–30.

[42] Dutta UK, Hossain MT, Das M, Siddique MP, Sarker CK, Rana M. Identification of microflora isolated from apparently healthy camels in Bangladesh. Int J BioRes 2010;2(7):01–6.

[43] Akhter J, Hossain MT, Islam MT, Siddique MP, Islam MA. Isolation and identification of microflora from apparently healthy caged parrots of Dhaka zoo of Bangladesh. Bangladesh J Vet Med 2010;8(1):05–10.

[44] Nooruddin GM, Hossain MT, Mohammad M, Rahman MM. Sero-epidemiology of avian influenza virus in native chicken in Bangladesh. Int J Poultry Sci 2006;5(11):1029–33. M Rahamatunnabi, MR Ali, MA Islam and Hossain MT (2005) Isolation and characterization of bacteria from keratoconjunctivitis affected cattle with their pathogrnicity and *in vitro* antibiotic sensitivity. Bangladesh Journal of Veterinary Medicine 3(1): 25-31.

[45] Collinge J. Prion diseases of humans and animals: their causes and molecular basis. Annu Rev Neurosci 2001; 24:519–50.

[46] Ellwanger JH, Chies JAB. Zoonotic spillover and emerging viral diseases - time to intensify zoonoses surveillance in Brazil. Braz J Infect Dis 2018;22(1):76–8.

[47] Dhama K, Sharun K, Tiwari R, Sircar S, Bhat S, Malik YS, Singh KP, Chaicumpa W, Bonilla-Aldana DK, Rodriguez-Morales AJ. Coronavirus disease 2019 – COVID-19. Preprints. 2020030001. 2020.

[48] Rodriguez-Morales AJ, Bonilla-Aldana DK, Balbin-Ramon GJ, Paniz-Mondolfi A, Rabaan A, Sah R, Esposito S. History is repeating itself, a probable zoonotic spillover as a cause of an epidemic: the case of 2019 novel coronavirus. Infezioni Med Le 2020;28(1): 3–5.

[49] https://www.who.int/emergencies/diseases/novel-coronavirus-2019/events-as-they-happen.

[50] World Health Organisation (WHO). WHO announces COVID-19 outbreak a pandemic. 2020. https://www.euro.who.int/en/health-topics/health-emergencies/coronavirus-covid-19/news/news/2020/3/who-announces-covid-19-outbreak-a-pandemic.

[51] Morawska L, Cao J. Airborne transmission of SARS-CoV-2: the world should face the reality. Environ Int 2020; 139:105730.

[52] Andersen KG, Rambaut A, Lipkin WI, Holmes EC, Garry RF. The proximal origin of SARS-CoV-2. Nat Med 2020;26:450–2.

[53] Mollentze N, Streicker DG. Viral zoonotic risk is homogenous among taxonomic orders of mammalian and avian reservoir hosts. Proc Natl Acad Sci U S A 2020; 117(17):9423–30.

[54] Olival KJ, Hosseini PR, Zambrana-Torrelio C, Ross N, Bogich TL, Daszak P. Host and viral traits predict zoonotic spillover from mammals. Nature 2017;546: 646−50.

[55] Letko M, Seifert SN, Olival KJ, Plowright RK, Munstar VJ. Bat-borne virus diversity, spillover and emergence. Nat Rev Microbiol 2020;18:461−71.

[56] Ohio State University. Evolutionary history of SARS supports bats as virus source. ScienceDaily ScienceDaily February 20, 2008. https://www.sciencedaily.com/releases/2008/02/080219150146.htm.

[57] Wang LF, Eaton BT. Bats, civets and the emergence of SARS. In: Childs JE, Mackenzie JS, Richt JA, editors. Wildlife and emerging zoonotic diseases: the biology, circumstances and consequences of cross-species transmission. Current topics in microbiology and immunology, vol. 315; 2007. p. 325−44.

[58] Li W, Wong SK, Li F, Kuhn JH, Huang IC, Choe H, Farzan M. Animal origins of the severe acute respiratory syndrome coronavirus: insight from ACE2-S-protein interactions. J Virol 2006;80(9):4211−9.

[59] Kan B, Wang M, Jing H, Xu H, Jiang X, Yan M, Liang W, Zheng H, Wan K, Liu Q, Cui B, Xu Y, Zhang E, Wang H, Ye J, Li G, Li M, Cui Z, Qi X, Chen K, Du L, Gao K, Zhao YT, Zou XZ, Feng YJ, Gao YF, Hai R, Yu D, Guan Y, Xu J. Molecular evolution analysis and geographic investigation of severe acute respiratory syndrome coronavirus-like virus in palm civets at an animal market and on farms. J Virol 2005;79(18):11892−900.

[60] Ries J. Here's how COVID-19 compares to past outbreaks. Healthline; 2020. https://www.healthline.com/health-news/how-deadly-is-the-coronavirus-compared-to-past-outbreaks.

[61] Watts J. Report details lessons from SARS outbreak. Lancet 2003;362(9391):1207.

[62] Hung LS. The SARS epidemic in Hong Kong: what lessons have we learned? J R Soc Med 2003;96(8): 374−8.

[63] World Health Organization (WHO). Summary of probable SARS cases with onset of illness from 1 November 2002 to 31 July 2003. 2003. https://www.who.int/csr/sars/country/table2004_04_21/en/.

[64] World Health Organization (WHO). Middle East respiratory syndrome coronavirus (MERS-CoV). 2019. https://www.who.int/emergencies/mers-cov/en/.

[65] Rahman M and Chakraborty A. Nipah virus outbreaks in Bangladesh: a deadly infectious disease. WHO South-East Asia J Public Health 2012;1:208−12.

[66] Banerjee S, Gupta N, Kodan P, Mittal A, Ray Y, Nischal N, Soneja M, Biswas A, Wig N. Nipah virus disease: a rare and intractable disease. Intract Rare Dis Res 2019;8:1−8.

[67] Luby SP, Rahman M, Hossain MJ, Blum LS, Husain MM, Gurley E, Khan R, Ahmed BN, Rahman S, Nahar N, Kenah E, Comer JA, Ksiazek and TG. Foodborne transmission of Nipah virus, Bangladesh. Emerg Infect Dis 2006;12(12):1888−94.

[68] Centers for disease control and prevention update: outbreak of Nipah virus–Malaysia and Singapore. MMWR Morb Mortal Wkly Rep 1999;48:335−7.

[69] Centers for Diseases Control and Prevention (CDC). Zoonotic diseases. 2017. https://www.cdc.gov/onehealth/basics/zoonotic-diseases.html.

[70] Salyer SJ, Silver R, Simone K, Behravesh CB. Prioritizing zoonoses for global health capacity building-themes from one health zoonotic disease workshops in 7 countries, 2014-2016. Emerg Infect Dis 2017;23(13): S55−64.

[71] Jones KE, Patel NG, Levy MA, Storeygard A, Balk D, Gittleman JL, et al. Global trends in emerging infectious diseases. Nature 2008;451:990−3. https://doi.org/10.1038/nature06536.

[72] Woolhouse ME, Gowtage-Sequeria S. Host range and emerging and reemerging pathogens. Emerg Infect Dis 2005;11:1842−7.

[73] Castro J. Bats host more than 60 human-infecting viruses. Livescience; 2013. https://www.livescience.com/26898-bats-host-human-infecting-viruses.html.

[74] Chomel BB, Stuckey MJ, Boulouis HJ, Aguilar-Setién A. Bat-related zoonoses. In: Zoonoses-infections affecting humans and animals: focus on public health aspects; 2015. p. 697−714.

[75] Wong S, Lau S, Woo P, Yuen KY. Bats as a continuing source of emerging infections in humans. Rev Med Virol 2007;17:67−91.

[76] Hayman DTS, Bowen RA, Cryan PM, McCracken GF, O'Shea TJ, Peel1 AJ, Gilbert A, Webb CT, Wood JLN. Ecology of zoonotic infectious diseases in bats: current knowledge and future directions. Zoonoses Public Health 2013;60:2−21.

[77] Rahman MA, Hossain MG, Singh AC, Islam MS, Islam MA. A retrospective analysis of influence of environmental/air temperature and relative humidity on Sars-CoV-2 outbreak. J Pure Appl Microbiol 2020; 14(3):1705−14.

[78] Lam TTY, Jia N, Zhang YW, Shum MHH, Jiang JF, Zhu HC, et al. Identifying SARS-CoV-2-related coronaviruses in Malayan pangolins. Nature 2020:1−4.

[79] Yeşilbağ K, Aytoğu G. Coronavirus host divergence and novel coronavirus (Sars-CoV-2) outbreak. Clin Exp Ocul Trauma Infect 2020;2(1):6−14.

[80] Zheng J. SARS-CoV-2: an emerging coronavirus that causes a global threat. Int J Biol Sci 2020;16(10): 1678−85.

[81] Liu P, Chen W, Chen JP. Viral metagenomics revealed sendai virus and coronavirus infection of malayan pangolins (*Manis javanica*). Viruses 2019;11:979.

[82] Zhou P, Yang XL, Wang XG, Hu B, Zhang L, Zhang W, Si HR, Zhu Y, Li B, Huang CL, Chen HD, Chen J, Luo Y, Guo H, Jiang RD, Liu MQ, Chen Y, Shen XR, Wang X, Zheng XS, Zhao K, Chen QJ, Deng F, Liu LL, Yan B, Zhan FX, Wang YY, Xiao GF, Shi ZL. A pneumonia outbreak associated with a new coronavirus of probable bat origin. Nature 2020;579:270−3.

[83] Decaro N, Martella V, Saif LJ, Buonavoglia C. COVID-19 from veterinary medicine and one health perspectives: what animal coronaviruses have taught us. Res Vet Sci 2020;131:21–3.
[84] Xia X. Extreme genomic CpG deficiency in SARS-CoV-2 and evasion of host antiviral defense. Mol Biol Evol 2020;37(9):2699–705.
[85] Kazi TA, Mitra S, Mukhopadhyay BC, Bhattacharya R, Mandal S, Biswas SR. COVID-19: probing the possibility of transmission in animal kingdom. Preprints, 2020050006. 2020.
[86] Definition of corona by Merriam-Webster. Merriam-Webster. Archived from the original on 2020-03-24. Retrieved 2020-03-24.
[87] Almeida JD, Berry DM, Cunningham CH, Hamre D, Hofstad MS, Mallucci L, McIntosh K, Tyrrell DA. Virology: coronaviruses. Nature 1968;220(5168):650.
[88] Sturman LS, Holmes KV, Lauffer MA, Maramorosch K. The molecular biology of coronaviruses. Adv Virus Res 1983;28:35–112.
[89] "ICTV Taxonomy history: Orthocoronavirinae". International committee on Taxonomy of viruses (ICTV). Retrieved 2020-01-24.
[90] Fan Y, Zhao K, Shi ZL, Zhou P. Bat coronaviruses in China. Viruses 2019;11(3):210.
[91] Paim FC, Bowman AS, Miller L, Feehan BJ, Marthaler D, Saif LJ, Vlasova AN. Epidemiology of deltacoronaviruses (δ-CoV) and gammacoronaviruses (γ-CoV) in wild birds in the United States. Viruses 2019;11:897.
[92] Carstens EB. Ratification vote on taxonomic proposals to the international committee on taxonomy of viruses (2009). Arch Virol 2010;155(1):133–46.
[93] International Committee on Taxonomy of Viruses (ICTV). talk.ictvonline.org. Retrieved 2020-09-14.
[94] Gui M, Liu X, Guo D, Zhang Z, Yin CC, Chen Y, Xiang Y. Electron microscopy studies of the coronavirus ribonucleoprotein complex. Protein Cell 2017;8(3):219–24.
[95] Paul S. The molecular biology of coronaviruses. Adv Virus Res 2006;66:193–292.
[96] Lalchhandama K. The chronicles of coronaviruses: the electron microscope, the doughnut, and the spike. Sci Vis 2020;20(2):78–92.
[97] Lai MM, Cavanagh D. The molecular biology of coronaviruses. Adv Virus Res 1997;48:1–100.
[98] de Groot RJ, Baker SC, Baric R, Enjuanes L, Gorbalenya AE, Holmes KV, Perlman S, Poon L, Rottier PJ, Talbot PJ, Woo PC, Ziebuhr J. FamilyCoronaviridae. In: King AM, Lefkowitz E, Adams MJ, Carstens EB, International Committee on Taxonomy of Viruses, International Union of Microbiological Societies, editors. Virology division. Oxford: Ninth Report of the International Committee on Taxonomy of Viruses; 2011. p. 806–28.
[99] Antonina N, Agnieszka D, Artur S, Aleksandra, Jasik KP, Pyrc K. Membrane protein of human coronavirus NL63 is responsible for interaction with the adhesion receptor. J Virol 2019;93(19).
[100] Tyrrell DAJ, Myint SH. Coronaviruses. In: Baron S, editor. Medical microbiology. 4th ed. Galveston (TX): University of Texas Medical Branch at Galveston; 1996 [Chapter 60] https://www.ncbi.nlm.nih.gov/books/NBK7782/.
[101] Yi Y, Lagniton P, Ye S, Li E, Xu RH. COVID-19: what has been learned and to be learned about the novel coronavirus disease. Int J Biol Sci 2020;16(10):1753–66.
[102] de Wit E, van Doremalen N, Falzarano D, Munster VJ. SARS and MERS: recent insights into emerging coronaviruses. Nat Rev Microbiol 2016;14:523–34.
[103] Liu P, Shi L, Zhang W, He J, Liu C, Zhao C, Kong SK, Loo JFC, Gu D, Hu L. Prevalence and genetic diversity analysis of human coronaviruses among cross-border children. Virol J 2017;14(1):230.
[104] Forgie S, Marrie TJ. Healthcare-associated atypical pneumonia. Semin Respir Crit Care Med 2009;30(1):67–85.
[105] Coronavirus-Wikipedia. https://en.wikipedia.org/wiki/Coronavirus.
[106] Monto AS, Cowling BJ, Peiris J. Coronaviruses. Viral Infect Hum Epidemiol Control 2014:199–223.
[107] Liu DX, Liang JQ, Fung TS. Human coronavirus-229e, -OC43, -NL63, and -HKU1. Ref Module Sci 2020. B978-0-12-809633-8.21501-X.
[108] Saif LJ, Jung K. Comparative pathogenesis of bovine and porcine respiratory coronaviruses in the animal host species and SARS-CoV-2 in humans. J Clin Microbiol 2020;58(8). e01355-20.
[109] Cavanagh D. Coronavirus avian infectious bronchitis virus. Vet Res 2007;38:281–97.
[110] Hernandez M, Abad D, Eiros JM, Rodriguez-Lazaro D. Are animals a neglected transmission route of SARA-CoV-2? Pathogens 2020;9:480–4.
[111] Gonultas S, Karabagli M, Bastug Y, Cilesiz NC, Kadioglu A. COVID-19 and animals: what do we know? Turk J Urol 2020;46:249–52.
[112] WHO. Coronavirus disease (COVID-19): how is it transmitted?. 2020. https://www.who.int/news-room/q-a-detail/coronavirus-disease-covid-19-how-is-it-transmitted.
[113] How does coronavirus spread? WebMD. https://www.webmd.com/lung/coronavirus-transmission-overview#2.
[114] Tang A, Tong Z, Wang H, Dai Y, Li K, Liu J, Wu W, Yuan C, Yu M, Li P, Yan J. Detection of novel coronavirus by RT-PCR in stool specimen from asymptomatic child, China. Emerg Infect Dis 2020;26(6):1337–9.
[115] AFCD. Detection of low level of COVID-19 virus in pet dog. 2020. https://www.afcd.gov.hk/english/publications/publications_press/pr2335.html.
[116] The Government of the Hong Kong Special Administrative Region - Press Releases. Pet dog tests positive for COVID-19 virus; 2020. https://www.info.gov.hk/gia/general/202003/19/P2020031900606.htm.
[117] The Brussels Times. Coronavirus: Belgian woman infected her cat. March 27 , 2020. https://www.brusselstimes.com/news/belgium-all-news/103003/coronavirus-belgian-woman-infected-her-cat/.

[118] Animal and Plant Health Inspection Service. USDA statement on the confirmation of COVID-19 in a Tiger in New York. May 13, 2020. https://www.aphis.usda.gov/aphis/newsroom/news/sa_by_date/sa-2020/ny-zoo-covid-19.
[119] Kim YI, Kim SG, Kim SM, Kim EH, Park SJ, Yu KM, Chang JH, et al. Infection and rapid transmission of SARS-CoV-2 in ferrets. Cell Host Microbe 2020;27:704–9.
[120] Lu R, Zhao X, Li J, Niu P, Yang B, Wu H, et al. Genomic characterisation and epidemiology of 2019 novel coronavirus: implications for virus origins and receptor binding. Lancet 2020;395:565–74.
[121] Martina BE, Haagmans BL, Kuiken T, Fouchier RAM, Rimmelzwaan GF, Van Amerongen G, et al. Virology: SARS virus infection of cats and ferrets. Nature 2003;425:915.
[122] Calisher CH, Childs JE, Field HE, Holmes KV, Schountz T. Bats: important reservoir hosts of emerging viruses. Clin Microbiol Rev 2006;19(3):531–45.
[123] Ji W, Wang W, Zhao X, Zai J, Li X. Cross-species transmission of the newly identified coronavirus 2019-nCoV. J Med Virol 2020;92(4):433–40.
[124] Li X, Song Y, Wong G, Cui J. Bat origin of a new human coronavirus: there and back again. Sci China Life Sci 2020;63(3):461–2.
[125] Salata C, Calistri A, Parolin C, Palu G. Coronaviruses: a paradigm of new emerging zoonotic diseases. Pathog Dis 2020;77(9):ftaa006.
[126] McIntosh K. Coronaviruses: a comparative review. In: Arber W, Haas R, Henle W, Hofschneider PH, Jerne NK, Koldovský P, Koprowski H, Maaløe O, Rott R, editors. Current Topics in Microbiology and Immunology/Ergebnisse der Mikrobiologie und Immunitätsforschung; 1974. p. 87.
[127] Dhama K, Patel SK, Pathak M, Yatoo MI, Tiwari R, Malik YS, Singh R, Sah R, Rabaan AA, Bonilla-Aldana DK, et al. An update on SARS-COV-2/COVID-19 with particular reference on its clinical pathology, pathogenesis, immunopathology and mitigation strategies – a review [Preprints] Travel Med Infect Dis 2020;37.
[128] Tiwari R, Dhama K, Sharun K, Yatoo MI, Malik YS, Singh R, Michalak I, Sah R, Bonilla-Aldana DK, Rodriguez-Morales AJ. COVID-19: animals, veterinary and zoonotic links. Vet Q 2020;40(1):169–82.
[129] Chapter 24: Coronaviridae. In: Fenner's veterinary virology. 5th ed. Academic Press; 2017. p. 435–61.
[130] Murphy FA, Gibbs EP, Horzinek MCand Studdart MJ. Veterinary virology. Boston: Academic Press; 1999. p. 495–508.
[131] Tirotta E, Carbajal KS, Schaumburg CS, Whitman L, Lane TE. Cell replacement therapies to promote remyelination in a viral model of demyelination. J Neuroimmunol 2010;224(1–2):101–7.
[132] Decaroa Nand Lorusso A. Novel human coronavirus (SARS-CoV-2): a lesson from animal coronaviruses. Vet Microbiol 2020;244:108693.
[133] Kasmi Y, Khataby K, Souiri A, Ennaji MM. Chapter 7: Coronaviridae: 100,000 years of emergence and reemergence, emerging and reemerging viral pathogens: fundamental and basic virology aspects of human. Anim Plant Pathog 2020;1:127–49.
[134] Almendros A. Can companion animals become infected with Covid-19? Vet Rec 2020;186(12).
[135] Mora-Díaz JC, Piñeyro PE, Houston E, Zimmerman J, Giménez-Lirola LG. Porcine hemagglutinating encephalomyelitis virus: a review. Front Vet Sci 2019;6:53.
[136] Wang Q, Vlasova AN, Kenney SP, Saif LJ. Emerging and re-emerging coronaviruses in pigs. Curr Opin Virol 2019;34:39–49.
[137] Hou Y, Wang Q. Emerging highly virulent porcine epidemic diarrhea virus: molecular mechanisms of attenuation and rational design of live attenuated vaccines. Int J Mol Sci 2019;20:E5478.
[138] Corman VM, Muth D, Niemeyer D, Drosten C. Hosts and sources of endemic human coronaviruses. Adv Virus Res 2018;100:163–88.
[139] Amer HM. Bovine-like coronaviruses in domestic and wild ruminants. Anim Health Res Rev 2018;19:113–24.
[140] Decaro N, Buonavoglia C. An update on canine coronaviruses: viral evolution and pathobiology. Vet Microbiol 2008;132:221–34.
[141] Merck Veterinary Manual. Merck veterinary manual. Retrieved 2020-06-08.
[142] Shrivastava N, Baxi P, Ratho RK, Saxena SK. Global trends in epidemiology of coronavirus disease 2019 (COVID-19). In: Saxena S, editor. Coronavirus disease 2019 (COVID-19). Medical virology: from pathogenesis to disease Control; 2020. p. 9–21.
[143] Autieri CR, Miller CL, Scott KE, Kilgore A, Papscoe VA, Garner MM, Haupt JL, Bakthavatchalu V, Muthupalani S, Fox JG. Systemic coronaviral disease in 5 ferrets. Comp Med 2015;65(6):508–16.
[144] Pusterla N, Vin R, Leutenegger CM, Mittel LD, Divers TJ. Enteric coronavirus infection in adult horses. Vet J 2018;231:13–8.
[145] Mihindukulasuriya KA, Wu G, St Leger J, Nordhausen RW, Wang D. Identification of a novel coronavirus from a beluga whale by using a panviral microarray. J Virol 2008;82(10):5084–8.
[146] Woo PC, Lau SK, Lam CS, Tsang AK, Hui SW, Fan RY, Martelli P, Yuen KY. Discovery of a novel bottlenose dolphin coronavirus reveals a distinct species of marine mammal coronavirus in Gammacoronavirus. J Virol 2014;88:1318–31.
[147] Chung JY, Kim HR, Bae YC, Lee OS, Oem JK. Detection and characterization of bovine-like coronaviruses from four species of zoo ruminants. Vet Microbiol 2011;148:396–401.
[148] Alekseev KP, Vlasova AN, Jung K, Hasoksuz M, Zhang X, Halpin R, Wang S, Ghedin E, Spiro D, Saif LJ. Bovine-like coronaviruses isolated from four species of captive wild ruminants are homologous to bovine coronaviruses, based on complete genomic sequences. J Virol 2008;82:12422–31.

[149] Hasoksuz M, Alekseev K, Vlasova A, Zhang X, Spiro D, Halpin R, Wang S, Ghedin E, Saif LJ. Biologic, antigenic, and full-length genomic characterization of a bovine-like coronavirus isolated from a giraffe. J Virol 2007; 81:4981–90.

[150] Rosa GM, Santos N, Grøndahl-Rosado R, Fonseca FP, Tavares L, Neto I, Cartaxeiro C, Duarte A. Unveiling patterns of viral pathogen infection in free-ranging carnivores of northern Portugal using a complementary methodological approach. Comp Immunol Microbiol Infect Dis 2020;69:101432.

[151] Alfano F, Dowgier G, Valentino MP, Galiero G, Tinelli A, Nicola D, Fusco G. Dentification of pantropic canine coronavirus in a wolf (*Canis lupus* italicus) in Italy. J Wildl Dis 2019;55:504–8.

[152] Goller KV, Fickel J, Hofer H, Beier S, East ML. Coronavirus genotype diversity and prevalence of infection in wild carnivores in the Serengeti National Park, Tanzania. Arch Virol 2013;158:729–34.

[153] Stephenson N, Swift P, Moeller RB, Worth SJ, Foley J. Feline infectious peritonitis in a mountain lion (Puma concolor), California, USA. J Wildl Dis 2013;49: 408–12.

[154] Kennedy M, Citino S, Dolorico T, McNabb AH, Moffat AS, Kania S. Detection of feline coronavirus infection in captive cheetahs (*Acinonyx jubatus*) by polymerase chain reaction. J Zoo Wildl Med 2001;32:25–30.

[155] Juan-Salles C, Domingo M, Herraez P, Fernandez A, Fernandez J. An outbreak of feline infectious peritonitis in captive servals (*Felis serval*): clinical, pathological, and immunohistochemical findings. In: Baer CK, editor. Proceedings of the American association of zoo veterinarians; 1997. p. 224–6.

[156] Watt NJ, MacIntyre NJ, McOrist S. An extended outbreak of infectious peritonitis in a closed colony of European wildcats (*Felis silvestris*). J Comp Pathol 1993;108:73–9.

[157] Dong BQ, Liu W, Fan XH, Vijaykrishna D, Tang XC, Gao F, Li LF, Li GJ, Zhang JX, Peiris JSM, Smith GJD, Chen H, Guan Y. Detection of a novel and highly divergent coronavirus from Asian leopard cats and Chinese ferret badgers in southern China. J Virol 2007;81: 6920–6.

[158] Suryaman GK, Soejoedono RD, Setiyono A, Poetri ON, Handharyani E. Isolation and characterization of avian coronavirus from healthy Eclectus parrots (*Eclectus roratus*) from Indonesia. Vet World 2019;12:1797–805.

[159] Wille M, Muradrasoli S, Nilsson A, Järhult JD. High prevalence and putative lineage maintenance of avian coronaviruses in Scandinavian waterfowl. PLoS One 2016;11:e01510198.

[160] Bande F, Arshad SS, Omar AR. Isolation and metagenomic identification of avian leukosis virus associated with mortality in broiler chicken. Adv Virol 2016; 2016:9058403.

[161] Papineau A, Berhane Y, Wylie TN, Wylie KM, Sharpe S, Lung O. Genome organization of Canada goose coronavirus, a novel species identified in a mass die-off of Canada geese. Sci Rep 2019;9:5954.

[162] Zhuang QY, Wang KC, Liu S, Hou GY, Jiang WM, Wang SC, Li JP, Yu JM, Chen JM. Genomic analysis and surveillance of the coronavirus dominant in ducks in China. PLoS One 2015;10:e0129256.

[163] Zhou H, Chen X, Hu T, Li J, Song H, Liu Y, Wang P, Liu D, Yang J, Holmes EC, Hughes AC, Bi Y, Shi W. A novel bat coronavirus closely related to SARS-CoV-2 contains natural insertions at the S1/S2 cleavage site of the spike protein. Curr Biol 2020;30(19):3896.

[164] Malik YS, Sircar S, Bhat S, Sharun K, Dhama K, Dadar M, Tiwari R, Chaicumpa W. Emerging novel Coronavirus (2019-nCoV)-current scenario, evolutionary perspective based on genome analysis and recent developments. Vet Q 2020;40(1):68–76.

[165] Chan JF, Kok KH, Zhu Z, Chu H, To KK, Yuan S, Yuen KY. Genomic characterization of the 2019 novel human-pathogenic coronavirus isolated from a patient with atypical pneumonia after visiting Wuhan. Emerg Microb Infect 2020;9(1):221–36.

[166] Hu D, Zhu C, Ai L, He T, Wang Y, Ye F, Yang L, Ding C, Zhu X, Lv R, et al. Genomic characterization and infectivity of a novel SARS-like coronavirus in Chinese bats. Emerg Microb Infect 2018;7(1):154.

[167] Hu B, Zeng LP, Yang XL, Ge XY, Zhang W, Li B, Xie JZ, Shen XR, Zhang YZ, Wang N, Luo DS, et al. Discovery of a rich gene pool of bat SARS-related coronaviruses provides new insights into the origin of SARS coronavirus. PLoS Pathog 2017;13(11):e1006698.

[168] Drexler JF, Corman VM, Drosten C. Ecology, evolution and classification of bat coronaviruses in the aftermath of SARS. Antivir Res 2014;101:45–56.

[169] Ren LL, Wang YM, Wu ZQ, Xiang ZC, Guo L, Xu T, Jiang YZ, Xiong Y, Li YJ, Li H, et al. Identification of a novel coronavirus causing severe pneumonia in human: a descriptive study. Chinese Med J 2020;133(9): 1015–24.

[170] Ramadan N, Shaib H. Middle East respiratory syndrome coronavirus (MERS-CoV): a review. Germs 2019;9(1): 35–42.

[171] Mohd HA, Al-Tawfiq JA, Memish ZA. Middle East Respiratory Syndrome Coronavirus (MERS-CoV) origin and animal reservoir. Virol J 2016;13:87.

[172] Wu A, Peng Y, Huang B, Ding X, Wang X, Niu P, Meng J, Zhu Z, Zhang Z, Wang J, et al. Genome composition and divergence of the novel coronavirus (2019- nCoV) originating in China. Cell Host Microbe 2020;27(3):325–8.

[173] Rossi GA, Sacco O, Mancino E, Cristiani L, Midulla F. Differences and similarities between SARS-CoV and SARS-CoV-2: spike receptor-binding domain recognition and host cell infection with support of cellular serine proteases. Infection 2020;48:665–9.

[174] Petersen E, Koopmans M, Go U, Hamer DH, Petrosillo N, Castelli F, Storgaard M, Khalili SA, Simonsen L. Comparing SARS-CoV-2 with SARS-CoV and influenza pandemics. Lancet Infect Dis 2020;20: e238–44.

[175] Lau S, Luk H, Wong A, Li K, Zhu L, He Z, Fung J, Chan T, Fung K, Woo P. Possible bat origin of severe acute

respiratory syndrome coronavirus 2. Emerg Infect Dis 2020;26(7):1542–7.
[176] Shabir O. How does the SARS virus genome compare to other viruses?. 2020. https://www.news-medical.net/health/How-Does-the-SARS-Virus-Genome-Compare-to-Other-Viruses.aspx.
[177] Yoo HS, Yoo D. COVID-19 and veterinarians for one health, zoonotic- and reverse-zoonotic transmissions. J Vet Sci 2020;21(3):e51.
[178] World Health Organization (WHO). WHO Statement regarding cluster of pneumonia cases in Wuhan, China. January 9, 2020. https://www.who.int/china/news/detail/09-01-2020-who-statement-regarding-cluster-of-pneumonia-cases-in-wuhan-china.
[179] World Health Organization (WHO). Laboratory testing of human suspected cases of novel coronavirus (nCoV) infection: interim guidance. 2020. https://apps.who.int/iris/handle/10665/330374.
[180] Dhama K, Patel SK, Sharun K. SARS-CoV-2 jumping the species barrier: zoonotic lessons from SARS, MERS and recent advances to combat this pandemic virus. Trav Med Infect Dis 2020;37.
[181] COVID-19 dashboard by the center for systems science and engineering (CSSE) at Johns Hopkins University (JHU). ArcGIS. Johns Hopkins University. Retrieved 2020-11-02. https://gisanddata.maps.arcgis.com/apps/opsdashboard/index.html#/bda7594740fd40299423467b48e9ecf6.
[182] Sarkar J, Guha R. Infectivity, virulence, pathogenicity, host-pathogen interactions of SARS and SARS-CoV-2 in experimental animals: a systematic review. Vet Res Commun 2020;10:1–10.

CHAPTER 3

Vaccine Development Through Reverse Vaccinology Using Artificial Intelligence and Machine Learning Approach

SWARNAV BHAKTA • SUVENDU CHOUDHURY • JOYDEEP PAUL • ARIJIT BHATTACHARYA

1 INTRODUCTION

Vaccinology involves science and technologic innovations to accomplish rapid development of safe and effective vaccines. The basics involve immunogens, the host immune response, delivery strategies and technologies, manufacturing, and above all laboratory and clinical evaluation. Alongside, safety, regulatory, ethical, and economic considerations are imperative for vaccine development and utilization. Historically, the vaccine research delved strategies for immunization against infectious agents, which include virus, pathogenic bacteria, protozoa, and fungi [1]. Scientific improvements in the past few decades include expression of genes and their delivery systems, introduction of synthetic adjuvants, development of nanoparticles [2] that help in maneuvering novel vaccine strategies targeting identification and expression of novel antigens, development of chimeric antibodies, and structural alterations of existing proteins. Barring infectious agents, vaccine development can also be exploited in controlling noninfectious disease such as cancer and developmental and neurodegenerative disorders. Veterinary vaccines are also of primary importance in pathogenic disease spread because of the serious involvement of various livestock and wild animals as potential zoonotic reservoirs for emerging pathogens [1,3].

There are around 1500 different known human pathogens; with more pathogens being discovered every year, the number of available vaccines is merely 23 and 24 more are in the pipeline (Table 3.1). The primary reason for such a paltry ratio is the plethora of challenges that a successful discovery of vaccine triumphs over. Over the years, the technology for vaccine development moved miles from the empirical "isolate, inactivate, and inject" paradigm, which involved the disease-causing entity to get isolated, inactivated, and properly administered to elicit a robust primary immune response and memory [4]. Series of effective vaccines have been developed through this strategy against pathogens that cause diseases such as diphtheria, tetanus, pertussis, and influenza [5]. A second approach involves administration of live-attenuated pathogens for the elimination of the potential disease-causing pathogens, which has also been quite successful against *Mycobacterium tuberculosis*, mumps, measles, and small pox [6]. Combination vaccines, evolved recently, attained the goal of conferring protection against more than one disease with a single composition [7]. This ensures fewer shots, minimizing the side effects and lowering cost. Albeit its tremendous impact on human health and survival, effective vaccine development against a number of pathogens such as *Plasmodium*, *Leishmania*, HIV, hookworm, dengue virus, and hepatitis C virus (HCV) remains unachieved owing to the existing impediments for vaccine discovery [7,8]. The key barriers can be summarized as follows

- i. generic variability of host and accumulation of variations in pathogens;
- ii. zoonotic pathogens and their evolution;
- iii. various environmental, physiologic, and geographic factors;
- iv. coinfection, particularly in immunocompromised conditions;
- v. lack of knowledge of systems immunology;
- vi. safety issue associated with novel vaccines.

To overcome such barriers, immunologic knowledge discovery has been amalgamated with computational biology to augment various strategies of which *reverse vaccinology* (RV) appears to be the most promising and systematic [9]. RV is a process in which vaccines are

COVID-19: Tackling Global Pandemics through Scientific and Social Tools. https://doi.org/10.1016/B978-0-323-85844-1.00006-4

TABLE 3.1
List of Approved Vaccines and Vaccine Candidates in Pipeline.

Available	Pipeline	RV-Implemented
Cholera	*Campylobacter jejuni*	GBS
Dengue	Chagas disease	Respiratory syncytial virus
Diphtheria	Chikungunya	HSV
Hepatitis	Dengue	Human cytomegalovirus
Haemophilus influenzae type b (Hib)	Enterotoxigenic *Escherichia coli*	*Bacillus anthracis*
Human papillomavirus	Enterovirus 71 (EV71)	*Streptococcus pneumoniae*
Influenza	GBS	*Mycobacterium tuberculosis*
Japanese encephalitis	HSV	*Cryptosporidium hominis*
Malaria	HIV-1	*Aeromonas hydrophila*
Measles	Human hookworm disease	*Nocardia asteroides*
Meningococcal meningitis	Leishmaniasis disease	Lassa virus
Mumps	Malaria	*Klebsiella pneumoniae*
Pertussis	Nipah virus	*Chlamydia pneumoniae*
Pneumococcal disease	Nontyphoidal Salmonella disease	
Poliomyelitis	Norovirus	
Rabies	Paratyphoid fever	
Rotavirus	Respiratory syncytial virus	
Tetanus	Schistosomiasis disease	
Tick-borne encephalitis	*Shigella*	
Tuberculosis	*Staphylococcus aureus*	
Typhoid	*S. pneumoniae*	
Varicella	*Streptococcus pyogenes*	
Yellow fever	Tuberculosis	
	Universal influenza vaccine	

Certain diseases for which vaccines are available are enlisted as *Available*, while vaccine candidates under development are enlisted as *Pipeline*. WHO's Product Development for Vaccines Advisory Committee (PDVAC) is overseeing the progress of pipeline vaccines.
GBS, group B Streptococcus; *HSV*, herpes simplex virus; *RV*, reverse vaccinology.
Based on: https://www.who.int/teams/immunization-vaccines-and-biologicals/diseases.

being developed or improved by using the bioinformatics approach. The main aim is to accelerate vaccine development process by precise prediction of pathogenic epitopes that can elicit a strong immune response in the host. Emerged as an offshoot of the genomics era, RV bloomed over the past decade as an effective strategy for rapid vaccine development exploiting deep learning algorithms to identify potential antigens and immunogenic regions [10]. Through RV, the major caveats and barriers of conventional vaccinology have significantly been addressed, with increased safety, reduced cost, less antigenic competition, and direct targeting to the site of infection [10,11].

The fundamental is to screen the whole genome of pathogens to identify specific genes that could elicit high antigenicity against host B cells and T cells [12]. Therefore the identified epitopes that are recognized by B cells and T_H cells or cytotoxic T lymphocytes (CTLs), itself can be combined to design novel vaccines [13,14]. Once these predicted candidates satisfy efficacy and safety after analyzing through prediction algorithms, they can be manufactured synthetically and are selected through infection in animal models [15]. The method is bolstered further with the advent of the machine learning (ML) and deep learning approach, with its robust potential to efficiently recognize biological patterns [16]. Introduction of ML revolutionized the RV process in the context of enhancing prediction accuracy and time management. Alongside it offered researchers to exploit bigger datasets and to compare with more number of biological parameters to add-in existing models of RV [17]. Through various user-friendly predicting ML tools, identification of immunodominant molecules and prediction of possible outcome in various genotypic background have radically improved in the recent years [18]. High-dimensional omics technologies supported by big-data analysis methods to generate datasets for ML-based prediction and utilization of the prediction outcome through RV can enhance success rates of the vaccine development endeavors [19]. The coming era of artificial intelligence (AI) through ML and deep learning will be the soul of next-generation medical research including vaccine designing.

One of the greatest lessons of the coronavirus disease 2019 (COVID-19) pandemic has been the earnest need of developing vaccine for emerging diseases as a crucial, if not the most important, step of pandemic preparedness. RV already had been proven quiet beneficial for predicting viral antigens, identifying their epitopes, and assessing their immunogenicity [20]. Moreover, this approach is also both time-saving and cost-effective. Furthermore, quick identification and isolation of the new corona virus strains are required for effective disease prevention and adequate control measures to hinder the community spread of severe acute respiratory syndrome coronavirus 2 (SARS-CoV-2) [21]. Here we discuss the detailed transformation of traditional vaccinology approaches to the RV approach and how it has proven beneficial to combat various pathogenic and nonpathogenic diseases, with special reference to some of the major RV-based vaccine development trials. Also, we have discussed about the fundamentals of ML algorithms used in developing different RV tools that are available to enhance prediction accuracy.

2 FROM TRADITIONAL VACCINOLOGY TO REVERSE VACCINOLOGY

Over the years, immunization efforts have become the major aspect of therapeutics, particularly against infective pathogens. In 1796, traditional vaccines were formally introduced into western medicine by Edward Jenner. He used infected materials isolated from cows to immunize against smallpox. But the systematic development of vaccines and establishment of the basic theory of vaccinology was first introduced by Louis Pasteur, who proposed "isolation, inactivation, and injection" as the steps to develop a vaccine; that is, one should identify and isolate a pathogen, inactivate it, and administer it to augment immunogenic memory against the pathogen [22]. Subsequent development of strategies for successful development of inactivated vaccines by Salk (chemical inactivation by formaldehyde) and attenuated vaccine by Sabine (serial passaging till attenuation was attained) against polio expedited vaccine research. Isolation and exploitation of purified cellular components from pathogens introduced a new dimension in vaccine research for diphtheria, tetanus, *Neisseria meningitidis*, and *Streptococcus pneumoniae* [22]. For pathogens that cannot be cultured in a laboratory, success was achieved by Buynak et al. in 1976 [23] by inactivating surface antigens present in the plasma of chronically infected individuals, the same subsequently developed into recombinant vaccines. Vaccines are considered one of the most important advances in modern medicine and have greatly improved our quality of life by reducing or eliminating many serious infectious diseases. Pasture's legacy has deeply impacted the sustainability of the human race, as today, 86% of the world's children receive essential, lifesaving vaccines [22]. However, for diseases such as malaria, AIDS, and other emerging diseases, no effective vaccine could be developed by the conventional approach so far.

To reciprocate the exigency, sequencing entire genomes and advances in bioinformatics have transformed traditional vaccinology with strategic exploitation of genomic data to discover novel immunorelevant information that had been perhaps beyond the scopes of conventional vaccinology. Since deciphering the first ever whole-genome draft of *Haemophilus influenzae* [24], generation of huge amount of sequencing data coupled to advent of robust in silico approaches offered a holistic purview on molecules as prospective vaccine candidates. Potential surface targets, assessment of variability (Shannon entropy), posttranslation modification, processing, presentation and nonspecific interaction can be assessed using various bioinformatics algorithms and help determine the targets or epitopes to develop novel and improved vaccines by using the information available in the genome of the pathogen. The approach enables systematic identification of all the potential antigens of a pathogen to develop a safe and efficient vaccine against known or emerging pathogens. Genome sequences serve as the seed data for RV followed by the use of algorithms to predict exposed/secreted proteins with immunogenic potential, the expression and purification of the identified species, and the immunization of laboratory animals for evaluation of the protective immune response and memory elicited by the antigen(s)/epitopes [25] (Fig. 3.1). Therefore the identified epitopes that are recognized by $CD4^+$ T cells or $CD8^+$ T cells can be used in "reverse" as a tool to identify new antigens [17]. This is how the term *reverse vaccinology* fulfills its activity. A basic workflow for the identification of a potential vaccine candidate is illustrated in Fig. 3.1. As there is no requirement for culturing bacteria in the laboratory, this revolutionized technique efficiently reduces the cost and increases the speed for developing vaccines [26]. This is the major advantage of RV over conventional vaccinology.

Interestingly, the RV approach has been adopted for vaccine development in 1989, even before the onset of the genome-sequencing era, against HCV. HCV RNA was isolated and sequenced and predicted to encode a number of envelope proteins [27]. When tested in primate models, the envelope proteins elicited immune responses and acted as candidate vaccines. Before the establishment of a standard culture system, candidate vaccines could be designed against a virus [28]. In a more extensive RV approach, meningococcus B (MenB) was targeted to prevent meningococcal meningitis. Among all the predicted open reading frames (ORFs) from the genome sequence, 600 proteins were analyzed for immunogenicity after expression in a bacterial system. Various surface antigens previously unknown were shown to raise serum immunoglobulin levels, with bacteriocidal action in murine models. Of such candidates, porA or fHbp emerged as the most potent antigen and vaccine candidate. After accomplishing successful preclinical studies, the MenB vaccine was tested for safety at population levels and approved for use in Europe in 2013 [29]. RV has been also applied to many other bacterial pathogens, such as group A and group B *Streptococcus* (GBS) [30], *Porphyromonas gingivalis* [31], and antibiotic-resistant *Staphylococcus aureus* [32].

RV has been implemented in designing vaccines against viral pathogens such as herpes simplex virus

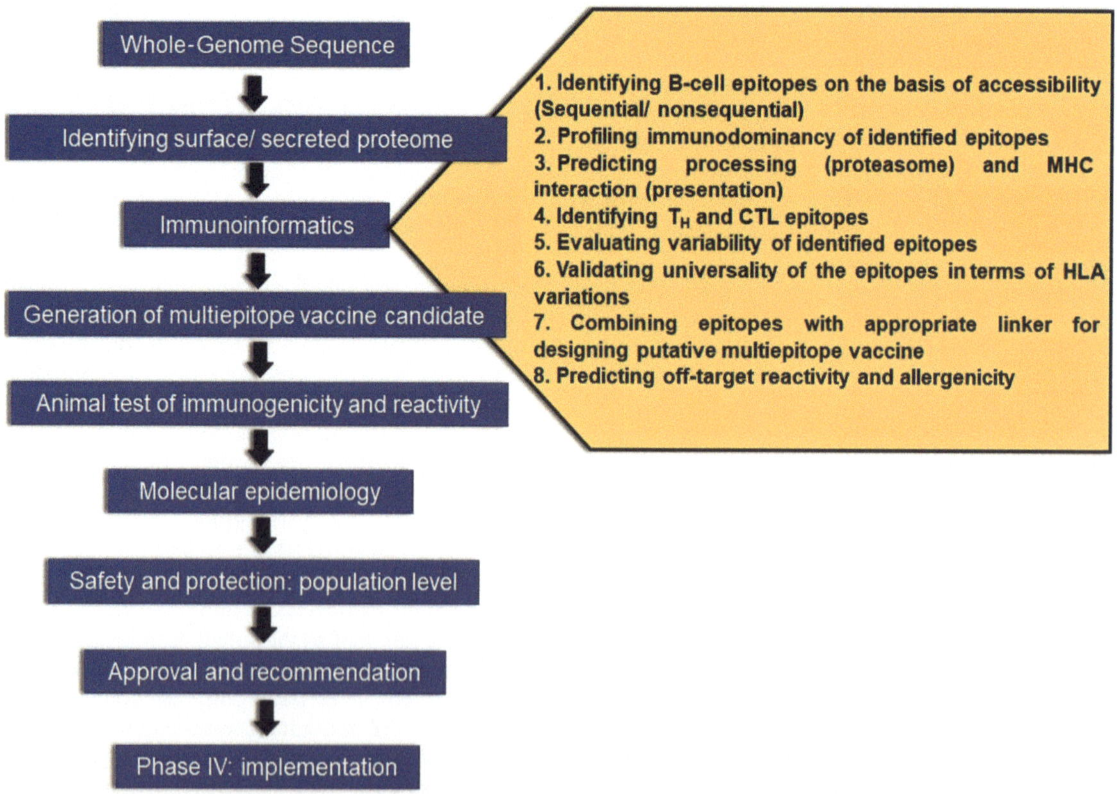

FIG. 3.1 The possible workflow for developing vaccines utilizing the reverse vaccinology approach. Steps for immunoinformatics analysis are further elaborated. *CTL*, cytotoxic T lymphocyte; *HLA*, human leukocyte antigen; *MHC*, major histocompatibility complex; T_H, helper T cell.

(HSV) and human cytomegalovirus (HCMV). HCMV, is a virus of the beta-herpes virus subfamily with a large DNA genome with 165 predicted ORFs. Out of the 165, 94 proteins were predicted to be transmembrane (TM) domains, and 74 of the 94 TM proteins were expressed in mammalian cells. A total of 36 proteins were screened to be potential antigens as revealed by analyzing clinical sera with selective immunoglobulins [33]. The RV strategy has also been implemented in predicting candidate subunit vaccines against Ebola and SARS-CoV-2 in the recent times.

Alongside the experimental success of RV approach, user-friendly and graphical user interface-assisted environments have been built for the rapid identification of cross-genome vaccine candidates. The first such package was provided by a software package for performing RV in silico studies, named *New Enhanced Reverse Vaccinology Environment* (NERVE) [34]. This software helps in silico identification of the best vaccine candidates from whole proteomes of infectious pathogens. This software platform utilizes bioinformatics algorithm for protein analysis and comparison. Recently, a new dimension to the RV approach has been inculcated which involve high-throughput screening of large numbers of antibody-secreting cells and identification of predominant immunoglobulins during infection followed by identification of potential antigens. This approach, assigned as RV 2.0, led to the identification of newer vaccine candidates from *N. meningitidis*, *M. tuberculosis*, and *S. aureus* [35].

Despite its tremendous potential in vaccine development, the RV approach does inherit limitations associated with the fundamentals of immune activations against different pathogens. Genomics-based approaches can predict putative protein products derived from predicted ORFs but cannot predict the expression levels and formation of protein complexes and is thereby not capable of evaluating single peptide as a component of avid immunogenicity. For example, for HCMV, the majority of neutralizing antibodies target the gH/gL/UL128/UL130/UL131 pentameric complex formed by the co-expression of each of the respective genes while none of the individual proteins can raise considerable neutralizing antibodies [36]. An additional layer of hindrance is the effective coding potential of the pathogen's genome, which often expands the range of polypeptides encoded beyond the ORFs predicted and annotated through established gene models [37]. Also, often the bioinformatically predicted subunits can bind preformed neutralizing antibodies and are unable to trigger neutralizing antibody production during in vivo immunization. This can be attributed to genetic variation and serum stability of the antigen or other unidentified factors. In this context the RV approach has to be pipelined with robust computation integrated with extensive screening-validation cycles.

3 ARTIFICIAL INTELLIGENCE-BASED ALGORITHMS IN REVERSE VACCINOLOGY

The backbone of RV approach is prediction, and the ML and deep learning-based approaches have enormous potential in predicting the efficacy or deleterious effects of vaccination. Inputs including direct experimental data derived from proteomics and metabolomics, structural properties, cytokine profiling, and mass cytometry, among others, can be processed for training by the ML module, which consequently establishes a classification/predictive features. The output of this rule predicts the status of immunogenicity (e.g., potential to cat as B-/T-cell epitopes, proteasomal processing, and major histocompatibility complex [MHC] presentation) or nonspecific reactogenicity (e.g., allergenicity). The four basic steps of employing AI in such prediction involve

A. preprocessing the data for training;
B. identifying vaccine candidates;
C. selecting a proper ML algorithm with desired selectivity, specificity, accuracy, and precision;
D. testing and validating the models.

Several algorithms are implemented for building up classifiers for making the predictions. A brief outline of the algorithms is discussed here.

Support vector machines (SVMs): SVM is an extremely elegant and robust prediction algorithm. Here data points are classified based on N number of features each of which defines a hyperplane on N-dimensional space. The key to classification is rooted in finding a plane to discriminate between two classes of data points. The distance between the classes is maximized so that future data points can be classified and prediction is possible. Hyperplanes serve as decision boundaries, and the dimension of the hyperplane depends on the number of features considered. Data points proximal to the hyperplane are *support vectors* that modulate margin and dimension. Such data points are relevant for the algorithm to build an SVM for prediction (https://towardsdatascience.com/support-vector-machine-introduction-to-machine-learning-algorithms-934a444fca47, https://towardsdatascience.com/support-vector- machine-support-vector-classifier-maximal-margin-classifier-22648a38ad9c). SVM can be applied by using the *polynomial kernel* where the sensitivity, specificity, and accuracy obtained were much better than the other kernel functions. In case of the polynomial kernel, the value of kernel parameter needs to be optimized to a specific value [38]. This optimization of the SVM model is a prerequisite to achieve high prediction accuracy at a specific cutoff score. Successful predictive models can be built with SVM when the number of observations is limited and the dimension of the training dataset is high. SVM can attain high level of accuracy, precision sensitivity, and specificity, as it avoids overfitting of training data. In RV, SVM is implemented by several tools to map immunogenic regions of proteins [39,40].

Artificial neural networks (ANNs): The deep learning methods using ANNs are not based on simplifying assumptions such as quantitative matrices; rather it requires larger amounts of complex relationships in the datasets. A "shallow" neural network comprises an input layer, one hidden layer, and an output layer [41]. A deep neural network includes two or more "hidden layers" that process inputs. This ML approach is very adaptive, effective with nonlinear problems, self-improving, and tolerant to a certain level of erroneous data [42]. Mimicking the biological neuronal system, ANN is constructed through various nodes, each of which can work on input data to perform simple operations resulting in an output (node value). The outputs from various nodes interact with each other, and the information can flow unidirectionally with the feedback loops to generate patterns or classifying features for learning [42]. For supervised learning, such features can be retrieved from some data that are fed to the network, whereas for unsupervised learning the algorithm searches for hidden patterns based on which clustering can be performed. For reinforcement learning, the sequence of decisions is made depending on the assessment of the environment and by maximizing the cumulative reward (https://arxiv.org/pdf/1709.06599.pdf). ANN is implemented in RV for predicting MHC-peptide binding, which involves degenerate motifs. Most of the tools implementing ANN attain high precision and accuracy for prediction [43].

Hidden Markov models (HMM): An HMM is a probabilistic graphical model in which the observed data are modeled as a series of outputs generated by one of the many hidden states; it is commonly used in pattern recognition for classification. The models contain a series of discrete states of first-order Markov chains with definite transition probabilities between the states alongside an initial probabilistic distribution [44]. For each of the states, there are discrete or continuous probability distributions for possible emissions or outputs. Transition probability is the probability of moving from one state to another, and emission probability is the probability of existing in a particular state. Hence, state-to-state transitions are guided by a set of transition and emission probabilities [42]. As for such propagation, the sequences of states through which the model traverses are hidden and cannot be observed individually, and hence it is mentioned as *hidden* Markov model. The predicted probability of any consequence is computed by multiplying the emission and transition probabilities along the trajectory. Various tools dedicated to the identification of T_H and CTL epitopes exploit HMM to accomplish an extremely high degree of prediction accuracy [45].

Random forest (RF): RF is a classifier algorithm. It consists of a large number of individual ensemble decision trees that allow the models to outperform individual models as the constituent models work to rectify errors in each model. Decision tree is a series of successive decisions made to reach a final result. The operation of RF algorithm is ensemble learning from randomly generated trees to generate a specific outcome. In vaccine development, particularly for epitope prediction, RF models are used alongside SVM to enhance the accuracy of predictions [46].

Logistic regression: It is the appropriate regression analysis to conduct when the dependent variable is binary. Like all regression analyses, the logistic regression is a predictive analysis [47]. Logistic regression is implemented for data description and to explain the relationship between one dependent binary variable and one or more nominal, ordinal, interval, or ratio-level independent variables [48]. This predictive regression often accompanies SVM and RF for epitope- and allergen-predicting tools.

k-Nearest neighbors (KNN) clustering: The KNN classifier is one of the predominantly used nonparametric classifier in which each of the test samples is classified based on its k-nearest neighbors [49]. The distance between test and training sample is first calculated to define the KNN attribute. Often, particularly while analyzing big data, distributed frameworks such as Hadoop are employed to find the KNN of training samples. The quality of the KNN classification depends on the detection of primary neighbors, which is affected by the quality of big-data processing. To expedite the execution of conventional KNN, random clustering (RC-KNN) and landmark spectral clustering (LC-KNN) [50], which apply the sequential KNN to test the input example with a limited set of examples, are implemented. KNN is used for identifying a potential immunogen/allergen when genome-wide identification is aimed.

Boosting: *Boosting* describes the combination of many weak learners into one very accurate prediction algorithm. A weak learner refers to a learning algorithm that only predicts slightly better than randomly. When looking at tree-based ensemble algorithms, a single *decision tree* would be the *weak learner* and a *combination* of these would result in the AdaBoost algorithm [51]. It is a boosting algorithm that allows decision trees to grow as weak learners. The AdaBoost algorithm is part of the family of boosting algorithms and was first introduced by Freund and Schapire in 1996. It is sequentially growing decision trees as weak learners. After each round of prediction, a penalty is imposed on incorrect callers. The final prediction is the weighted median. Thus AdaBoost is a metalearning method with iterative approach to boost weak learners and thereby enhances the performance of an ML algorithm.

Chen and Guestrin introduced the XGBoost (eXtreme Gradient Boosting) algorithm in 2016. The algorithm utilizes the concept of *gradient tree boosting*. The aim of increasing the speed and performance and reducing overfitting was accomplished. Here the gradient-boosted trees use classification and regression trees in a sequential learning process as weak learners. Although organizationwise the regression trees are similar to decision trees, they differ in continuous score assignment in each ultimate node. The score is summed up for the final prediction [52].

4 IMMUNOINFORMATICS TOOLS USING ARTIFICIAL INTELLIGENCE-BASED ALGORITHMS AND CLASSIFIERS

Vaccine designing is strategically complex, and to accomplish a successful vaccine, various factors are needed to be considered, including the intrinsic physicochemical property of the antigen to various host factors responsible for an adequate humoral response and memory. By exploiting the available algorithms and statistical packages, a number of databases (Table 3.2) and tools (Table 3.3) have been developed in the past two - decades. The tools predict B-cell and T-cell epitopes, MHC binding, and the allergenicity of candidate immunodominant molecules. There are various tools dedicated to pathogen-specific vaccine designing, such as Mosaic Vaccine Designer, epigraph, etc. [53]. The following discussion would be restricted to more generic and popular tools.

4.1 B-Cell Epitope Prediction

BepiPred-2.0: BepiPred-2.0, a sequential B-cell epitope predictor, is trained on epitopes derived from antibody-antigen complex structures through an RF algorithm [54]. Apart from the solved three-dimensional (3D) structures, prediction is also based on a large collection

TABLE 3.2
A Brief List of Useful Resources (Databases) on B-Cell Epitopes, T-Cell Epitopes, Allergens, and Evolutionary Information of Immune System Components.

Prediction	Names	URLs	Refs
B-cell epitopes	IEDB	http://www.immuneepitope.org/	[100]
	Bcipep	http://www.imtech.res.in/raghava/bcipep	[101]
	CED	http://immunet.cn/ced/	[102]
	Epitome	http://cubic.bioc.columbia.edu/services/epitome/	[103]
	IMGT	http://imgt.cines.fr	[104]
T-cell epitopes	IEDB	http://www.immuneepitope.org/	[100]
	JenPep	http://www.darrenflower.info/jenpep/	[105]
	Syfpeithi	http://www.syfpeithi.de	[106]
	FRED	http://www-bs.informatik.uni-tuebingen.de/Software/FRED	[107]
	IMGT	http://imgt.cines.fr	[104]
Allergen	AllergenPro	http://www.niab.go.kr/nabic/	[108]
	Database of IUIS	http://www.allergen.org	[109]
	SDAP	http://fermi.utmb.edu/SDAP/	[110]
Evolutionary information of immune system components	ImmTree	http://bioinf.uta.fi/ImmTree	[111]
	ImmunomeBase	http://bioinf.uta.fi/ImmunomeBase	[112]
	Immunome database	http://bioinf.uta.fi/Immunome/	[113]
	Immunome knowledge base	http://bioinf.uta.fi/IKB/	[114]

TABLE 3.3
A List of Popular Web Servers and Tools for Immunoinformatics Analysis.

Web Servers and Tools	Names	URLs	Refs
B-cell epitope prediction	BepiPred	http://www.cbs.dtu.dk/services/BepiPred	[115]
	IEDB	http://www.immuneepitope.org/	[100]
	DiscoTope	http://www.cbs.dtu.dk/services/DiscoTope/	[116]
	ABCpred	http://www.imtech.res.in/raghava/abcpred	[55]
	BEPITOPE	jlpellequer@cea.fr	[117]
	COBEpro	http://scratch.proteomics.ics.uci.edu/	[118]
	IMGT	http://imgt.cines.fr	[104]
	CEP	http://bioinfo.ernet.in/cep.htm	[119]
	MIMOP	Request from franck.molina@cpbs.univ-montp1.fr	[120]
	MIMOX	http://web.kuicr.kyoto-u.ac.jp/hjian/mimox	[121]
	AgAbDb	http://202.41.70.51:8080/agabdb2/	[122]
	Pepitope	http://pepitope.tau.ac.il/	[123]
	3DEX	http://www.schreiber-abc.com/3dex/	[124]
T-cell epitope prediction	IEDB	http://www.immuneepitope.org/	[100]
	IMGT	http://imgt.cines.fr	[104]
	MHCPred	http://www.darrenflower.info/mhcpred/	[125]
	Propred	http://www.imtech.res.in/raghava/proped	[60]
	Syfpeithi	http://www.syfpeithi.de	[106]
	EpiToolKit	http://www.epitoolkit.org	[126]
	ElliPro	http://tools.immuneepitope.org/tools/ElliPro	[127]
	Pcleavage	http://www.imtech.res.in/raghava/pcleavage/	[62]
	TAPPred	http://www.imtech.res.in/raghava/tappred/	[128]
	NetCTL	http://www.cbs.dtu.dk/services/NetCTL/	[63]
	NetMHC 3.0	http://www.cbs.dtu.dk/services/NetMHC	[129]
	MMBPred	http://www.imtech.res.in/raghava/mmbpred/	[130]
Allergy prediction	AllerTOP v.2	https://www.ddg-pharmfac.net/AllerTOP/	[71]
	AllergenFP	https://ddg-pharmfac.net/AllergenFP/	[72]
	AllerTool	http://research.i2r.a-star.edu.sg/AllerTool/	[131]
	AllerHunter	http://tiger.dbs.nus.edu.sg/AllerHunter	[131]
	AlgPred	http://www.imtech.res.in/raghava/algpred	[73]
	APPEL	http://jing.cz3.nus.edu.sg/cgi-bin/APPEL	[132]
	Allermatch	http://www.allermatch.org	[133]
	EVALLER	http://bioinformatics.bmc.uu.se/evaller.html	[134]
In silico vaccination	VaxiJen	http://www.darrenflower.info/VaxiJen/	[74]
	Vaxign	http://www.violinet.org/vaxign/	[135]
	VIOLIN	http://www.violinet.org	[75]
	NERVE	http://www.bio.unipd.it/molbinfo	[34]
	DyNAVacs	http://miracle.igib.res.in/dynavac/	[136]

of linear epitopes downloaded from the Immune Epitope Database (IEDB). The interface for input data and output displays is convenient and tractable; hence BepiPred-2.0 is becoming a popular tool in the bioinformatics and immunology community.

ABCpred: The first web tool developed based on recurrent neural network for predicting B-cell epitopes was ABCpred server [55]. The tool uses ANN with fixed-length patterns for training and testing with a dataset consisting 700 B-cell epitopes and 700 non-B-cell epitopes (random peptides) of a maximum length of 20 residues. An accuracy of 65.93% could be achieved in predicting B-cell and non-B-cell epitopes.

IgPred: IgPred is the first classifier web-server-based tool to predict B-cell epitopes with the potential to induce different classes of antibodies [56]. By implementing models developed on experimentally validated nonredundant datasets and reciprocated with cross-validation, the tools attain an accuracy of 80% in prediction. The tool scans immunogenic regions on proteins and candidate epitopes and also identifies specific antibody-binding motifs by the MEME/MAST software and similarity searches using the Smith-Waterman algorithm against experimentally validated epitopes [56].

LBtope: It was developed for predicting linear B-cell epitopes [57]. Primary models were built on SVM and IBk to analyze 12,063 epitopes and 20,589 nonepitopes obtained from IEDB. The tool overcame the limitations of earlier predictions based on small datasets.

SVMTriP: It includes "realistic prediction" of protein surface area preferentially recognized by immunoglobulins. Yao et al. [58] applied SVM in combination with tri-peptide similarity and propensity scores (SVMTriP) in order to accomplish better prediction accuracy. The performance for linear B-cell epitope prediction can be improved by the combination of similarity and propensity of tripeptide subsequences [59]. The tool is particularly helpful for vaccine machinery and immunodiagnostic reagents.

4.2 Major Histocompatibility Complex-Binding Peptide and T-Cell Epitope Prediction

MHCPred: Here allele QSAR models were generated using partial least squares. The binding affinity of MHC class I and II molecules and their transport through transporter associated with processing (TAP) are predicted using such models [60].

EpiToolKit (ETK): ETK is a virtual workbench hosting an array of immunoinformatics tools including MHC genotyping, epitope and neoepitope prediction, epitope selection for vaccine design, and epitope assembly [61]. Here immunogenicity of an epitope is considered to be a sum over the immunogenicity of each epitope-MHC allele pair, weighted by the probability of an MHC allele to appear in the target population or person. It attains superior performance compared with analogous tools by integrating a next-generation sequencing (NGS)-based genotyping approach called OptiType, which can integrate MHC allele data with RNAseq, Exomeseq, and whole-genome sequencing reads. Such approach enables ETK use in polymorphic epitope prediction and personalized design approaches.

Pcleavage: It is another SVM-based prediction tool to detect the immunoproteasome (20S) cleavage sites in antigenic sequences [62]. Using conventional sparse coding, each amino acid is assigned with a value based on which patterns are generated for SVM classifiers. MAPPP is another tool for predicting 20S proteasomal cleavage sites.

NetCTL 1.2 Server: NetCTL [63] is an ANN-based prediction tool for CTL epitopes in proteins. Prediction of peptide MHC class I binding, proteasomal C terminal cleavage, and TAP transport efficiency is integrated to rank epitopes. The server allows for predictions of CTL epitopes restricted to 12 MHC class I supertypes. MHC class I binding and proteasomal cleavage is performed using ANN. TAP transport efficiency is predicted using weight matrix.

CTLPred: It is a direct method for prediction of CTL epitopes in subunit vaccine design. In direct methods the information or patterns of T-cell epitopes instead of MHC binders were used for the development of methods. The methods are based on elegant ML techniques such as ANN and SVM. The methods also allow the *consensus and combined* prediction based on the two methods [40].

NetMHC4.0: NetMHC4.0 is an ANN-based prediction tool developed by Andreatta and Nielsen et al. Here ANNs have been trained for 81 different human MHC alleles including human leukocyte antigen (HLA)-A, B, C, and E. Although most HLA have preference for binding to nonamers, predictions can be made for peptides of any length [64]. NetMHCpan-4.1 is a similar tool that predicts binding through ANN trained on a combination of more than 850,000 quantitative binding affinity and mass-spectrometry-eluted ligand peptides covering 201 MHC molecules from humans (HLA-A, B, C, and E), mice (H-2), cattle (BoLA), primates (Patr, Mamu, Gogo), swine (SLA), and equine (Eqca) for binding affinity and 289 MHC molecules from humans (HLA-A, B, C, and E), mice (H-2), cattle (BoLA), primates (Patr, Mamu, Gogo), swine (SLA), equine (Eqca), and dogs (DLA) for eluted ligands.

SVMHC: SVMHC [65] is applied for the prediction of both MHC class I and class II binding peptides. An SVM approach is implemented for predicting MHC class I binding peptides, while prediction of MHC class II binding peptides is based on the matrices published by Sturniolo et al. SVMHC conducts analysis for a broad collection of alleles, identifying prospective binding stretches in a protein sequence and investigating the effects of single-nucleotide variations in such interactions [66].

Rankpep: It is server designed to predict probable binders to MHC class I and class II molecules from alignments using Position-Specific Scoring Matrices (PSSMs) [67]. FOR MHC class I, PSSMs were obtained from alignments of MHC ligands using PROFILE-WEIGHT that imposes branch-proportional sequence weights. For MHC class II, BLK2PSSM with position-based weights is implemented. RANKPEP includes a selection of 102 and 80 PSSMs for the prediction of peptide-binding MHC class I and class II molecules, respectively. The immunodominance filter used is an SVM-based classifier trained on residue properties and amino acid sequence to discriminate between immunogenic and nonimmunogenic peptides.

POPI: It is another prediction server of MHC class I and class II. This SVM-based algorithm is used with physicochemical properties of MHC and ligand peptides as the training dataset [68].

TEpredict: TEpredict is a tool for the T-cell epitope detection. The predictions are based on MHC class I binders, MHC class II binders, proteasomal/immunoproteasomal processing of antigens, and peptide-TAP binding, excluding peptides sharing local similarity with human proteins from the set of predicted epitopes. As a classifier for the parameters, Sparse Partial Least Squares regression is used. The models are capable of generating predictions of both high sensitivity (0.50–0.80) and high specificity (0.75–0.99) [69].

4.3 Allergen Prediction Tools

Allergen-associated protein motifs had been identified by analyzing sets of allergen and nonallergen protein sequences. Zorzet et al. [70] introduced a computational approach for classifying the amino acid sequences in allergens and nonallergens. Subsequently several prediction tools were developed to predict the allergenicity of proteins/peptides.

AllerTopV2: ML methods such as logistic regression, decision tree, naïve Bayes, RF, multilayer perceptron, and KNN have been implemented while developing this allergen-predicting tool. The best performing method was KNN with models resulting in 85.3% accuracy at fivefold cross-validation [71].

AllergenFP: AllergenFP implements a four-step algorithm to accurately predict allergen motifs [72]. Protein sequences describe the principal properties of amino acids and the strings of diverse lengths generated are converted into vectors of identical length by auto- and cross-covariance (ACC). For evaluation the vectors are transformed into binary fingerprints for comparison.

AlgPred: This similarity-based prediction software can be applied to identify stretch of polypeptides with similarity to reported allergens through server search for MEME/MAST allergen motifs. It includes IgE epitope mapping to pinpoint the position of the allergenic regions on proteins. It also adopts a hybrid approach to combine two or more such strategies [73].

4.4 Tools for In Silico Vaccination

VaxiJen: RV is implemented in Vaxign [74], which was the first server for alignment-independent prediction of protective antigens. The usual limitations of alignment-based approach are dealt with ACC transformation of sequences into uniform vectors of principal amino acid properties. Whole-protein antigenicity models are built on bacterial, viral, and tumor datasets. The accuracy of the models was calculated to vary between 70% and 89%.

Vaccine Investigation and Online Information Network (VIOLIN): VIOLIN assembles vaccine research data with optimal curtain and analysis [75]. Alongside VIOLIN, a bunch of independent programs have been developed. Such programs include Vaxign, which was the first RV-based vaccine candidate prediction program. It also hosts the Ontology of Vaccine Adverse Events for the systematic representation of adverse effects of licensed human vaccines.

PEPVAC: Promiscuous EPitope-based VACcine predicts multiepitope vaccines with a broad population coverage [76]. It aims identifying promiscuous peptide binders to five distinct HLA class I supertypes (A2, A3, B7, A24, and B15) for which the population frequency is estimated to be 95%, regardless of ethnicity.

This optimization is accomplished through the prediction of peptides that bind to several HLA molecules with similar peptide-binding specificity (supertypes). PSSMs, derived from block alignments of MHC class I, are used for prediction.

4.5 Commercial Software and Tools

There are several convenient commercial software for huge data handing and high-throughput prediction.

DNASTAR's Epitope Prediction Software: Lasergene Structural Biology Suite from DNASTAR have different software and tools for epitope prediction to identify B-cell epitopes based on protein sequence data [77]. The identification is combined with visualization software. Protean 3D, part of Lasergene protein, uses an ML approach that improves the ability to perform B-cell epitope prediction using only sequence-based information, providing better overall predictive accuracy than other leading epitope prediction methods.

EpiQuest: EpiQuest is a package for prediction of linear B-cell and T-cell epitopes (http://www.aptumbio.com/software). It can predict potential epitopes based on not only their binding property but also their overall functionality. The software can be used for developing combination vaccines, specific antibodies for immunodetection, and biomimetic immunomodulator designing [78].

TEPITOPEpan: TEPITOPEpan predicts T-cell epitopes by identifying ligands for HLA-DR molecules. Principally it deals with estimating similarity between various HLA-DR pockets by residues in binding pockets for the HLA molecules that are in contact with core residues. A huge library of 11 PSSMs with 1 PSSM is a 20 × 9 matrix where 9 binding specificity vectors correspond to 9 pockets, building the backbone of TEPITOPE [79]. TEPITOPEpan extracts this library to generate a new PSSM. In a generated PSSM, each vector is a weighted average of binding specificity vectors of the corresponding pocket over 11 DRB alleles. The weight can be computed by pocket sequence similarity.

For the majority of the tools, estimation of immunogenicity or allergenicity of specific proteins is based on only primary sequences. As it is generally assumed that T lymphocytes recognize short linear fragments, 3D accessibility is of limited effect on epitope prediction for T cells, while B-cell receptors recognize nonsequential epitopes and structural consideration should be an integral component for predicting nonsequential B-cell epitopes. Other important caveats include posttranslation modifications in epitopes that are not predicted [80]. In this scenario, mass spectroscopy-based sequencing prediction methods have been considered to be one of the most effective techniques to map specific epitopes, particularly PSSM combined with deep learning can be of immense applicability for this purpose. Therefore RV through deep learning provides the unique opportunity to optimize vaccines by making available the entire landscape of immune recognition and activation.

5 IMPLEMENTING REVERSE VACCINOLOGY IN VACCINE DEVELOPMENT ENDEAVORS

The concept of RV is applicable to any pathogen whose genome is known and is accessible for the discovery of protective antigens. One of the most successful examples of the RV approach applied to a bacterial pathogen is Bexsero, a novel vaccine against meningococcus B with broad strain coverage and being approved in more than 30 countries for use in all age groups including infants who are the most susceptible to meningococcal disease [81]. Along with other bacterial pathogens, this vaccine development approach was explored for a vaccine against *Chlamydia* [82]. Although several RV approaches were successful in the context of bacterial pathogens (Table 3.1), at present, RV remains an overseen approach for many other classes of pathogens such as viruses, protozoa, and helminths.

Recent works using NGS highlight viral dynamics and genomic diversity even within the host, when the virus is subject to selective pressure [83]. In recent years, success rate and speed in the development of challenging viral vaccines, such as dengue virus, respiratory syncytial virus (RSV), HSV, or HCMV, have increased for the NGS-based diversity profiling [84–86]. Especially in case of dengue virus, a systematic bioinformatics approach has been applied to define the antigenic diversity of viral strains by analyzing more than 9000 dengue protein sequences and identifying peptide combinations sufficient to capture the antigenic diversity of T-cell epitopes [87]. Using the RV approach, a multiepitope-based subunit vaccine (MEV) was developed against RSV, through epitope mapping from two highly antigenic and conserved proteins of RSV (glycoprotein and fusion protein). The outcomes of development of the MEV proposed that the vaccine construct will be a significant candidate against RSV to control and prevent RSV-related disorders if further investigated experimentally [88]. For HSV, three potential antigenic envelope glycoproteins (glycoproteins E, B, and D) were selected from nine envelope glycoproteins for the development of an efficient vaccine construct. Through the immunoinformatics approach, potential antigenic epitopes were identified in such a way that these epitopes should be nonallergenic, nontoxic, nonhomologous (to the human proteome) and maximally conserved across the selected strains and species. This vaccine construct was designed for mainly multiple strains of HSV-1 and HSV-2 [89]. For HCMV, which is responsible for congenital infection worldwide, attempts through live-attenuated viruses, glycoprotein subunit formulations, viral vectors, and single/bivalent DNA plasmids were unable to achieve desired success. Identification of novel antigens for HCMV vaccine development through RV rejuvenated the possibility of preventing congenital cytomegalovirus infection through NGS [33]. Combining sequencing data with deep learning inducted the 'pan-genome' concept in RV [90]. Pan-genome RV is emerging as a powerful tool to identify unique vaccine candidates in antigenically distinct species [91] and its first application was for the development of a universal vaccine against GBS [30]. This approach covers a wide range of species and the entire gene set of all strains of a species.

The recent outbreak of COVID-19 is caused by SARS-CoV-2. The four main human coronaviruses are HCoV-HKU1, HCoV-OC43, HCoV-229E, and HCoV-NL63, which are circulating in the human population worldwide and cause mild symptoms [92]. Like other coronaviruses, SARS-CoV-2 also encodes structural proteins, including the spike (S) protein, the envelope (E) protein, the membrane (M) protein, and the nucleocapsid (N) protein. The RV approach has been adopted to explore the potential of the in silico examination of the viral proteome to hunt for antigenic and surface-exposed proteins. In fact, a number of reports have already come up using the abovementioned immunoinformatics approach to explore candidate vaccines for COVID-19 [93–96]. Proteins such as spike glycoprotein, nucleocapsid phosphoprotein, membrane glycoprotein, and ORF3a accessory protein, which contain more efficient antigenic epitope-rich domains, have been identified for the development of epitope-based peptide vaccines. Among them, the spike glycoprotein has been extensively used for designing multiepitope vaccines. After selecting the target protein, multiple CTL, T_H-, and B-cell epitopes of the target protein are predicted using different immunoinformatics tools such as ProPred-1, ProPred, and ABCPred servers, respectively. Following this, the T-cell and B-cell epitopes are again selected based on the antigenicity, toxicity, allergenicity, and cross-reactivity with human proteomes. Afterward, the selected epitopes were joined together using the appropriate linker to construct the main core of the multiepitope vaccine. The vaccine construct should have appropriate physicochemical properties such as stability at room temperature, more antigenic capability, and should not induce a hypersensitive or an acute inflammatory reaction. The antigenicity, allergenicity, and physiochemical properties of the designed vaccine are being estimated using the VaxiJen v2.0 server, AllergenFP v.1.0 server, and ProteinSol web server, respectively. Finally, the affinity of the vaccine construct to the human immune receptors, such as toll-like receptor (TLR)3, TLR5, TLR8, or purinergic receptors, should be investigated through docking, followed by the molecular dynamics simulation of docking complexes. Predicting such interactions would evaluate the potential of the vaccine candidate to induce a proinflammatory response and also augment the vaccine adjuvant designing [97].

6 CONCLUSION

Technical and conceptual advances in bioinformatics have led to a better exposition of the immune system through structural immunoinformatics. Comprehensive meta-analyses of the immune epitope data help researchers to understand the immune system pathway in infectious disease pathogenesis through computational framework. Implication of AI in bioinformatics represents a huge advantage in faster outputs of complex comparative data. Enhancements for discriminating antigens from nonantigens are a prior step to design an effective vaccine. In identifying surface proteins during host-pathogen interaction and selecting potential vaccine candidates, different omics technologies has also been implemented in this field. Immunomics has been used as a complementary approach to proteomics for the critical assessment of epitope prediction [98]. Designing of multiepitope vaccines can generate a wide range of immune responses, which effectively stimulate the immune components to fight against an infectious pathogen. Although the development of multiepitope peptide vaccines is an ardent task, AI helps assess the standard benchmark datasets very accurately [99]. But in order to obtain consensus predictions, the process of combining the results of several servers sometimes gets complicated because of the lack of a unified output format, which needs to be sorted. Development of standardized data representations is still a need in data science. Hence, it is necessary to develop metaservers for the evaluation of different prediction methods and for combining all the predicted data after proper comparison. Sometime two different servers give two contradictory outputs, which creates a problem in the overall analysis. Another major issue that needs to be taken care is achieving sufficient coverage of MHC polymorphism [25]. The amount of data regarding immune recognition and genomics, transcriptomics, and proteomics information for various pathogens is increasing day by day. An ever-growing widespread application of RV needs dynamic and timely retrieval of data associated with infectious pathogens. For this purpose, development of big databases for both immune-reactivity data and pathogen data and proper curation is a prerequisite. Further improvements in the databases and tools will be critical to advance in vaccine development. Addressing the apparent lacuna associated with the vaccine development through RV can immensely augment future vaccine discovery efforts.

CONFLICT OF INTEREST

The authors declare no conflict of interest.

REFERENCES

[1] Barrett ADT. Vaccinology in the twenty-first century. NPJ Vaccines 2016;1:16009.

[2] Wallis J, Shenton DP, Carlisle RC. Novel approaches for the design, delivery and administration of vaccine technologies. Clin Exp Immunol 2019;196(2):189–204.

[3] Meeusen EN, Walker J, Peters A, Pastoret PP, Jungersen G. Current status of veterinary vaccines. Clin Microbiol Rev 2007;20(3):489–510 [table of contents].

[4] Doolan DL, Apte SH, Proietti C. Genome-based vaccine design: the promise for malaria and other infectious diseases. Int J Parasitol 2014;44(12):901–13.

[5] Amanna IJ, Slifka MK. Successful vaccines. Curr Top Microbiol Immunol 2020;428:1–30.

[6] Greenwood B. The contribution of vaccination to global health: past, present and future. Philos Trans R Soc Lond B Biol Sci 2014;369(1645):20130433.
[7] Delany I, Rappuoli R, De Gregorio E. Vaccines for the 21st century. EMBO Mol Med 2014;6(6):708–20.
[8] De Rycker M, Baragana B, Duce SL, Gilbert IH. Challenges and recent progress in drug discovery for tropical diseases. Nature 2018;559(7715):498–506.
[9] Rappuoli R. Reverse vaccinology, a genome-based approach to vaccine development. Vaccine 2001; 19(17–19):2688–91.
[10] Vivona S, Gardy JL, Ramachandran S, Brinkman FS, Raghava GP, Flower DR, et al. Computer-aided biotechnology: from immuno-informatics to reverse vaccinology. Trends Biotechnol 2008;26(4):190–200.
[11] Babiuk LA. Broadening the approaches to developing more effective vaccines. Vaccine 1999;17(13–14): 1587–95.
[12] Oli AN, Obialor WO, Ifeanyichukwu MO, Odimegwu DC, Okoyeh JN, Emechebe GO, et al. Immunoinformatics and vaccine development: an overview. ImmunoTargets Ther 2020;9:13–30.
[13] Doolan DL, Southwood S, Freilich DA, Sidney J, Graber NL, Shatney L, et al. Identification of Plasmodium falciparum antigens by antigenic analysis of genomic and proteomic data. Proc Natl Acad Sci U S A 2003;100(17):9952–7.
[14] Lauemoller SL, Kesmir C, Corbet SL, Fomsgaard A, Holm A, Claesson MH, et al. Identifying cytotoxic T cell epitopes from genomic and proteomic information: "The human MHC project". Rev Immunogenet 2000;2(4):477–91.
[15] Kennedy RB, Ovsyannikova IG, Palese P, Poland GA. Current challenges in vaccinology. Front Immunol 2020;11:1181.
[16] Han Y, Kim D. Deep convolutional neural networks for pan-specific peptide-MHC class I binding prediction. BMC Bioinf 2017;18(1):585.
[17] Heinson AI, Ewing RM, Holloway JW, Woelk CH, Niranjan M. An evaluation of different classification algorithms for protein sequence-based reverse vaccinology prediction. PLoS One 2019;14(12):e0226256.
[18] Gallion J, Koire A, Katsonis P, Schoenegge AM, Bouvier M, Lichtarge O. Predicting phenotype from genotype: improving accuracy through more robust experimental and computational modeling. Hum Mutat 2017;38(5):569–80.
[19] Bragazzi NL, Gianfredi V, Villarini M, Rosselli R, Nasr A, Hussein A, et al. Vaccines meet big data: state-of-the-art and future prospects. From the classical 3Is ("Isolate-Inactivate-Inject") vaccinology 1.0 to vaccinology 3.0, vaccinomics, and beyond: a historical overview. Front Public Health 2018;6:62.
[20] Ong E, Wong MU, Huffman A, He Y. COVID-19 coronavirus vaccine design using reverse vaccinology and machine learning. Front Immunol 2020;11:1581. https://doi.org/10.3389/fimmu.2020.01581. PMID: 32719684; PMCID: PMC7350702.
[21] Yuen KS, Ye ZW, Fung SY, Chan CP, Jin DY. SARS-CoV-2 and COVID-19: the most important research questions. Cell Biosci 2020;10:40.
[22] Rappuoli R. From pasteur to genomics: progress and challenges in infectious diseases. Nat Med 2004; 10(11):1177–85.
[23] Buynak EB, Roehm RR, Tytell AA, Bertland 2nd AU, Lampson GP, Hilleman MR. Development and chimpanzee testing of a vaccine against human hepatitis B. Proc Soc Exp Biol Med 1976;151(4):694–700.
[24] Fleischmann RD, Adams MD, White O, Clayton RA, Kirkness EF, Kerlavage AR, et al. Whole-genome random sequencing and assembly of Haemophilus influenzae Rd. Science 1995;269(5223):496–512.
[25] Backert L, Kohlbacher O. Immunoinformatics and epitope prediction in the age of genomic medicine. Genome Med 2015;7:119.
[26] Adu-Bobie J, Capecchi B, Serruto D, Rappuoli R, Pizza M. Two years into reverse vaccinology. Vaccine 2003;21(7–8):605–10.
[27] Choo QL, Kuo G, Weiner AJ, Overby LR, Bradley DW, Houghton M. Isolation of a cDNA clone derived from a blood-borne non-A, non-B viral hepatitis genome. Science 1989;244(4902):359–62.
[28] Bukh J. The history of hepatitis C virus (HCV): basic research reveals unique features in phylogeny, evolution and the viral life cycle with new perspectives for epidemic control. J Hepatol 2016;65(1 Suppl. l):S2–21.
[29] Rivero-Calle I, Raguindin PF, Gomez-Rial J, Rodriguez-Tenreiro C, Martinon-Torres F. Meningococcal group B vaccine for the prevention of invasive meningococcal disease caused by Neisseria meningitidis serogroup B. Infect Drug Resist 2019;12:3169–88.
[30] Maione D, Margarit I, Rinaudo CD, Masignani V, Mora M, Scarselli M, et al. Identification of a universal Group B streptococcus vaccine by multiple genome screen. Science 2005;309(5731):148–50.
[31] Ross BC, Czajkowski L, Hocking D, Margetts M, Webb E, Rothel L, et al. Identification of vaccine candidate antigens from a genomic analysis of Porphyromonas gingivalis. Vaccine 2001;19(30):4135–42.
[32] Etz H, Minh DB, Henics T, Dryla A, Winkler B, Triska C, et al. Identification of in vivo expressed vaccine candidate antigens from *Staphylococcus aureus*. Proc Natl Acad Sci U S A 2002;99(10):6573–8.
[33] Nelson CS, Herold BC, Permar SR. A new era in cytomegalovirus vaccinology: considerations for rational design of next-generation vaccines to prevent congenital cytomegalovirus infection. NPJ Vaccines 2018;3:38.
[34] Vivona S, Bernante F, Filippini F. NERVE: new enhanced reverse vaccinology environment. BMC Biotechnol 2006;6:35.
[35] Rappuoli R, Bottomley MJ, D'Oro U, Finco O, De Gregorio E. Reverse vaccinology 2.0: human immunology instructs vaccine antigen design. J Exp Med 2016;213(4):469–81.
[36] Fouts AE, Chan P, Stephan JP, Vandlen R, Feierbach B. Antibodies against the gH/gL/UL128/UL130/UL131

complex comprise the majority of the anti-cytomegalovirus (anti-CMV) neutralizing antibody response in CMV hyperimmune globulin. J Virol 2012;86(13):7444–7.

[37] Wu A, Peng Y, Huang B, Ding X, Wang X, Niu P, et al. Genome composition and divergence of the novel coronavirus (2019-nCoV) originating in China. Cell Host Microbe 2020;27(3):325–8.

[38] Savas C, Dovis F. The impact of different kernel functions on the performance of scintillation detection based on support vector machines. Sensors 2019;19(23).

[39] Zhao Y, Zeng D, Rush AJ, Kosorok MR. Estimating individualized treatment rules using outcome weighted learning. J Am Stat Assoc 2012;107(449):1106–18.

[40] Bhasin M, Raghava GP. Prediction of CTL epitopes using QM, SVM and ANN techniques. Vaccine 2004;22(23–24):3195–204.

[41] Wilentzik Muller R, Gat-Viks I. Exploring neural networks and related visualization techniques in gene expression data. Front Genet 2020;11:402.

[42] Shahid N, Rappon T, Berta W. Applications of artificial neural networks in health care organizational decision-making: a scoping review. PLoS One 2019;14(2):e0212356.

[43] Zhao Y, Pinilla C, Valmori D, Martin R, Simon R. Application of support vector machines for T-cell epitopes prediction. Bioinformatics 2003;19(15):1978–84.

[44] Titsias MK, Holmes CC, Yau C. Statistical inference in hidden Markov models using k-segment constraints. J Am Stat Assoc 2016;111(513):200–15.

[45] Yoon BJ. Hidden Markov models and their applications in biological sequence analysis. Curr Genomics 2009;10(6):402–15.

[46] Javadi A, Khamesipour A, Monajemi F, Ghazisaeedi M. Computational modeling and analysis to predict intracellular parasite epitope characteristics using random forest technique. Iran J Public Health 2020;49(1):125–33.

[47] Bewick V, Cheek L, Ball J. Statistics review 14: logistic regression. Crit Care 2005;9(1):112–8.

[48] Sperandei S. Understanding logistic regression analysis. Biochem Med 2014;24(1):12–8.

[49] Wang X. A fast exact k-nearest neighbors algorithm for high dimensional search using k-means clustering and triangle inequality. Proc Int Jt Conf Neural Netw 2012;43(6):2351–8.

[50] Hassanat ABA. Two-point-based binary search trees for accelerating big data classification using KNN. PLoS One 2018;13(11):e0207772.

[51] Luna JM, Gennatas ED, Ungar LH, Eaton E, Diffenderfer ES, Jensen ST, et al. Building more accurate decision trees with the additive tree. Proc Natl Acad Sci U S A 2019;116(40):19887–93.

[52] Chang W, Liu Y, Xiao Y, Yuan X, Xu X, Zhang S, et al. A machine-learning-based prediction method for hypertension outcomes based on medical data. Diagnostics 2019;9(4).

[53] Theiler J, Yoon H, Yusim K, Picker LJ, Fruh K, Korber B. Epigraph: a vaccine design tool applied to an HIV therapeutic vaccine and a pan-filovirus vaccine. Sci Rep 2016;6:33987.

[54] Jespersen MC, Peters B, Nielsen M, Marcatili P. BepiPred-2.0: improving sequence-based B-cell epitope prediction using conformational epitopes. Nucleic Acids Res 2017;45(W1):W24–9.

[55] Saha S, Raghava GP. Prediction of continuous B-cell epitopes in an antigen using recurrent neural network. Proteins 2006;65(1):40–8.

[56] Gupta S, Ansari HR, Gautam A, Open Source Drug Discovery C, Raghava GP. Identification of B-cell epitopes in an antigen for inducing specific class of antibodies. Biol Direct 2013;8:27.

[57] Singh H, Ansari HR, Raghava GP. Improved method for linear B-cell epitope prediction using antigen's primary sequence. PLoS One 2013;8(5):e62216.

[58] Yao B, Zhang L, Liang S, Zhang C. SVMTriP: a method to predict antigenic epitopes using support vector machine to integrate tri-peptide similarity and propensity. PLoS One 2012;7(9):e45152.

[59] Yao B, Zheng D, Liang S, Zhang C. SVMTriP: a method to predict B-cell linear antigenic epitopes. Methods Mol Biol 2020;2131:299–307.

[60] Guan P, Doytchinova IA, Zygouri C, Flower DR. MHCPred: bringing a quantitative dimension to the online prediction of MHC binding. Appl Bioinf 2003;2(1):63–6.

[61] Schubert B, Brachvogel HP, Jurges C, Kohlbacher O. EpiToolKit–a web-based workbench for vaccine design. Bioinformatics 2015;31(13):2211–3.

[62] Bhasin M, Raghava GP. Pcleavage: an SVM based method for prediction of constitutive proteasome and immunoproteasome cleavage sites in antigenic sequences. Nucleic Acids Res 2005;33(Web Server issue):W202–7.

[63] Larsen MV, Lundegaard C, Lamberth K, Buus S, Lund O, Nielsen M. Large-scale validation of methods for cytotoxic T-lymphocyte epitope prediction. BMC Bioinf 2007;8:424.

[64] Nielsen M, Lundegaard C, Worning P, Lauemoller SL, Lamberth K, Buus S, et al. Reliable prediction of T-cell epitopes using neural networks with novel sequence representations. Protein Sci 2003;12(5):1007–17.

[65] Donnes P, Kohlbacher O. SVMHC: a server for prediction of MHC-binding peptides. Nucleic Acids Res 2006;34:W194–7. Web Server Issue.

[66] Sturniolo T, Bono E, Ding J, Raddrizzani L, Tuereci O, Sahin U, et al. Generation of tissue-specific and promiscuous HLA ligand databases using DNA microarrays and virtual HLA class II matrices. Nat Biotechnol 1999;17(6):555–61.

[67] Reche PA, Glutting JP, Zhang H, Reinherz EL. Enhancement to the RANKPEP resource for the prediction of peptide binding to MHC molecules using profiles. Immunogenetics 2004;56(6):405–19.

[68] Tung CW, Ho SY. POPI: predicting immunogenicity of MHC class I binding peptides by mining informative physicochemical properties. Bioinformatics 2007;23(8):942–9.
[69] Antonets DV, Maksiutov AZ. [TEpredict: software for T-cell epitope prediction]. Mol Biol (Mosk) 2010;44(1): 130–9.
[70] Zorzet A, Gustafsson M, Hammerling U. Prediction of food protein allergenicity: a bioinformatic learning systems approach. Silico Biol 2002;2(4):525–34.
[71] Dimitrov I, Bangov I, Flower DR, Doytchinova I. AllerTOP v.2-a server for in silico prediction of allergens. J Mol Model 2014;20(6):2278.
[72] Dimitrov I, Naneva L, Doytchinova I, Bangov I. AllergenFP: allergenicity prediction by descriptor fingerprints. Bioinformatics 2014;30(6):846–51.
[73] Saha S, Raghava GP. AlgPred: prediction of allergenic proteins and mapping of IgE epitopes. Nucleic Acids Res 2006;34(Web Server issue):W202–9.
[74] Doytchinova IA, Flower DR. VaxiJen: a server for prediction of protective antigens, tumour antigens and subunit vaccines. BMC Bioinf 2007;8:4.
[75] Xiang Z, Todd T, Ku KP, Kovacic BL, Larson CB, Chen F, et al. VIOLIN: vaccine investigation and online information network. Nucleic Acids Res 2008;36(Database issue):D923–8.
[76] Reche PA, Reinherz EL. PEPVAC: a web server for multi-epitope vaccine development based on the prediction of supertypic MHC ligands. Nucleic Acids Res 2005; 33(Web Server issue):W138–42.
[77] Burland TG. DNASTAR's Lasergene sequence analysis software. Methods Mol Biol 2000;132:71–91.
[78] Sima M, Ferencova B, Bhattacharyya T, Miles MA, Litvinov SV, Hailu A, et al. Synthetic peptides as a novel approach for detecting antibodies against sand fly saliva. PLoS Negl Trop Dis 2019;13(1):e0007078.
[79] Zhang L, Chen Y, Wong HS, Zhou S, Mamitsuka H, Zhu S. TEPITOPEpan: extending TEPITOPE for peptide binding prediction covering over 700 HLA-DR molecules. PLoS One 2012;7(2):e30483.
[80] Sette A, Rappuoli R. Reverse vaccinology: developing vaccines in the era of genomics. Immunity 2010;33(4): 530–41.
[81] Esposito S, Castellazzi L, Bosco A, Musio A, Stoddard J. Use of a multicomponent, recombinant, meningococcal serogroup B vaccine (4CMenB) for bacterial meningitis prevention. Immunotherapy 2014;6(4):395–408.
[82] Thorpe C, Edwards L, Snelgrove R, Finco O, Rae A, Grandi G, et al. Discovery of a vaccine antigen that protects mice from Chlamydia pneumoniae infection. Vaccine 2007;25(12):2252–60.
[83] Chabria SB, Gupta S, Kozal MJ. Deep sequencing of HIV: clinical and research applications. Annu Rev Genomics Hum Genet 2014;15:295–325.
[84] Renzette N, Gibson L, Jensen JD, Kowalik TF. Human cytomegalovirus intrahost evolution-a new avenue for understanding and controlling herpesvirus infections. Curr Opin Virol 2014;8:109–15.
[85] McLellan JS, Chen M, Joyce MG, Sastry M, Stewart-Jones GB, Yang Y, et al. Structure-based design of a fusion glycoprotein vaccine for respiratory syncytial virus. Science 2013;342(6158):592–8.
[86] Van Damme E, Van Loock M. Functional annotation of human cytomegalovirus gene products: an update. Front Microbiol 2014;5:218.
[87] Khan AM, Heiny AT, Lee KX, Srinivasan KN, Tan TW, August JT, et al. Large-scale analysis of antigenic diversity of T-cell epitopes in dengue virus. BMC Bioinf 2006; 7(Suppl. 5):S4.
[88] Tahir Ul Qamar M, Shokat Z, Muneer I, Ashfaq UA, Javed H, Anwar F, et al. Multiepitope-based subunit vaccine design and evaluation against respiratory syncytial virus using reverse vaccinology approach. Vaccines (Basel). 2020;8(2).
[89] Sarkar B, Ullah MA, Araf Y, Das S, Rahman MH, Moin AT. Designing novel epitope-based polyvalent vaccines against herpes simplex virus-1 and 2 exploiting the immunoinformatics approach. J Biomol Struct Dyn 2020:1–21.
[90] Tettelin H, Masignani V, Cieslewicz MJ, Donati C, Medini D, Ward NL, et al. Genome analysis of multiple pathogenic isolates of Streptococcus agalactiae: implications for the microbial "pan-genome". Proc Natl Acad Sci U S A 2005;102(39):13950–5.
[91] Tettelin H. The bacterial pan-genome and reverse vaccinology. Genome Dyn 2009;6:35–47.
[92] Cabeca TK, Granato C, Bellei N. Epidemiological and clinical features of human coronavirus infections among different subsets of patients. Influenza Other Respir Viruses 2013;7(6):1040–7.
[93] Enayatkhani M, Hasaniazad M, Faezi S, Guklani H, Davoodian P, Ahmadi N, et al. Reverse vaccinology approach to design a novel multi-epitope vaccine candidate against COVID-19: an in silico study. J Biomol Struct Dyn 2020:1–16.
[94] Yazdani Z, Rafiei A, Yazdani M, Valadan R. Design an efficient multi-epitope peptide vaccine candidate against SARS-CoV-2: an in silico analysis. Infect Drug Resist 2020;13:3007–22.
[95] Sanami S, Zandi M, Pourhossein B, Mobini GR, Safaei M, Abed A, et al. Design of a multi-epitope vaccine against SARS-CoV-2 using immunoinformatics approach. Int J Biol Macromol 2020;164:871–83.
[96] Tahir Ul Qamar M, Shahid F, Aslam S, Ashfaq UA, Aslam S, Fatima I, et al. Reverse vaccinology assisted designing of multiepitope-based subunit vaccine against SARS-CoV-2. Infect Dis Poverty 2020;9(1):132.
[97] Mosaheb MM, Reiser ML, Wetzler LM. Toll-like receptor ligand-based vaccine adjuvants require intact MyD88 signaling in antigen-presenting cells for germinal center formation and antibody production. Front Immunol 2017;8:225.
[98] De Groot AS. Immunomics: discovering new targets for vaccines and therapeutics. Drug Discov Today 2006; 11(5–6):203–9.

[99] Zhang L. Multi-epitope vaccines: a promising strategy against tumors and viral infections. Cell Mol Immunol 2018;15(2):182–4.
[100] Peters B, Sidney J, Bourne P, Bui HH, Buus S, Doh G, et al. The immune epitope database and analysis resource: from vision to blueprint. PLoS Biol 2005;3(3):e91.
[101] Saha S, Bhasin M, Raghava GP. Bcipep: a database of B-cell epitopes. BMC Genomics 2005;6:79.
[102] Huang J, Honda WCED. A conformational epitope database. BMC Immunol 2006;7:7.
[103] Schlessinger A, Ofran Y, Yachdav G, Rost B. Epitome: database of structure-inferred antigenic epitopes. Nucleic Acids Res 2006;34(Database issue):D777–80.
[104] Lefranc MP, Giudicelli V, Ginestoux C, Jabado-Michaloud J, Folch G, Bellahcene F, et al. IMGT, the international ImMunoGeneTics information system. Nucleic Acids Res 2009;37(Database issue):D1006–12.
[105] Blythe MJ, Doytchinova IA, Flower DR. JenPep: a database of quantitative functional peptide data for immunology. Bioinformatics 2002;18(3):434–9.
[106] Rammensee H, Bachmann J, Emmerich NP, Bachor OA, Stevanovic S. SYFPEITHI: database for MHC ligands and peptide motifs. Immunogenetics 1999;50(3–4):213–9.
[107] Feldhahn M, Donnes P, Thiel P, Kohlbacher O. FRED–a framework for T-cell epitope detection. Bioinformatics 2009;25(20):2758–9.
[108] Kim C, Kwon S, Lee G, Lee H, Choi J, Kim Y, et al. A database for allergenic proteins and tools for allergenicity prediction. Bioinformation 2009;3(8):344–5.
[109] Allergen nomenclature. IUIS/WHO allergen nomenclature subcommittee. Bull World Health Organ 1994; 72(5):797–806.
[110] Ivanciuc O, Schein CH, Braun W. SDAP: database and computational tools for allergenic proteins. Nucleic Acids Res 2003;31(1):359–62.
[111] Ortutay C, Siermala M, Vihinen M. ImmTree: database of evolutionary relationships of genes and proteins in the human immune system. Immunome Res 2007;3:4.
[112] Rannikko K, Ortutay C, Vihinen M. Immunity genes and their orthologs: a multi-species database. Int Immunol 2007;19(12):1361–70.
[113] Ortutay C, Vihinen M. Immunome: a reference set of genes and proteins for systems biology of the human immune system. Cell Immunol 2006;244(2):87–9.
[114] Ortutay C, Vihinen M. Immunome knowledge base (IKB): an integrated service for immunome research. BMC Immunol 2009;10:3.
[115] Larsen JE, Lund O, Nielsen M. Improved method for predicting linear B-cell epitopes. Immunome Res 2006;2:2.
[116] Haste Andersen P, Nielsen M, Lund O. Prediction of residues in discontinuous B-cell epitopes using protein 3D structures. Protein Sci 2006;15(11):2558–67.
[117] Odorico M, Pellequer JL. BEPITOPE: predicting the location of continuous epitopes and patterns in proteins. J Mol Recognit 2003;16(1):20–2.
[118] Sweredoski MJ, Baldi P. COBEpro: a novel system for predicting continuous B-cell epitopes. Protein Eng Des Sel 2009;22(3):113–20.
[119] Kulkarni-Kale U, Bhosle S, Kolaskar AS. CEP: a conformational epitope prediction server. Nucleic Acids Res 2005;33(Web Server issue):W168–71.
[120] Moreau V, Granier C, Villard S, Laune D, Molina F. Discontinuous epitope prediction based on mimotope analysis. Bioinformatics 2006;22(9):1088–95.
[121] Huang J, Gutteridge A, Honda W, Kanehisa M. MIMOX: a web tool for phage display based epitope mapping. BMC Bioinf 2006;7:451.
[122] Ghate AD, Bhagwat BU, Bhosle SG, Gadepalli SM, Kulkarni-Kale UD. Characterization of antibody-binding sites on proteins: development of a knowledgebase and its applications in improving epitope prediction. Protein Pept Lett 2007;14(6):531–5.
[123] Mayrose I, Penn O, Erez E, Rubinstein ND, Shlomi T, Freund NT, et al. Pepitope: epitope mapping from affinity-selected peptides. Bioinformatics 2007;23(23): 3244–6.
[124] Schreiber A, Humbert M, Benz A, Dietrich U. 3D-Epitope-Explorer (3DEX): localization of conformational epitopes within three-dimensional structures of proteins. J Comput Chem 2005;26(9):879–87.
[125] Guan P, Doytchinova IA, Zygouri C, Flower DR. MHCPred: a server for quantitative prediction of peptide-MHC binding. Nucleic Acids Res 2003;31(13): 3621–4.
[126] Feldhahn M, Thiel P, Schuler MM, Hillen N, Stevanovic S, Rammensee HG, et al. EpiToolKit–a web server for computational immunomics. Nucleic Acids Res 2008;36(Web Server issue):W519–22.
[127] Ponomarenko J, Bui HH, Li W, Fusseder N, Bourne PE, Sette A, et al. ElliPro: a new structure-based tool for the prediction of antibody epitopes. BMC Bioinf 2008; 9:514.
[128] Bhasin M, Raghava GP. Analysis and prediction of affinity of TAP binding peptides using cascade SVM. Protein Sci 2004;13(3):596–607.
[129] Lundegaard C, Lamberth K, Harndahl M, Buus S, Lund O, Nielsen M. NetMHC-3.0: accurate web accessible predictions of human, mouse and monkey MHC class I affinities for peptides of length 8-11. Nucleic Acids Res 2008;36(Web Server issue):W509–12.
[130] Bhasin M, Raghava GP. Prediction of promiscuous and high-affinity mutated MHC binders. Hybrid Hybridomics 2003;22(4):229–34.
[131] Zhang ZH, Koh JL, Zhang GL, Choo KH, Tammi MT, Tong JC. AllerTool: a web server for predicting allergenicity and allergic cross-reactivity in proteins. Bioinformatics 2007;23(4):504–6.
[132] Cui J, Han LY, Li H, Ung CY, Tang ZQ, Zheng CJ, et al. Computer prediction of allergen proteins from sequence-derived protein structural and physicochemical properties. Mol Immunol 2007;44(4):514–20.

[133] Fiers MW, Kleter GA, Nijland H, Peijnenburg AA, Nap JP, van Ham RC. Allermatch, a webtool for the prediction of potential allergenicity according to current FAO/WHO Codex alimentarius guidelines. BMC Bioinf 2004;5:133.
[134] Martinez Barrio A, Soeria-Atmadja D, Nister A, Gustafsson MG, Hammerling U, Bongcam-Rudloff E. EVALLER: a web server for in silico assessment of potential protein allergenicity. Nucleic Acids Res 2007; 35(Web Server issue):W694−700.
[135] He Y, Xiang Z, Mobley HL. Vaxign: the first web-based vaccine design program for reverse vaccinology and applications for vaccine development. J Biomed Biotechnol 2010;2010:297505.
[136] Harish N, Gupta R, Agarwal P, Scaria V, Pillai B. DyNAVacS: an integrative tool for optimized DNA vaccine design. Nucleic Acids Res 2006;34(Web Server issue): W264−6.

CHAPTER 4

Riddle of Herd Immunity in SARS-CoV-2-Induced Viral Terrorism: Science to Society

ASHA SHELLY • DIBYAJNAN CHAKRABORTY • AMULYA K. PANDA • TANMAY MAJUMDAR

1 HERD IMMUNITY

Herd immunity or society immunity is essentially a conception of obtaining protection through natural infection or mass vaccination in a specific cohort. The purpose of individual vaccination is to establish stable immunity in the body to prevent, or reduce, the chances of recurrent infection of a particular disease. Nonetheless, in the public health, the purpose of herd immunity is to escalate immunization efficacy to control or eliminate the infection in a particular society. It is evidently a sequential and continuous process of immunization among individuals because natural or vaccine-based immunity is lost over time through the waning of vaccinated individuals due to death and the arrival of newly susceptible individuals due to birth or migration. Thus to endure herd immunity in a particular cohort, it is imperative to vaccinate new arrivals at regular intervals [1]. Herd immunity recuperates the protection against pathogens that are transmissible in nature [2–4]. The successful eradication of several viral diseases such as smallpox [5] and poliovirus, the reduction in the transmission of pertussis, and the protection against influenza, pneumococcal disease, cholera [6], and rotavirus [7] were potentially accomplished only through expanded immunization programs that have been the pertinent and essential factors for evolving herd immunity [8]. In addition, the level of vaccination needed to be achieved through herd immunity varies depending upon the frequencies of secondary infections [3,9]. For example, in measles, a highly infectious viral disease, one person can infect up to 18 individuals. Thus 95% of the people are required to be immune in order to achieve herd immunity. The new severe acute respiratory syndrome coronavirus 2 (SARS-CoV-2) has a lower infection rate than measles. On average, each infected person can pass the virus onto two or three new people [10]. This means that herd immunity should be accomplished when around 60% of a specific population becomes exposed to coronavirus disease 2019 (COVID-19) [11].

Here, we have discussed several cardinal factors that can directly or indirectly alter the acquisition of immunity. Among those factors, the gut microbiome is an indispensable component that has the capacity to influence individual immunity and thereby mapping the population immunity. Gut microbiota, along with nutrients and environmental factors, play an indispensable and cumulative role in evolving herd immunity against any infectious disease. Several eminent peers recommended that heterologous immunity is the furthermost important curriculum contributing protection against SARS-CoV-2 infection in India. It has been proposed that the robust oral poliovirus vaccination program developed herd immunity. This same method has the capability to deliver the protection individually against coronavirus infection. The herd immunity against COVID-19 may be possible through rigorous immunization process with SARS-CoV-2-specific candidate vaccine in several communities at a large scale and in a repetitive method.

2 NETWORK AMONG THE MICROBIOME, NUTRIENTS, AND ENVIRONMENTAL FACTORS INFLUENCES IMMUNE SYSTEM DEVELOPMENT

2.1 Impact of Microbiome—The Second Brain to Mature the Immune System

To generate competent herd immunity, the maturation of the immune system is an indispensable and cardinal

COVID-19: Tackling Global Pandemics through Scientific and Social Tools. https://doi.org/10.1016/B978-0-323-85844-1.00005-2

factor. Alterations of the microbiota by any means could refute impact on the immune system's development, which thwarts the vaccination program and subsequently herd immunity [12,13]. Collectively, the microbial community residing in our body is termed as the microbiota. Microbiota has been popularly known as the 'second brain' of the body owing to its fundamental essentiality for achieving protective immunity. Microbiome in totality outnumbers the total cells in the human body by several fold; 99% of the total genome of humans comes from microbiota and 1% comes from 23,000 genes of the body's own cells [14–16]. The gut microbiota dysbiosis significantly impedes the development and function of both innate and adaptive immunity [17,18]. The human microbiome synchronizes several functions of the body such as nutrient metabolism, intestinal barrier functions, shaping of the immune system, etc. It supports the prevention of several diseases and contributes to cultivating the genetic diversity in the population [19]. The absence of normal microbiota exacerbates the maturation of the immune system, worsening both structure and functions as intimated by studies in germ-free (GF) mice [20,21]. However, by reestablishing microbiota symbiosis, all these effects can be reinstated.

In addition, the microbiome seems to postulate a reasonable justification about the disparity in vaccine responses at both individual and community levels. Gut microbiota have specific immunomodulatory properties [22]. *Bacteroides fragilis*, a gram-negative bacteria (phylum: Bacteroidetes) regulates mucosal tolerance to self-antigens by maintaining T-cell homeostasis, preventing T-helper (Th) 1/Th17 balance [23], and also induces suppressive forkhead box p3^{+} (Foxp3^{+}) T-regulatory (Treg) cell functions by bolstering anti-inflammatory cytokines such as interleukin 10 and transforming growth factor β [24]. Similar preferment of Treg cell differentiation has been shown with the presence of *Clostridia* sp. in the colon, but not in the small intestine in maintaining the immune cell homeostasis [25]. Furthermore, differentiation of Th17 cells and mucosal immunoglobulin A (IgA) secretion are possible only with the colonization of segmented filamentous bacteria (SFB) [26]. Several mucosal pathogens such as bacteria, virus, and fungi uphold both Th17 and IgA secretion. The ecology of the microbiome is crucial from birth, for achieving the highest efficacy of vaccination because the immunization process is mostly been accomplished during childhood (Fig. 4.1). An infant's microbiome is determined by the maternal-offspring exchange of microbiota [27]. It becomes similar to the adult microbiome by 3 years of age. Until then, it is highly variable [28] and grows uninterrupted, along with the development of the immune system [20]. The total microbiome composition of infants is mostly dominated by phyla such as Firmicutes, Tenericutes, Proteobacteria, Bacteroidetes, and Fusobacteria based on the commensal microbiota harbored in the placenta [29–31]. The variation of commensals is also related to the mode of infant delivery. The intestinal flora of neonates born by vaginal delivery resemble the maternal vaginal flora. They have higher frequencies of both *Lactobacillus* sp. and *Prevotella* sp. [32]. In contrast, vaginal delivery was associated with a higher prevalence of facultative anaerobic organisms such as *Escherichia coli*, *Staphylococcus* sp., *Streptococcus* sp., and other Enterobacteriaceae (phylum: Proteobacteria) for the first few days of birth [27,33]. Nonetheless, cesarean babies are predominantly colonized with maternal-skin microbiota such as *Clostridium*, *Staphylococcus*, *Propionibacterium*, and *Corynebacterium* [34,35]. After birth, milk feeding has the most significant impact on altering the microbiome composition of children. Breast milk provides augmented existence of *Bifidobacterium* sp. (phylum: Actinobacteria) in the intestinal tract of infants. Infants also have a higher abundance of *Staphylococcus* sp. *Streptococcus* sp., *Lactobacillus* sp. (phylum: Firmicutes), *Serratia* sp., *Ralstonia* sp. (phylum: Proteobacteria), *Corynebacterium* sp. (phylum: Actinobacteria), etc. in their gut [36]. In contrast, there is a richness of aerobic bacteria and lower frequencies of *Bifidobacterium* sp. in the guts of bottle milk-fed infants [37–39]. The neonatal gut microbiota not only programs the metabolic function but also educates the naïve immune system, without which achieving the success of vaccination will be precluded.

2.2 Effect of Nutrients

The balance among nutrient metabolism, microbiota, and vaccination efficiency is delimited by leptin signaling [40], which synchronizes the hunger threshold, as well as cellular immunity by maintaining Th1/Th2 balance [41] and suppressing Treg cell differentiation [42]. In addition, GF mice have reduced leptin expression, suggesting a connection between microbiota and leptin-mediated immune response in the vaccination process [43]. Furthermore, several metabolites are critical—short-chain fatty acids (SCFAs) are an example of how nutrient processing by microbiota shapes the maturation of the immune system [44]. SCFAs are generated exclusively by microbiota. Important SCFAs (butyrate and acetate) help maintain intestinal epithelial integrity [45,46]. Vitamin A (retinoic acid) deficiency in the diet has also been linked to the augmented frequency of *E. coli*, causing enteric

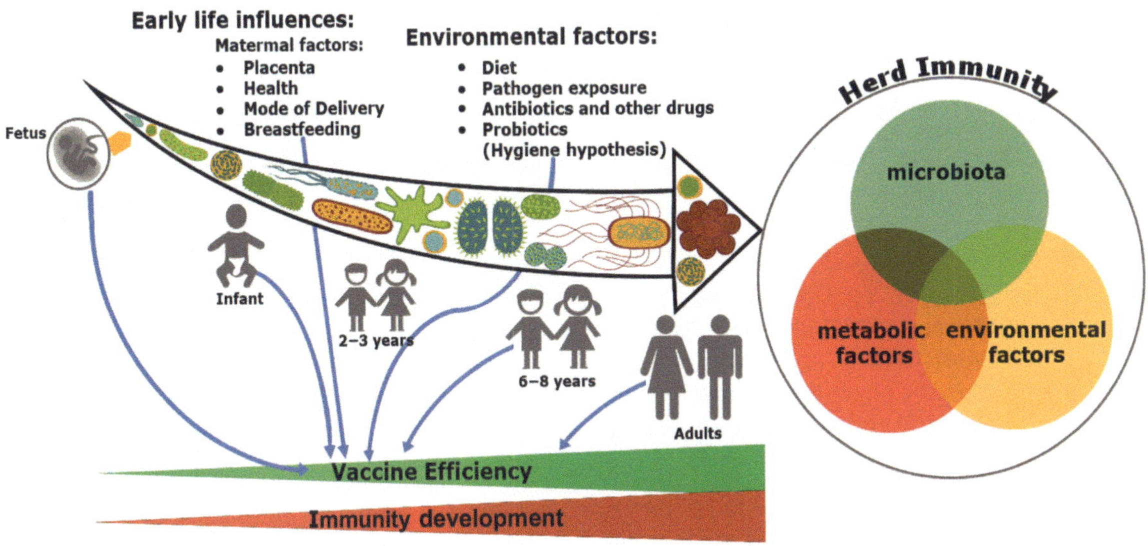

FIG. 4.1 **The Coevolution of Immune System and Microbiota Determines Herd Immunity.** The immune system and microbiota coevolve together in a symbiotic relationship. The impact of microbiome on the immune system development cannot be ignored. From fetus to adulthood, microbiomes are synchronized by maternal transfer and environmental factors. Early maternal factors such as mode of delivery, breastfeeding, antibiotics, and diets all influence the immune system. Hence, all have an impact on subsequent immunologic responses to many vaccines. Development of herd immunity in a community against any infection is the result of a complex outcome of host-specific factors such as microbiota, metabolism, and environmental conditions. Malnutrition and repeated gastrointestinal infections reduce many vaccines' efficacy. Microbial dysbiosis, along with environmental enteropathy, influences undernourishment. It impairs the immune system development and decreases the efficiency of vaccines in a community, thereby compromising herd immunity.

infections [47]. Vitamin A can endorse the balance between Treg and Th17 subsets. Imbalance in vitamin A levels causes enrichment of Th17 subsets in the small intestine, which is associated with ameliorated frequencies of SFB [48,49]. Thus microbiota frequencies are the indispensable factors for efficacious immunization and the influence of nutrients to support the microbiota symbiosis can be the preeminent factor for reaching the avidity of immunity (Fig. 4.1).

2.3 Environmental Factors, Microbiome, and Vaccine Efficacy

The aforementioned reports suggest that both nutrition metabolism and the composition of microbiota have a cardinal impact on the expansion of innate and adaptive immune systems (Fig. 4.1). According to the Centers for Disease Control, every 1 in 10 children in the United States and 1 in 4 children in Europe have various allergic disorders [50], bringing allergic diseases to an epidemic condition [51,52]. This is strongly aggravated in urban environments involving various pollutants. Exposure to various pollutants interrupts the frequencies of the microbiome in the body. In contrast, the concept of the hygiene hypothesis advocates that exposure to certain viral infections in early life and large family size thwart the risk of suffering from hay fever and allergic rhinitis [53]. The hygiene hypothesis basically takes into account the "critical period of immune system development" or the "window of opportunities" in early childhood, during which the immune system of the child is shaped for the adulthood by several intrinsic or extrinsic factors [54] (Fig. 4.1). The shaping of an ideal immune system decides the accomplishment of the vaccination. Dietary changes and environmental factors can account for up to 57% of gut microbiota changes, affecting the immune system [55]. Consequently, westerners have reduced frequencies of microbial diversity in the gut because of high-fat and low-fiber diets, which unwittingly reduces the microbiota enrichments [56,57]. In a similar study, Filippo et al. observed that the microbiota of rural African children fluctuates vividly from that of city-dwelling European children. African children are predominantly occupied by more anti-inflammatory commensal bacteria such as *Bacteroides* sp., *Prevotella* sp., *Faecalibacterium prausnitzii*, *Xylanibacter* sp., etc. They had lower frequencies of both Firmicutes and Enterobacteriaceae, and this is obligatory for the synthesis of SCFAs, providing anti-inflammatory responses in the gut. These studies furthermore implied that higher frequencies of anti-inflammatory commensal microbiota boosted the immunization adeptness to protect children from certain pathogens (including enteropathogens). In contrast, malnutrition encumbered vaccine efficacy in a community where microbiome dysbiosis is a preponderant factor due to environmental enteropathy (EE) [12,58].

3 VACCINE CANDIDATE AND ROUTE OF IMMUNIZATION DETERMINE THE FATE OF HERD IMMUNITY

Vaccine is the imperative prophylactic intervention preventing the burden of several infectious diseases [59,60]. Oral vaccines are used mostly in developing countries, although significant discrepancies exist in the competency of the oral vaccines based on the geographic distributions [61–63]. Many vaccines with both living or nonliving agents, such as B-subunit-inactivated whole-cell combination vaccine [64], poliovirus vaccine, SC602 live *Shigella flexneri* 2a vaccine [65], oral rotavirus vaccine (RVV) [66], and CVD 103-HgR live cholera vaccine [67,68], have also demonstrated differential immunogenic responses based on their route of immunization. In the case of oral poliovirus vaccine (OPV), live-attenuated poliovirus is given orally in India from birth. OPV can spread efficiently and triggers the innate immunity followed by T-cell and B-cell activation because live microbes can persist for longer periods than dead organisms. Thus significant T-memory cells and B-memory cells are generated, which are key factors for the competence of herd immunity. OPV can generate antibodies that limit the proliferation of the challenged wild poliovirus and thwart person-to-person transmission. Nonetheless, in the United States, formalin-fixed inactivated poliovirus vaccine (IPV) is given to newborn babies. IPV cannot spread to different organs and can activate only B cells for antibody production. The dead vaccines sometimes escape T-cell activation because the antigens do not persist in the immune system for a substantial period. To activate T cells, vaccine candidates must endure in the body long enough for antigen presentation (Fig. 4.2A and B). In the case of IPVs, dead vaccine candidates are directly processed by B cells to produce antibodies for neutralizing the pathogens.

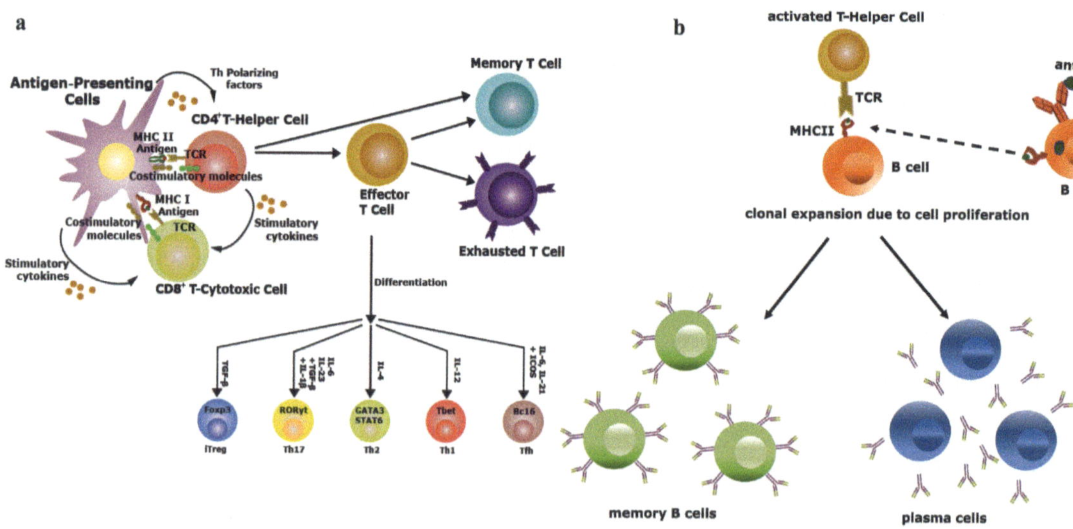

FIG. 4.2 **The Antigenic Diversity of T Cells and B Cells Defines the Fate of Vaccination. (A)** Upon contact with antigen-presenting cells (APCs) through the T-cell receptor (TCR)-antigen-major histocompatibility complex (MHC) complex, naïve T cells are differentiated into effector and memory T cells. Based on the antigenic property, effector T cells are of two types, i.e., $CD4^+$ T-helper (Th) or $CD8^+$ T-cytotoxic (Tc) cells. Antigens exposed by MHC class II (MHCII) bind with TCRs of Th cells, and antigens exposed by MHC class I (MHCI) bind with TCRs of Tc cells. Th cells have the potential to further differentiate into Th1, Th2, Th17, and induced regulatory T cells (iTreg), a process controlled by the lineage-specific transcription factors and effector cytokines produced by APCs. The effectiveness of memory T cells decides the efficacy of vaccination. **(B)** B cells (APCs) recognize antigens by their B-cell receptors (BCRs), followed by internalization of antigens and presentation to Th cells, which are specific to the same antigen. The TCRs of Th cells interact with antigens exposed by MHCII of B cells. Then the activated B cells trigger their own proliferation and differentiate into antibody-secreting plasma cells and memory B cells. Activation and class switching of B cells is the key factor for the success of immunization. *IL*, interleukin; *TGF-β*, transforming growth factor β.

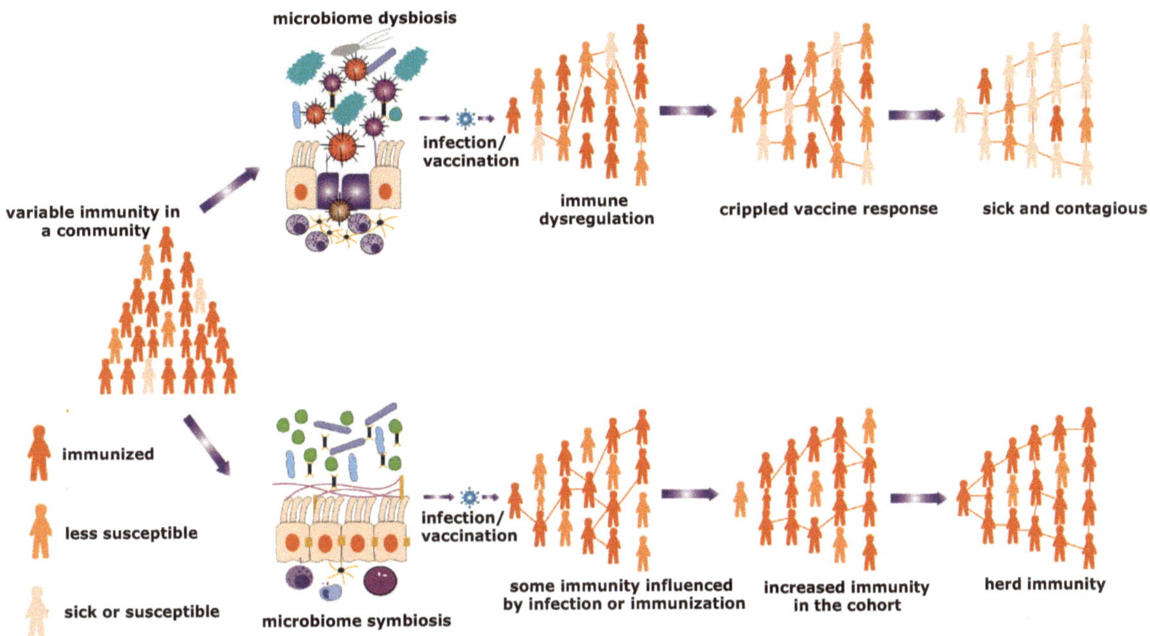

FIG. 4.3 **Microbiota, Vaccination, and Herd Immunity.** Maturity of herd immunity against any infectious disease is a significant means of acquiring protection during an epidemic or a pandemic. A spatial community is occupied by people having a variable degree of resistance against a particular infection. Microbial symbiosis regulates pathogen transmission dynamics and the efficacy of vaccination among different individuals in a population, promoting the development of herd immunity. In contrast, dysbiosis of the microbiome causes immune dysregulation, generating a suboptimal immune response and negative impacts on the vaccination program. Therefore the number of sick individuals increases in a particular cohort. This reduces the chances of evolving herd immunity and increases the number of sick and contagious patients in a community.

Thus IPV poses only individual protection but cannot prevent the spread of wild poliovirus in the community. Hence, the chances of herd immunity against IPV are very low. In different oral vaccines, the microbiome plays a pivotal role in providing mucosal immunity [69]. The enhancement of mucosal immunity is the symphony of the immunization process. Similarly, in OPV, the intestinal microbiome has a significant influence on mucosal immunity-mediated adequate vaccine responses. This is deficient in IPV, in which the route of immunization is parenteral [70]. Thus the microbiome seems to give plausible justification for developing herd immunity with OPV (Fig. 4.3).

4 POTENTIAL ENGROSSMENT OF THE MICROBIOME FROM INDIVIDUAL VACCINATION TO COMMUNITY IMMUNITY

4.1 Variable Expression of Microbiota Defines the Outcome of Vaccination

Vaccines are the utmost significant protective interventions that reduce the burden of diverse infectious ailments [59,60]. The microbiome not only plays a crucial role in shaping the immune system but also determines the efficacy of several vaccines, thereby develops protective immune responses against diseases. Previous studies have proposed that differences in even normal residential intestinal microbiota between individuals of a population can affect the efficacy of many vaccines remarkably both locally and systemically [71] (Table 4.1). Intestinal dysbiosis causes different residential intestinal microbiota amid different individuals. Previously, it has been observed that gut colonization with human *Bifidobacterium* sp. in GF mice led to enhanced immune response against rotaviruses by increasing antirotavirus IgA secretion. Hence, *Bifidobacterium* sp. engrossment is essential for priming an infant's adaptive immune system activation [61,72–75]. Conversely, a reduced vaccine response has been correlated with increased frequencies of "Enterobacteriales", "Pseudomonadales," and "Clostridiales" in both oral and parenteral vaccines [61]. This variance in immunization response decreases the probability of developing herd immunity in a particular cohort. Separately, a positive correlation has been observed between the

TABLE 4.1
The Effect of Differential Species Richness of Intestinal Microbiome on Immune Responses and Vaccination.

Vaccine Type	Microbial Richness Responsible for Vaccine Effectiveness	Tested Population (Ref)
1. INTESTINAL DYSBIOSIS: DIFFERENCES IN MICROBIOTA COMPOSITION OF VACCINE RESPONDERS AND NONRESPONDERS AMONG DIFFERENT POPULATIONS		
Pentavalent diphtheria-tetanus-acellular pertussis-inactivated poliomyelitis-*Haemophilus influenzae* type B vaccine (DTaP-IPV-*Hib*)	Existence of intestinal *Bifidobacterium* sp. especially *Bifidobacterium longum*, *Bifidobacterium infantis*, and *Bifidobacterium breve* under the phylum Actinobacteria enhanced antipoliovirus IgA antibodies.	Infants (France) [76]
OPV, parenteral TT, BCG, and HBV	Abundance of *B. longum* of the phylum Actinobacteria augmented oral and parenteral vaccine avidities among infants. Amplified frequencies of Enterobacteriales and Pseudomonadales encumbered vaccine-specific immune responses.	Infants (Bangladesh) [61]
RVV	*Streptococcus bovis* under the phylum Firmicutes correlated with oral RVV response. Increased *Bacteroides* sp. and *Prevotella* sp. of the phylum Bacteroidetes were associated with lack of RVV response.	Infants (Ghana) [74]
	Abundance of bacteria related to *Clostridium* cluster IX of the phylum Firmicutes and *Serratia* sp. and *Escherichia coli* of the phylum Proteobacteria augmented RVV efficiency in a population.	Infants (Pakistan) [73]
TIV and OPV	Activation of TLR5 enriched antibody response against TIV and OPV. Impaired antibody response in both GF mice and antibiotic-treated mice was restored by oral reconstitution with the flagellated strain of *E. coli*.	Mice [96,97]
Live attenuated oral typhoid (Ty21a) vaccine	Cell-mediated immune response was found to be associated with SFB under the phylum Firmicutes, while humoral response was independent of any microbial dysbiosis.	Healthy adults [72]
2. EE: SBBO DISTURBING THE NORMAL GUT MICROBIOTA IS RESPONSIBLE FOR SUBOPTIMAL RESPONSE OF MANY VACCINES IN LOW-INCOME SETTINGS AND DEVELOPING COUNTRIES		
Live cholera vaccine CVD-103-HgR	SBBO could blunt the immunologic response to live cholera vaccine candidate CVD 103-HgR.	Children from high- and low-income countries [60,89]

(*continued*)

TABLE 4.1
The Effect of Differential Species Richness of Intestinal Microbiome on Immune Responses and Vaccination.—cont'd

Vaccine Type	Microbial Richness Responsible for Vaccine Effectiveness	Tested Population (Ref)
Live oral *Shigella flexneri* 2a candidate SC602 vaccine	Lower dose of 10^4 CFU was adequate to elicit a protective response in North American cohort, while the ingestion of 10^4, 10^5, or 10^6 CFU of SC602 was not adequate for Bangladeshi toddlers because of SBBO.	[91,92]
Oral RVV, OPV	EE was found to be associated with refuted efficacy of oral RVV but had no impact on the parenteral administered vaccines—tetanus, pertussis, diphtheria, *Haemophilus influenzae* type B, and measles.	[93,94]
	Enhanced frequencies of *Campylobacter* sp. under the phylum Proteobacteria and enteroviruses at the time of immunization negatively correlated with the immunogenicity of OPV and diminished rotavirus immunoglobulin A titer (RVI).	Urban slum areas of Bangladesh [95]
3. PROBIOTICS AND PREBIOTICS: SEVERAL STUDIES MENTIONED ABOUT THE ROLE OF BOTH PROBIOTICS AND PREBIOTICS ON VACCINE EFFICACY, WHICH FURTHER SUBSTANTIATES THE SIGNIFICANCE OF RESIDENTIAL GUT MICROBIOTA ON VACCINE EFFECTIVENESS		
The impact of several probiotic strains on the efficacy of 17 different vaccines, e.g., DTP, whole-cell DTP *vaccine* (DTwP), diphtheria, tetanus, acellular pertussis, with *H. influenzae* type B (DTaP-Hib), DTaP-IPV-Hib, HAV, HBV, Hib, LAIV, attenuated virus MMR vaccine with chickenpox vaccine or varicella vaccine (MMRV), OCV, OPV, ORV, pneumococcal conjugate *vaccine* (PCV7), pneumococcal polysaccharide vaccine (PPV23), polio, TIV, and Ty21a	*Lactobacillus* sp., *Bifidobacterium* sp., and *Saccharomyces boulardii* were the most frequently used microorganisms in probiotics. The beneficial effects of probiotics were found to be strongest in both oral and parenteral vaccines.	[71]
Salmonella typhimurium SL1479 vaccination	The administration of a prebiotic (a nondigestible food component that promotes the growth of beneficial microorganisms), e.g., fructo-oligosaccharide/inulin mix, was shown to enhance the efficacy of *S. typhimurium* SL1479 vaccine.	Mice [98]

(continued)

TABLE 4.1
The Effect of Differential Species Richness of Intestinal Microbiome on Immune Responses and Vaccination.—cont'd

Vaccine Type	Microbial Richness Responsible for Vaccine Effectiveness	Tested Population (Ref)
4. ANTIBIOTICS: SEVERAL STUDIES IN BOTH MICE AND HUMANS HAVE SHOWN THE EFFECT OF ANTIBIOTIC-MEDIATED DYSBIOSIS ON VACCINE RESPONSES		
BCG vaccine; Bexsero meningococcal serogroup B vaccine (MenB); the meningococcal serogroup C vaccine-NeisVac-C (MenC); the Prevenar 13-valent pneumococcal conjugate vaccine (PVC13); the hexavalent combination vaccine against hepatitis B, diphtheria, tetanus, pertussis; *H. influenzae* type B; inactivated poliomyelitis virus (INFANRIX Hexa)	Antibiotic-mediated dysbiosis in early life impaired antibody response against these vaccines. Restoration of commensal microbiota retrieved vaccine efficiencies.	Newborn mice [13]
RVV	Positive correlation was observed in antibiotic-driven microbiota modulation and increased immunologic response.	Adult cohort [73]

BCG, bacille Calmette-Guérin; *DTP*, diphtheria and tetanus toxoids and acellular pertussis; *EE*, environmental enteropathy; *GF*, germ-free; *HAV*, hepatitis A vaccine; *HBV*, hepatitis B vaccine; *LAIV*, live attenuated influenza vaccine; *OCV*, oral cholera vaccine; *OPV*, oral poliovirus vaccine; *ORV*, oral rabies vaccine; *RVV*, rotavirus vaccine; *SBBO*, small bowel bacterial overgrowth; *SFB*, segmented filamentous bacteria; *TIV*, trivalent inactivated influenza vaccine; *TLR5*, toll-like receptor 5; *TT*, tetanus toxoid.

Bifidobacterium sp. (*Bifidobacterium longum, Bifidobacterium infantis,* or *Bifidobacterium breve*) in the intestine and enhanced antipoliovirus IgA antibody production followed by immunization with OPV with a pentavalent diphtheria-tetanus-acellular pertussis-inactivated poliomyelitis-*Haemophilus influenzae* type B vaccine [76]. Thus certain specific microbiome frequencies are more wanted because they remain a source of natural adjuvant that facilitates the efficacy of vaccine responses and helps in developing herd immunity against pathogens [77].

4.2 Exacerbation of the Vaccination Process Because of Environmental Enteropathy

Metabolic dysfunction is connected with a feeble response to many traditional oral vaccines. Poor economic conditions have compromised the accessibility of a balanced diet in various populations. This impairs the efficacy of many oral vaccines, including cholera vaccine, poliovirus vaccine, and RVV. Briefly, nutrient imbalance supports EE, leading to small bowel bacterial overgrowth (SBBO) among different individuals [78]. In addition, underprivileged children in slum areas commonly have excessive bacterial colonization in the proximal small intestine. This impairs the architecture of the intestine, with blunted villi and abnormal crypt-to-villus ratio [71,79–81]. The excessive bacterial growth augments the number of intraepithelial lymphocytes [82,83], leading to a marked increase in lymphocyte numbers, which is a signature factor of allergic diseases [84–86]. The regular ingestion of both contaminated food and water containing fecal-oral bacteria aggravates plasmacytic infiltration in the lamina propria of intestinal mucosa [87]. The concurrent enteric infections with enteropathogens cause enhanced small intestinal inflammation, resulting in intestinal barrier dysfunction and reducing the absorption of nutrients [88]. This results in a blunted immunization process, as observed with the live cholera vaccine CVD-103-HgR in an underdeveloped community. The success of oral vaccines depends upon both diet and hygiene conditions. High seroconversion rates in Indonesian children living in poor conditions required 10-fold higher dose, i.e., 5×10^9 colony-forming units (CFUs) of CVD-103-HgR rather than the 5×10^8 CFU dose, which is constantly immunogenic in North Americans and Europeans [67,68,89] who generally take healthy diets. People with SBBO typically have an immunologically activated small intestine [87], with a proinflammatory state in the small intestine. Thus for any live or attenuated vaccine material, the small intestine microenvironment becomes hostile, impeding the vaccine. Poor induction of innate and adaptive immune responses occurs, and the chances of herd immunity become futile [90]. Similarly, differential response of the live oral *S. flexneri* 2a candidate SC602 vaccine with respect to diet has also been observed where it showed a protective response in North American volunteers with a strong immune response [91]. In contrast, in Bangladeshi toddlers who had poor diets, the ingestion of SC602 did not produce any immunized efficiency [92]. In a separate study, infants from the slum area of Bangladesh also showed reduced efficacy of both oral RVV and OPV due to intestinal injury, whereas no impact was observed on the parenterally administered vaccines such as for tetanus, pertussis, diphtheria, *H. influenzae* type B, and measles [93,94]. Because of microbiome-dysbiosis-induced nutritional imbalance, people in the slum areas of Bangladesh had increased frequencies of both *Campylobacter* sp. and enterovirus in the gut. This was negatively connected with the immunogenicity of OPV and also diminished the IgA titer of rotavirus (Table 4.1) [95].

4.3 Counterintuition Between Probiotics and Antibiotics on the Efficacy of Vaccination

Probiotics are live microorganisms that, when consumed orally in sufficient amounts, are beneficial to the vaccine effectiveness of the host [71,99]. Their beneficial effects have been found to be the strongest in both oral and parenteral influenza vaccines. Additionally, the administration of a prebiotic-like fructo-oligosaccharide/inulin mix augmented the efficacy of *Salmonella typhimurium* SL1479 vaccine in a murine model [98], signifying that the resident gastrointestinal microbiota can be modulated by prebiotics as well as probiotics, which may confer improved immunologic response toward vaccines.

Earlier studies have further supplemented the role of microbiota in vaccine efficiency. Rigorous exposure to antibiotics exacerbates microbiome dysbiosis and impairs vaccine responses producing long-lasting deleterious effects [84]. Infants exposed to antibiotics, given early in life, have been shown to be more prone to many diseases such as asthma, obesity [100,101], and metabolic syndrome [102]. Furthermore, Lynn et al. reported that early-life antibiotic-mediated dysbiosis caused impaired antibody response against many vaccines. Thus both probiotics and antibiotics associated with microbiome symbiosis are important factors that help in sustaining the success of vaccines (Table 4.1).

5 INNATE IMMUNITY SETS THE FOOTPRINTS FOR ANTIBODY TITER

Innate immunity is the first line of defense, which encompasses different pathogen recognition receptors (PRRs). PRRs are majorly responsible for the sensing of foreign microbes. They recognize specific microbial structures called pathogen-associated molecular patterns (PAMPs) and also the endogenous cellular products produced in response to cell stress (damage-associated molecular pattern [DAMPs]). PRRs are majorly classified into four different classes. Examples of PRRs include C-reactive protein, transmembrane different toll-like receptors (TLRs), and C-type lectin receptors, as well as cytoplasmic nucleotide binding oligomerization domain-like receptor (NLR), retinoic acid-inducible gene I (RIG-I)-like receptors, melanoma differentiation-associated gene 5 (MDA5), stimulator of interferon gene (STING), etc. [103,104] (Fig. 4.4). For a successful immune response against any pathogen, its recognition is a critical practice that must competently differentiate between self and nonself in order to facilitate a specific immune response and to circumvent collateral damage rendered by autoimmunity. For RNA viruses, the RNA genome and its replication and metabolic products represent the major non-self-products harboring PAMPs which are recognized by several PRRs such as TLR3, RIG-I, or MDA5 [105] (Fig. 4.4).

6 INNATE IMMUNITY: FUNDAMENTAL FACTOR FOR SPECIFICITY OF HERD IMMUNITY

Under an ideal infectious stage, the recognition of different PAMPs of microbes first activates the innate immune system (Fig. 4.4) followed by activation of both the arms of the adaptive immunity, which later generate the memory response (Fig. 4.2) [106]. For herd immunity generation, this phenomenon should happen in mass fraction of the population to develop memory response against different antigens. However, each antigen type has both benefits and difficulties that can perturb the activation of the immune system and limit the chances of developing herd immunity in the community [107]. For immune system activation, pathogens enter the body and the naked DNA/RNA of these microbes starts producing antigens. These antigens are processed by the innate immune components displaying intracellular PRRs, e.g., TLR3, MDA5, STING (Fig. 4.4) [103]. The activated innate immunity further gets involved in T-cell differentiation through antigen-presenting cells (APCs) and develops memory T cells (Fig. 4.2A), which is followed by the activation of B cells (Fig. 4.2B). Activated plasma cells produce both a strong antibody response to neutralize the antigen and memory B cells, which provide protection against the same or similar type of infection in the future

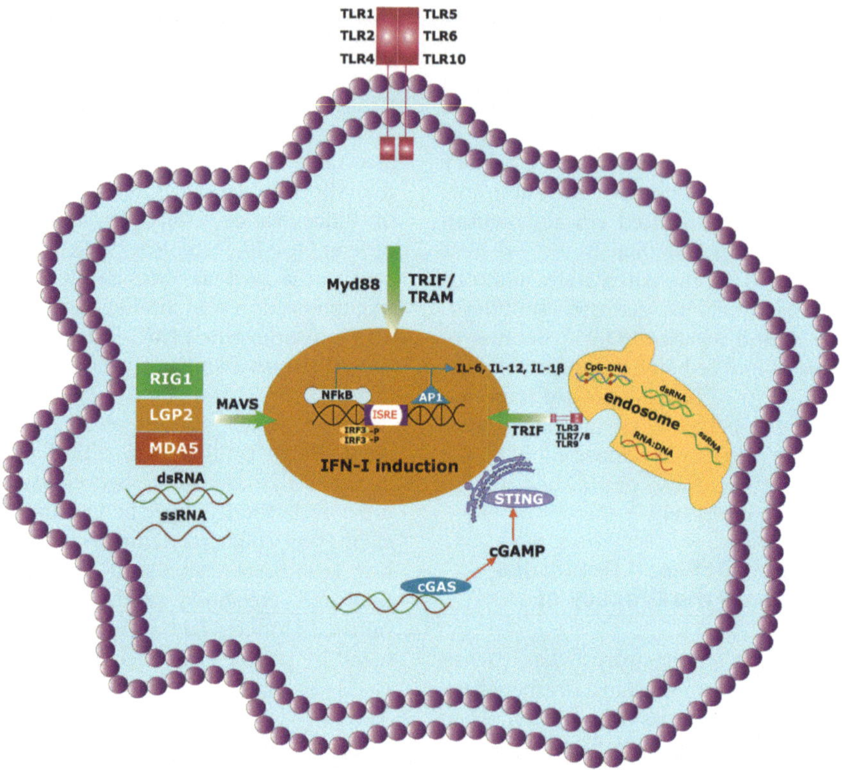

FIG. 4.4 Innate Immunity Sets the Fate for Immune Response to Any Pathogen. Innate immunity is activated by germline-encoded pathogen recognition receptors (PRRs) that recognize pathogen-associated molecular patterns and damage-associated molecular patterns of the microbes. Specific PRRs mainly include membrane-bound several toll-like receptors (TLRs), cytosolic DNA sensor, e.g., stimulator of interferon gene (STING), RIG-I-like receptor (RLR), MDA5, NOD-like receptor, all of which coordinate with the host's innate immune responses through the activation of the nuclear factor κB (NF-κB), activator protein 1 (AP1), and interferon regulatory factor (IRF)-signaling pathway, which triggers the antiviral interferon type I (IFN-I) pathway. Activated innate immunity can commence the adaptive immunity by triggering both the T cells and B cells. *dsRNA*, double-stranded RNA; *IL*, interleukin; *ssRNA*, single-stranded RNA.

(Fig. 4.2B) [108]. During pandemic, epidemic, or endemic (repeated infections by similar pathogens in an area) situations, the generated memory responses act as a natural vaccine and help in developing herd immunity against the microbes. Antibody titer is the major criterion for long-term herd immunity. Nakaya et al. investigated the antibody-generation sequence in response to trivalent inactivated influenza vaccine (TIV) by systematic analysis. The authors observed a positive correlation between early expression of TLR5—an important PRR for bacterial flagella of the microbiome population—and antibody titer [96,109]. Consistently, it was also found that both GF and antibiotic-treated mice could not provide immunity to TIV, which suggested that the intestinal microbiota were a source of TLR5 and helped in providing immunity against TIV. Nonetheless, antibody titer was reestablished by oral reconstitution with flagellated *E. coli*. This observation also reveals a prominent role of the intestinal microbiota in controlling immunity to parenteral vaccines [97]. Several earlier reports have also shown that abrogation of gut microbiota symbiosis exacerbates the host's susceptibility to several viral infections (Fig. 4.3) [110].

7 MONARCHY OF SOCIETY IMMUNITY CAN BE FOUNDED BY HETEROLOGOUS IMMUNITY

Society immunity or herd immunity is important for the establishment of enduring immunologic memory responses. Immunologic memory can be of two types—

memory responses generated by confronting the pathogens by the innate immune system, which are later followed by activation of adaptive immune responses referred to as homologous memory response. However, herd immunity can also be instigated in response to different pathogens or antigens in a particular group as a result of the action of memory response generated previously, known as heterologous immunity [111]. Theoretically, the cross-reactivity (or poly-specificity) of lymphocytes in antigen (or epitope) recognition is the essential factor of heterologous herd immunity (Fig. 4.5A). Cross-reactive memory-T-cell responses and memory-B-cell responses are engaged at the evolutionarily conserved sites among many virus groups, such as influenza or dengue virus, or among different members of the same virus group, such as arenaviruses, flaviviruses, and hantaviruses [112]. Nonetheless, heterologous immune response among absolutely isolated viruses has also been found for example between human papillomavirus and coronavirus [113], lymphocytic choriomeningitis virus and vaccinia virus [116,117], influenza virus and hepatitis C virus [114], influenza virus and Epstein-Barr virus [118] and influenza virus and HIV [115].

In these infections, immune responses facilitated by cross-reactive antibodies are often observed due to the existence of the heterologous, but comparable, antigens. After immunization with different vaccines, our body has a mixture of various antibodies. Among these different antibodies, one or more can recognize the antigens of second pathogens (Fig. 4.5A). This heterologous immunity is supportive to provoke herd immunity against the new pathogen in any community [119,120]. However, heterologous herd immunity can also be observed in circumstances like "coinfections," where infections of two unrelated pathogens occur concurrently or within a short period before the systemic dissemination of the first pathogen in the host, or "superinfections," where a second pathogen enters after the first pathogen is well established.

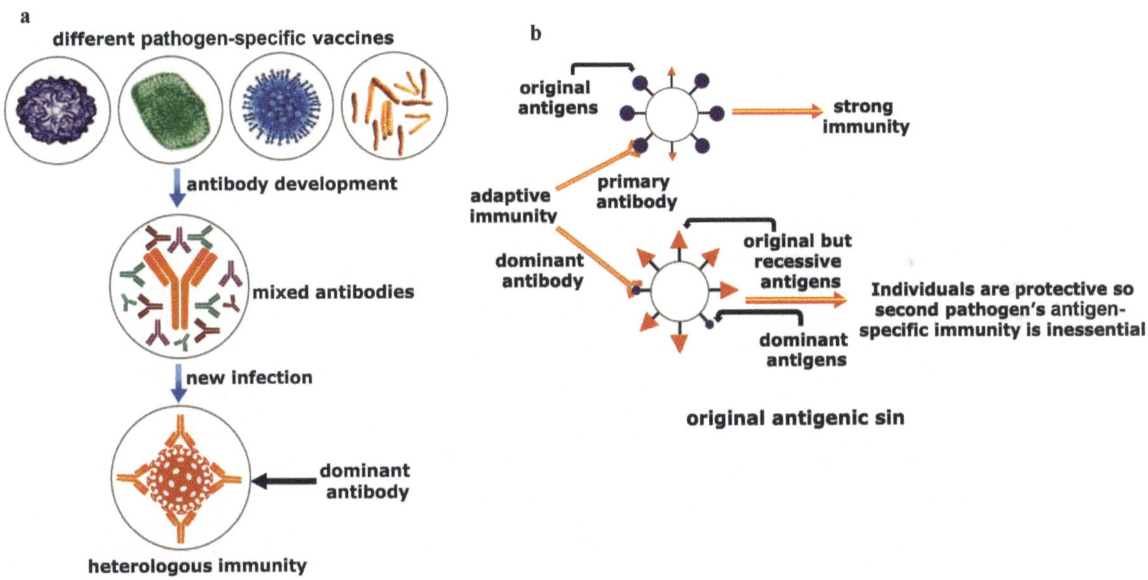

FIG. 4.5 **Heterologous Immunity: Yin and Yang.** **(A)** Heterologous immunity develops from the opinion that the antibody of one pathogen may protect an individual from a similar pathogen or in some cases even from phylogenetically distinct invaders. Vigorous immunization programs with several vaccine candidates are required to improve protective response against any new incoming pathogen having comparable antigens. Both humoral and cellular cross-reactive heterologous memory responses may influence the foundation of the natural course of herd immunity against any new infection. **(B)** Original antigenic sin refers to a phenomenon in which the development of immunity against pathogens or antigens is shaped by the first exposure to related pathogens. The memory antibodies generated as a result of first exposure will interact with similar antigens of the second pathogen for neutralizing it. This anamnestic recall against the second pathogen dampens the probability of immune system activation against original antigens of the second pathogen. This phenomenon is reasonably applicable to severe acute respiratory syndrome coronavirus 2 (SARS-CoV-2) infection where its infection rate is very mild.

8 THE PREVALENCE OF HETEROLOGOUS IMMUNITY CORRELATES WELL WITH LOW SARS-COV-2 INFECTION RATE IN INDIA

8.1 Influence of Microbiota on Community Immunity in Coronavirus Infection

The inception of herd immunity in the current SARS-CoV-2 pandemic still has a long road to cover. The precise herd immunity threshold for the SARS-CoV-2 infection is not yet evident. Nevertheless, several experts suggest that if infection spreads in more than 60% of a population, then herd immunity can be achieved against this virus [11]. Several studies have observed that the percentage of people infected so far with the virus is still in a single digit in contrast to the broad size of the tested population. This occurred because of the government-induced stringent lockdowns in various parts of the world. Similar efforts of limited lockdowns for generation of herd immunity have also been experimented by several countries notably Sweden and briefly in Britain. There are also only 7%–17% of infected people. In May 2020, the SARS-CoV-2 infection spread to as many as 20% of the total city dwellers in New York City, USA. This is the highest number for an outbreak in a particular cohort in the United States. Both the density of city dwellers in an area and the physical contacts among individuals are critical factors in reaching the threshold of herd immunity [121,122]. Thus if the disease spreads faster than is currently believed, the herd immunity generation could be achieved at a faster speed. The differences in immunity, nutrition, and epigenetic modifications among residents may also drive herd immunity generation, even downward [123]. Several previously common infectious diseases among children, e.g., measles and chickenpox, are now extremely infrequent in the United States because vaccine-induced herd immunity obstructs the outbreaks [2,124]. COVID-19 is potentially a much more detrimental disease than any other pneumonia-causing pathogen. SARS-CoV-2 infects mostly those who are at risk of getting sick because of underlying comorbidities. In several countries, e.g., the United States, many people already have threshold immunity, either because they had been sick with a similar viral strain of pneumonia in the past or because they have received a shot of variable vaccines. Among these vaccines, one or a few may have a good match for the version of the virus they encountered. This number is not high enough for reaching herd immunity, and those viruses still can circulate in every year. However, the existence of partial immunity in a population helps lower the risk of death by any new pathogen in that population. COVID-19, unlike influenza, is a brand-new disease. It has the potential to kill many more people because the SARS-CoV-2-antigen-specific immunity is still poor among variable populations. This partial immunity may not have the vigor to develop herd immunity against SARS-CoV-2.

Nonetheless, SARS-CoV-2 is definitely the cause of a pandemic, and mass infection during pandemic can be a potential solution against infection by developing herd immunity. It is essential to consider the impact of microbiome in developing herd immunity against SARS-CoV-2. Virulence factors of the virus trigger the cytokine storm, a hyperinflammatory response generated by the body's immune system that which is responsible for many complications and deaths [125]. The gut microbiome is the leading factor that controls the cytokine storm [126]. Specific microbiota composition controls the inflammatory responses that may predict the predisposition of COVID-19 [127] (Table 4.2). MRx-4DP0004, a strain of the bacterium *B. breve* in the phylum Actinobacteria, originally developed for asthma [128], inhibited the hyperinflammatory response by reducing the expression of angiotensin-converting enzyme 2 (ACE2) receptors for maintaining the potential antiviral response [129]. The predominance of the ACE2, SARS-CoV-2 receptor, is high in microbiome-enriched gut enterocytes and colonocytes [70,130]. Thereby, in the current crisis, the impact of microbiome cannot be ignored. The "microbial dark matter" inside the body and around us indeed possesses the potential to provide us with the tools that help us to develop immunity with limited spread of multidrug-resistant pathogens.

8.2 Microbiome: Novel Song in an Old Melody

The frequencies of microbiome in the lung-gut axis are the crucial and comprehensive biomarkers for viral diseases. Several studies revealed that higher frequencies of *Leptotrichia buccalis*, *Capnocytophaga gingivalis*, *Veillonella* sp., *Veillonella parvula*, and *Prevotella melaninogenica* in the bronchoalveolar lavage fluid worsened COVID-19 in affected patients [131]. These microbiome populations augmented the synthesis of both nucleotides and amino acids and enhanced the metabolism rates of carbohydrates, which is the precursor of a worsening infection. The respiratory tract and lungs are enriched with *Streptococcus infantis* during SARS-CoV-2 infection [132]. Conversely, *Fusobacterium periodonticum*, a prevalent component of the lung microbiota, has inhibited the severity of COVID-19 [133]. The imbalance of

TABLE 4.2
Disparity in Expression of Commensal Microbiome Between SARS-CoV-2-Sensitive and SARS-CoV-2-Resistant Patients.

ALTERATION OF MICROBIOME FREQUENCIES DUE TO SARS-COV-2 INFECTION	
Sensitive Patients	**Resistant Patients**
PHYLUM BACTEROIDETES	
Capnocytophaga gingivalis	*Alistipes onderdonkii*
Prevotella melaninogenica	*Parabacteroides merdae*
Bacteroides nordi	*Bacteroides stercoris*
Capnocytophaga sp.	*A. onderdonkii*
	Bacteroides ovatus
	Bacteroides dorei
	Bacteroides thetaiotaomicron
	Bacteroides massiliensis
	B. stercoris
PHYLUM FIRMICUTES	
Clostridium ramosum	*Faecalibacterium prausnitzii*
Clostridium hathewayi	Lachnospiraceae bacteria
Erysipelotrichaceae bacteria	*Eubacterium rectale*
Ruthenibacterium lactatiformans	*Ruminococcus obeum*
Veillonella sp.	*Dorea formicigenerans*
Veillonella parvula	*Lactobacillus* sp.
Streptococcus infantis	Lachnospiraceae bacteria
Ruminococcus gnavus	
Enterococcus avium	
PHYLUM ACTINOBACTERIA	
Actinomyces viscosus	*Bifidobacterium* sp.
Corynebacterium sp.	
Collinsella aerofaciens	
Collinsella tanakaei	
PHYLUM FUSOBACTERIA	
Leptotrichia buccalis	*Fusobacterium periodonticum*

(continued)

TABLE 4.2
Disparity in Expression of Commensal Microbiome Between SARS-CoV-2-Sensitive and SARS-CoV-2-Resistant Patients.—cont'd

ALTERATION OF MICROBIOME FREQUENCIES DUE TO SARS-COV-2 INFECTION	
Sensitive Patients	**Resistant Patients**
PHYLUM PROTEOBACTERIA	
Acinetobacter baumannii	
Klebsiella pneumoniae	
Morganella morganii	

SARS-CoV-2, severe acute respiratory syndrome coronavirus 2.

microbiota frequencies in gut also correlates with the variable expression of ACE2, which also increases the severity of pathogenesis in SARS-CoV-2 infection. There are certain bacteria of microbiome population that help in maintaining the ACE2 expression, which further accelerates the infection rate in the host. *Collinsella aerofaciens, Collinsella tanakaei, S. infantis,* and *Morganella morganii* are prominent commensals found in the gut microbiome of severely infected COVID-19 patients [132]. Several opportunistic pathogens have been persistently associated with exaggerated infection in patients with COVID-19 (e.g., *Ruminococcus gnavus, Clostridium hathewayi, Enterococcus avium, Alistipes onderdonkii, C. aerofaciens,* and *M. morganii*). In contrast, a few bacteria such as *Parabacteroides merdae, Bacteroides stercoris,* and *Lachnospiraceae* bacteria of the intestinal microbiota have the competence to reduce ACE2 expression in the gut epithelial cells, thereby influencing the calibrated immune system to impede the pathogenicity of COVID-19. These commensal bacteria provide the nutritional factors in the gut. This is essential for enhancing the immune system and is cardinal for protection against the SARS-CoV-2 infection. SCFAs (butyrate and acetate) are synthesized by *P. merdae, B. stercoris, A. onderdonkii,* and a *Lachnospiraceae* bacterium; an abundance of these bacteria is imperative for providing prophylactic efficacy against COVID-19. The metabolism of tryptophan to melatonin by *A. onderdonkii* is necessary to reduce the infection. These bacteria have a salutary role in combating the SARS-CoV-2 infection [134,135] (Table 4.2). Other than bacteria, COVID-19 patients also have higher frequencies of *Aspergillus flavus, Candida glabrata,* and *Candida albicans,* which are also pertinent commensals of the human gut microbiome [136].

8.3 Herd Immunity: Friends or Foes for COVID-19

The diversity of human leukocyte antigen (HLA) molecules is responsible for ensuring the diversified antigenic peptides that individuals across a population can present. It ensures the chances that individuals of a particular population may survive against the emergent diseases. However, the first exposure to a pathogen establishes a 'footprint' of the immune response. It is mainly the reexposure to the same or a similar antigen of the second pathogen that stimulates and cross-reacts with memory T cells that are specific for antigens of the first pathogen or vaccination (Fig. 4.2A). Thereby, a whole immune repertoire of memory T cells have been preferentially reexpanded, which refuted the clonal expansion of new antigen-specific T cells. Subsequently, chances of production of a new type of memory T cells for the second pathogen will be thwarted unwittingly by a process known as 'original antigenic sin' (Fig. 4.5B) [137]. Thus T-cell cross-reactivity and HLA diversity to past pandemics can persuade a suboptimal protective immune response to the second pathogen owing to 'original antigenic sin' (Fig. 4.5B). A similar mechanism has been proposed for B-cell responses (Fig. 4.2B). The antigenic sin can extend beyond low sensitivity to the second antigen to a situation in which the original antigen has established a Th1-, Th2-, or Th17-type response, which is inappropriate as well as attritional for the second pathogen. This dampens the concept of heterologous herd immunity, which is required for evolving the community vaccination [138].

8.4 Poliovirus Vaccine May Provide Heterologous Protection Against COVID-19

In the current pandemic, the entire world is facing a severe infection with SARS-CoV-2. To date, there is a very low rate of death of infected children under the age of 10 years, as well as for those not having any underlying comorbidity (like any type of autoimmunity). This observation is highly significant because it suggests children who have received regular immunization of different vaccines may have heterologous protection against COVID-19. In pan-India, the continuum of mass vaccinations with OPV, bacille Calmette-Guérin (BCG), measles vaccine, etc. has established memory responses against many types of variable antigens in the body. A single type or multiple types of antibodies from the pool of the antibody, present in the body, can recognize the novel pathogens carrying the same or similar types of antigens. This phenomenon is considered the major reason for the unexpectedly low rate of COVID-19 occurrence in India with a faster recovery rate. The infection rate until September 15, 2020, with SARS-CoV-2 is only 8.45% of the tested population. The total tested samples are almost 58 million, whereas the sum of the total infected patients including dead, recovered, and active is 4.9 million in India. If we compare the total infected samples (4.9 million) with the total population of India (1300 million), then the percentage of infected patients is very negligible (0.37%), which validates that SARS-CoV-2 infection has been cleared by the community comfortably. The total mortality—including comorbidity deaths due to COVID-19—is 80,808, which is actually 0.006% of the pan-India population (1300 million). This higher percentage of protection has been possible because of the heterologous memory immune response against COVID-19. Dominant antibodies, generated from previous vaccinations, have the aptitude to clear the SARS-CoV-2 infection comfortably, which very well explains the low infection rate observed in Indian population. Innate and adaptive memory responses in the course of poliovirus vaccination collude to develop heterologous immunity against the SARS-CoV-2 infection and are helping slow down the spread of COVID-19 [139]. Among different vaccines, two candidate vaccines are of paramount importance and are also gradually becoming the focus of repurposing of vaccines for COVID-19: the poliovirus vaccine and the BCG vaccine [139,140]. The BCG vaccine was originally administered to provide protection against *Mycobacterium tuberculosis* (TB) infection. BCG has also been shown to provide heterologous protection against other unrelated infections, as in yellow fever viral infection by inducing epigenetic reprogramming of monocytes [141] and reducing child mortality rate [142]. BCG also showed heterologous immunologic effects in low-birth-weight infants [143]. There has been comparatively less number of SARS-CoV-2-infection-related deaths in countries that have continued with the usage of OPV than in those countries that have switched to IPV [140]. This correlation of OPV usage and the reduced COVID-19 mortality rate is urging scientists to explore the possibility of using poliovirus vaccine to help lower the severity of COVID-19. Furthermore, the similarity between the SARS-CoV-2 encoded 3C-like protease and the 3CPro of picornaviruses, which has the role in viral pathogenesis through viral protein maturation, has also been used to design drugs against the 3C-like protease of SARS-CoV-2 to impede the infection [144]. These available vaccines can provide protection against COVID-19 possibly by antigen similarity with SARS-CoV-2, or through antigen-independent innate and adaptive memory [145]. We discuss the first possibility in the following section.

8.5 *In silico* Comparison of SARS-CoV-2 With OPV and BCG

To explore the chances of antigen sharing, we ran a preliminary analysis by comparing the SARS-CoV-2 epitopes with OPV and BCG epitopes at the sequence level using the protein basic local alignment search tool (BLAST). For this comparison, the information of efficient epitopes of the three organisms was extracted from the Immune Epitope Database and Analysis Resource (IEDB) [146] which is a database for experimentally known epitopes. We downloaded the sequences of SARS-CoV-2 (ID: 2697049), all three types of poliovirus (ID: 12080, 12083, and 12086), and BCG (ID: 33892) with only "linear epitopes" and "human host" using both as filters from IEDB, which provided 321, 88, and 464 epitope sequences for the three organisms, respectively. As in the majority of epitopes, these epitopes were 9–12 amino acids long. Then the protein BLAST was run for each of these 321 SARS-CoV-2 epitope protein sequences against the database of the downloaded polio and BCG epitope sequences using the National Center for Biotechnology Information (NCBI) offline protein BLAST [147]. Considering the usual length range of epitopes (9–12 residues), an E-value cutoff of 10e-2 and minimum alignment length of eight residues with no gaps were used to filter the BLAST hits. Upon performing this analysis, the SARS-CoV-2 open reading frame 7a (Orf7a) protein epitope (residues 72–81 with the sequence 'KHVYQLRARS') showed similarity (80%) with the human poliovirus type 3 Sabin strain epitope (residues 62–71 with the sequence 'RHVVQRRSRS') of VP1 protein. Similarly, the SARS-CoV-2 nucleocapsid epitope (residues 201–209 with the sequence 'SSRGTSPAR') shared similarity (87%) with part of the human poliovirus type 1 Mahoney epitope (residues 186–193 with the sequence "TYGTAPAR"). No such direct matching of epitope sequence was found between BCG and SARS-CoV-2. However, this is only the first step to analyze antigen similarity and requires other important factors such as the three-dimensional structure of epitopes, their subcellular localization, antisera profile, etc. to be taken into account to understand the complete scenario of cross-reactivity of antibodies.

9 PERSPECTIVES AND FUTURE OPPORTUNITIES

The concept of herd immunity—or community immunity—has triggered an intense debate concerning whether it would prevent the pandemic of COVID-19, and if so, how much herd immunity would be required to effectively impede its spread? Along with this, an important unanswered question in the current pandemic situation is the discovery of a future probable vaccine against SARS-CoV-2, against which the propensity of herd immunity expansion is rather uncertain. Based on the theory of original antigenic sin, the second infection from an unlikely pathogen activates the memory immune response that was developed from the first infection. The first pathogen-specific antibodies can neutralize new dominant antigens of the second pathogen. The effective antibodies against those dominant antigens coming from the first infection are thus able to eradicate the second infection. The immune system needs a reasonable time to recognize the new antigens through APCs, followed by T-cell activation, generation of memory T cells, and triggering of B cells for antibody secretion. Nonetheless, if dominant antibodies encounter comparable antigens from the second pathogen that are enough to eliminate the second infection, then the rest of the different antigens of the second pathogen will be presumed to be recessive antigens. The immune system will only generate the former dominant antibodies against this new second pathogen, failing to produce new antibodies against the first pathogen. This phenomenon results in an inefficient and a weak immunity, which has been expected against SARS-CoV-2 infection in India because of the regular poliovirus vaccinations. Thus the probability of attaining herd immunity against COVID-19 is very limited, given the current context of its low infectivity rate. Discovery of a future probable vaccine against SARS-CoV-2 is very uncertain. Despite the existence of heterologous immunity and original antigenic sin, induction of herd immunity by vaccination could be a potential counterpandemic measure against COVID-19, especially for vulnerable groups including individuals with immunodeficiency or a weakened immune system.

10 CONCLUSIONS

1. The critical role of herd immunity has been recognized with the current pandemic state due to the transmissible infection of SARS-CoV-2.
2. Herd immunity is the most critical prophylactic intervention, which has delivered protective immunity against several infectious diseases such as smallpox, polio, measles, etc. in the past. Thus the new paradigm of evolving herd immunity during a pandemic situation cannot be disregarded.
3. Microbiota, metabolism, and environmental factors coordinate with each other for developing an efficient immune system, which is a prerequisite for evolving herd immunity generation.

4. In India, the infection rate of SARS-CoV-2 is unexpectedly very low, i.e., only 0.37% of the total population. The prevalence of heterologous immunity due to rigorous vaccination programs at the grass-root level may provide protection against the SARS-CoV-2 pandemic.
5. Comparisons between SARS-CoV-2 Orf7a protein epitope (KHVYQLRARS) and the human poliovirus type 3 Sabin strain epitope (RHVVQRRSRS) from VP1 protein offer great insights into the concept of heterologous immunity. Similarly, the SARS-CoV-2 nucleocapsid epitope (SSRGTSPAR) shared resemblance (87%) with part of the human poliovirus type 1 Mahoney epitope (TYGTAPAR), which can provide an alternative prophylactic intervention against both poliovirus and SARS-CoV-2.

CONFLICT OF INTEREST

The authors declare no conflict of interest.

REFERENCES

[1] Nokes DJ, Anderson RM. The use of mathematical models in the epidemiological study of infectious diseases and in the design of mass immunization programmes. Epidemiol Infect 1988;101:1–20. https://doi.org/10.1017/s0950268800029186.
[2] Anderson RM, May RM. Vaccination and herd immunity to infectious diseases. Nature 1985;318:323–9. https://doi.org/10.1038/318323a0.
[3] Fine P, Eames K, Heymann DL. "Herd immunity": a rough guide. Clin Infect Dis Off Public Infect Dis Soc Am 2011;52:911–6. https://doi.org/10.1093/cid/cir007.
[4] John TJ, Samuel R. Herd immunity and herd effect: new insights and definitions. Eur J Epidemiol 2000;16: 601–6. https://doi.org/10.1023/a:1007626510002.
[5] Lane JM. Mass vaccination and surveillance/containment in the eradication of smallpox. Curr Top Microbiol Immunol 2006;304:17–29. https://doi.org/10.1007/3-540-36583-4_2.
[6] Ali M, Emch M, von Seidlein L, Yunus M, Sack DA, Rao M, Holmgren J, Clemens JD. Herd immunity conferred by killed oral cholera vaccines in Bangladesh: a reanalysis. Lancet 2005;366:44–9. https://doi.org/10.1016/s0140-6736(05)66550-6.
[7] Mast TC, Wang FT, Su S, Seeger JD. Evidence of herd immunity and sustained impact of rotavirus vaccination on the reduction of rotavirus-related medical encounters among infants from 2006 through 2011 in the United States. Pediatr Infect Dis J 2015;34:615–20. https://doi.org/10.1097/inf.0000000000000702.
[8] Kim TH, Johnstone J, Loeb M. Vaccine herd effect. Scand J Infect Dis 2011;43:683–9. https://doi.org/10.3109/00365548.2011.582247.
[9] Metcalf CJE, Ferrari M, Graham AL, Grenfell BT. Understanding herd immunity. Trends Immunol 2015;36: 753–5. https://doi.org/10.1016/j.it.2015.10.004.
[10] Liu Y, Gayle AA, Wilder-Smith A, Rocklöv J. The reproductive number of COVID-19 is higher compared to SARS coronavirus. J Travel Med 2020;27. https://doi.org/10.1093/jtm/taaa021.
[11] Kwok KO, Lai F, Wei WI, Wong SYS, Tang JWT. Herd immunity—estimating the level required to halt the COVID-19 epidemics in affected countries. J Infect 2020; 80:e32–3. https://doi.org/10.1016/j.jinf.2020.03.027.
[12] Ferreira RB, Antunes LC, Finlay BB. Should the human microbiome be considered when developing vaccines? PLoS Pathog 2010;6:e1001190. https://doi.org/10.1371/journal.ppat.1001190.
[13] Lynn DJ, Pulendran B. The potential of the microbiota to influence vaccine responses. J Leukoc Biol 2018;103: 225–31. https://doi.org/10.1189/jlb.5MR0617-216R.
[14] Guttinger S. The anti-vaccination debate and the microbiome: how paradigm shifts in the life sciences create new challenges for the vaccination debate. EMBO Rep 2019;20. https://doi.org/10.15252/embr.201947709.
[15] Lederberg J, McCray AT. Ome SweetOmics—a genealogical treasury of words. Scientist 2001;15:8.
[16] Ley RE, Peterson DA, Gordon JI. Ecological and evolutionary forces shaping microbial diversity in the human intestine. Cell 2006;124:837–48. https://doi.org/10.1016/j.cell.2006.02.017.
[17] Cerf-Bensussan N, Gaboriau-Routhiau V. The immune system and the gut microbiota: friends or foes? Nat Rev Immunol 2010;10:735–44. https://doi.org/10.1038/nri2850.
[18] Wu HJ, Wu E. The role of gut microbiota in immune homeostasis and autoimmunity. Gut Microb 2012;3:4–14. https://doi.org/10.4161/gmic.19320.
[19] Grice EA, Segre JA. The human microbiome: our second genome. Annu Rev Genom Hum Genet 2012;13:151–70. https://doi.org/10.1146/annurev-genom-090711-163814.
[20] Gensollen T, Iyer SS, Kasper DL, Blumberg RS. How colonization by microbiota in early life shapes the immune system. Science 2016;352:539–44. https://doi.org/10.1126/science.aad9378.
[21] Magwira CA, Taylor MB. Composition of gut microbiota and its influence on the immunogenicity of oral rotavirus vaccines. Vaccine 2018;36:3427–33. https://doi.org/10.1016/j.vaccine.2018.04.091.
[22] Ivanov II, Honda K. Intestinal commensal microbes as immune modulators. Cell Host Microbe 2012;12: 496–508. https://doi.org/10.1016/j.chom.2012.09.009.
[23] Mazmanian SK, Liu CH, Tzianabos AO, Kasper DL. An immunomodulatory molecule of symbiotic bacteria directs maturation of the host immune system. Cell 2005; 122:107–18. https://doi.org/10.1016/j.cell.2005.05.007.
[24] Round JL, Mazmanian SK. Inducible Foxp3^{+} regulatory T-cell development by *a* commensal bacterium of the intestinal microbiota. Proc Natl Acad Sci USA 2010;107: 12204–9. https://doi.org/10.1073/pnas.0909122107.

[25] Atarashi K, Tanoue T, Shima T, Imaoka A, Kuwahara T, Momose Y, Cheng G, Yamasaki S, Saito T, Ohba Y, et al. Induction of colonic regulatory T cells by indigenous Clostridium species. Science 2011;331:337–41. https://doi.org/10.1126/science.1198469.

[26] Gaboriau-Routhiau V, Rakotobe S, Lécuyer E, Mulder I, Lan A, Bridonneau C, Rochet V, Pisi A, De Paepe M, Brandi G, et al. The key role of segmented filamentous bacteria in the coordinated maturation of gut helper T cell responses. Immunity 2009;31:677–89. https://doi.org/10.1016/j.immuni.2009.08.020.

[27] Mueller NT, Bakacs E, Combellick J, Grigoryan Z, Dominguez-Bello MG. The infant microbiome development: mom matters. Trends Mol Med 2015;21:109–17. https://doi.org/10.1016/j.molmed.2014.12.002.

[28] Penders J, Thijs C, Vink C, Stelma FF, Snijders B, Kummeling I, van den Brandt PA, Stobberingh EE. Factors influencing the composition of the intestinal microbiota in early infancy. Pediatrics 2006;118:511–21. https://doi.org/10.1542/peds.2005-2824.

[29] Aagaard K, Ma J, Antony KM, Ganu R, Petrosino J, Versalovic J. The placenta harbors a unique microbiome. Sci Transl Med 2014;6. https://doi.org/10.1126/scitranslmed.3008599. 237ra265.

[30] Gritz EC, Bhandari V. The human neonatal gut microbiome: a brief review. Front Pediatr 2015;3:17. https://doi.org/10.3389/fped.2015.00017.

[31] Sood R, Zehnder JL, Druzin ML, Brown PO. Gene expression patterns in human placenta. Proc Natl Acad Sci USA 2006;103:5478–83. https://doi.org/10.1073/pnas.0508035103.

[32] Dominguez-Bello MG, Costello EK, Contreras M, Magris M, Hidalgo G, Fierer N, Knight R. Delivery mode shapes the acquisition and structure of the initial microbiota across multiple body habitats in newborns. Proc Natl Acad Sci USA 2010;107:11971–5. https://doi.org/10.1073/pnas.1002601107.

[33] Pantoja-Feliciano IG, Clemente JC, Costello EK, Perez ME, Blaser MJ, Knight R, Dominguez-Bello MG. Biphasic assembly of the murine intestinal microbiota during early development. ISME J 2013;7:1112–5. https://doi.org/10.1038/ismej.2013.15.

[34] Biasucci G, Rubini M, Riboni S, Morelli L, Bessi E, Retetangos C. Mode of delivery affects the bacterial community in the newborn gut. Early Hum Dev 2010;86(Suppl. S1):13–5. https://doi.org/10.1016/j.earlhumdev.2010.01.004.

[35] Madan JC, Farzan SF, Hibberd PL, Karagas MR. Normal neonatal microbiome variation in relation to environmental factors, infection and allergy. Curr Opin Pediatr 2012;24:753–9. https://doi.org/10.1097/MOP.0b013e32835a1ac8.

[36] Hunt KM, Foster JA, Forney LJ, Schütte UM, Beck DL, Abdo Z, Fox LK, Williams JE, McGuire MK, McGuire MA. Characterization of the diversity and temporal stability of bacterial communities in human milk. PLoS One 2011;6:e21313. https://doi.org/10.1371/journal.pone.0021313.

[37] Balmer SE, Wharton BA. Diet and faecal flora in the newborn: breast milk and infant formula. Arch Dis Child 1989;64:1672–7. https://doi.org/10.1136/adc.64.12.1672.

[38] Harmsen HJ, Wildeboer-Veloo AC, Raangs GC, Wagendorp AA, Klijn N, Bindels JG, Welling GW. Analysis of intestinal flora development in breast-fed and formula-fed infants by using molecular identification and detection methods. J Pediatr Gastroenterol Nutr 2000;30:61–7. https://doi.org/10.1097/00005176-200001000-00019.

[39] Yoshioka H, Iseki K, Fujita K. Development and differences of intestinal flora in the neonatal period in breast-fed and bottle-fed infants. Pediatrics 1983;72:317–21.

[40] Pérez-Pérez A, Vilariño-García T, Fernández-Riejos P, Martín-González J, Segura-Egea JJ, Sánchez-Margalet V. Role of leptin as a link between metabolism and the immune system. Cytokine Growth Factor Rev 2017;35:71–84. https://doi.org/10.1016/j.cytogfr.2017.03.001.

[41] La Cava A, Matarese G. The weight of leptin in immunity. Nat Rev Immunol 2004;4:371–9. https://doi.org/10.1038/nri1350.

[42] De Rosa V, Procaccini C, Calì G, Pirozzi G, Fontana S, Zappacosta S, La Cava A, Matarese G. A key role of leptin in the control of regulatory T cell proliferation. Immunity 2007;26:241–55. https://doi.org/10.1016/j.immuni.2007.01.011.

[43] Bäckhed F, Ding H, Wang T, Hooper LV, Koh GY, Nagy A, Semenkovich CF, Gordon JI. The gut microbiota as an environmental factor that regulates fat storage. Proc Natl Acad Sci USA 2004;101:15718–23. https://doi.org/10.1073/pnas.0407076101.

[44] Kim CH, Park J, Kim M. Gut microbiota-derived short-chain Fatty acids, T cells, and inflammation. Immune Netw 2014;14:277–88. https://doi.org/10.4110/in.2014.14.6.277.

[45] Fukuda S, Toh H, Hase K, Oshima K, Nakanishi Y, Yoshimura K, Tobe T, Clarke JM, Topping DL, Suzuki T, et al. Bifidobacteria can protect from enteropathogenic infection through production of acetate. Nature 2011;469:543–7. https://doi.org/10.1038/nature09646.

[46] Peng L, He Z, Chen W, Holzman IR, Lin J. Effects of butyrate on intestinal barrier function in a Caco-2 cell monolayer model of intestinal barrier. Pediatr Res 2007;61:37–41. https://doi.org/10.1203/01.pdr.0000250014.92242.f3.

[47] Sommer A, Tarwotjo I, Katz J. Increased risk of xerophthalmia following diarrhea and respiratory disease. Am J Clin Nutr 1987;45:977–80. https://doi.org/10.1093/ajcn/45.5.977.

[48] Cha HR, Chang SY, Chang JH, Kim JO, Yang JY, Kim CH, Kweon MN. Downregulation of Th17 cells in the small intestine by disruption of gut flora in the absence of retinoic acid. J Immunol 2010;184:6799–806. https://doi.org/10.4049/jimmunol.0902944.

[49] Shapiro H, Thaiss CA, Levy M, Elinav E. The cross talk between microbiota and the immune system: metabolites take center stage. Curr Opin Immunol 2014;30:54–62. https://doi.org/10.1016/j.coi.2014.07.003.

[50] Pawankar R, Canonica GW, Holgate ST, Lockey RF, Blaiss MS. WAO white book on allergy, vol. 3. Milwaukee, WI, USA: World Allergy Organization; 2011.
[51] Platts-Mills TA. The allergy epidemics: 1870–2010. J Allergy Clin Immunol 2015;136:3–13. https://doi.org/10.1016/j.jaci.2015.03.048.
[52] Ritchel M. Your environment is cleaner. Your immune system has never been so unprepared. The New York Times; 2019. D, 1.
[53] Strachan DP. Hay fever, hygiene, and household size. BMJ Clin Res Ed 1989;299:1259–60. https://doi.org/10.1136/bmj.299.6710.1259.
[54] Apostol AC, Jensen KDC, Beaudin AE. Training the fetal immune system through maternal inflammation-A layered hygiene hypothesis. Front Immunol 2020;11:123. https://doi.org/10.3389/fimmu.2020.00123.
[55] Zhang C, Zhang M, Wang S, Han R, Cao Y, Hua W, Mao Y, Zhang X, Pang X, Wei C, et al. Interactions between gut microbiota, host genetics and diet relevant to development of metabolic syndromes in mice. ISME J 2010;4:232–41. https://doi.org/10.1038/ismej.2009.112.
[56] Deehan EC, Walter J. The fiber gap and the disappearing gut microbiome: implications for human nutrition. Trends Endocrinol Metab 2016;27:239–42. https://doi.org/10.1016/j.tem.2016.03.001.
[57] Sonnenburg ED, Smits SA, Tikhonov M, Higginbottom SK, Wingreen NS, Sonnenburg JL. Diet-induced extinctions in the gut microbiota compound over generations. Nature 2016;529:212–5. https://doi.org/10.1038/nature16504.
[58] De Filippo C, Cavalieri D, Di Paola M, Ramazzotti M, Poullet JB, Massart S, Collini S, Pieraccini G, Lionetti P. Impact of diet in shaping gut microbiota revealed by a comparative study in children from Europe and rural Africa. Proc Natl Acad Sci USA 2010;107:14691–6. https://doi.org/10.1073/pnas.1005963107.
[59] Andre FE, Booy R, Bock HL, Clemens J, Datta SK, John TJ, Lee BW, Lolekha S, Peltola H, Ruff TA, et al. Vaccination greatly reduces disease, disability, death and inequity worldwide. Bull World Health Organ 2008;86:140–6. https://doi.org/10.2471/blt.07.040089.
[60] Levine MM. Immunogenicity and efficacy of oral vaccines in developing countries: lessons from a live cholera vaccine. BMC Biol 2010;8:129. https://doi.org/10.1186/1741-7007-8-129.
[61] Huda MN, Lewis Z, Kalanetra KM, Rashid M, Ahmad SM, Raqib R, Qadri F, Underwood MA, Mills DA, Stephensen CB. Stool microbiota and vaccine responses of infants. Pediatrics 2014;134:e362–72. https://doi.org/10.1542/peds.2013-3937.
[62] John TJ, Jayabal P. Oral polio vaccination of children in the tropics: I. The poor seroconversion rates and the absence of viral interference. Am J Epidemiol 1972;96:263–9. https://doi.org/10.1093/oxfordjournals.aje.a121457.
[63] Patriarca PA, Wright PF, John TJ. Factors affecting the immunogenicity of oral poliovirus vaccine in developing countries: Review. Rev Infect Dis 1991;13:926–39. https://doi.org/10.1093/clinids/13.5.926.
[64] Hallander HO, Paniagua M, Espinoza F, Askelöf P, Corrales E, Ringman M, Storsaeter J. Calibrated serological techniques demonstrate significant different serum response rates to an oral killed cholera vaccine between Swedish and Nicaraguan children. Vaccine 2002;21:138–45. https://doi.org/10.1016/s0264-410x(02)00348-1.
[65] Levine MM, Kotloff KL, Barry EM, Pasetti MF, Sztein MB. Clinical trials of Shigella vaccines: two steps forward and one step back on a long, hard road. Nat Rev Microbiol 2007;5:540–53. https://doi.org/10.1038/nrmicro1662.
[66] Lanata CF, Midthun K, Black RE, Butron B, Huapaya A, Penny ME, Ventura G, Gil A, Jett-Goheen M, Davidson BL. Safety, immunogenicity, and protective efficacy of one and three doses of the tetravalent rhesus rotavirus vaccine in infants in Lima, Peru. J Infect Dis 1996;174:268–75. https://doi.org/10.1093/infdis/174.2.268.
[67] Gotuzzo E, Butron B, Seas C, Penny M, Ruiz R, Losonsky G, Lanata CF, Wasserman SS, Salazar E, Kaper JB, et al. Safety, immunogenicity, and excretion pattern of single-dose live oral cholera vaccine CVD 103-HgR in Peruvian adults of high and low socioeconomic levels. Infect Immun 1993;61:3994–7. https://doi.org/10.1128/iai.61.9.3994-3997.1993.
[68] Simanjuntak C, Witham N, Punjabi N, Heppner DG, Losonsky G, Totosudirjo H, Rifai AR, Clemens J, Lim YL, et al. Safety and immunogenicity of single-dose live oral cholera vaccine CVD 103-HgR in 5–9-year-old Indonesian children. Lancet 1992;340:689–94. https://doi.org/10.1016/0140-6736(92)92231-4.
[69] McDermott AJ, Huffnagle GB. The microbiome and regulation of mucosal immunity. Immunology 2014;142:24–31. https://doi.org/10.1111/imm.12231.
[70] Wang J, Zhao S, Liu M, Zhao Z, Xu Y, Wang P, Lin M, Xu Y, Huang B, Zuo X, et al. ACE2 expression by colonic epithelial cells is associated with viral infection, immunity and energy metabolism. medRxiv 2020. https://doi.org/10.1101/2020.02.05.20020545.
[71] Zimmermann P, Curtis N. The influence of the intestinal microbiome on vaccine responses. Vaccine 2018;36:4433–9. https://doi.org/10.1016/j.vaccine.2018.04.066.
[72] Eloe-Fadrosh EA, McArthur MA, Seekatz AM, Drabek EF, Rasko DA, Sztein MB, Fraser CM. Impact of oral typhoid vaccination on the human gut microbiota and correlations with s. Typhi-specific immunological responses. PLoS One 2013;8:e62026. https://doi.org/10.1371/journal.pone.0062026.
[73] Harris V, Ali A, Fuentes S, Korpela K, Kazi M, Tate J, Parashar U, Wiersinga WJ, Giaquinto C, de Weerth C, et al. Rotavirus vaccine response correlates with the infant gut microbiota composition in Pakistan. Gut Microb 2018;9:93–101. https://doi.org/10.1080/19490976.2017.1376162.
[74] Harris VC, Armah G, Fuentes S, Korpela KE, Parashar U, Victor JC, Tate J, de Weerth C, Giaquinto C, Wiersinga WJ, et al. Significant correlation between the

infant gut microbiome and rotavirus vaccine response in rural Ghana. J Infect Dis 2017;215:34–41. https://doi.org/10.1093/infdis/jiw518.

[75] Moreau MC, Routhiau V, Dubuquoy C, Bisetti N, Bouley C, Prévoteau H. Modulating properties of intestinal bacterial strains, *Escherichia coli* and bifidobacterium, on two specific immune responses generated by the gut, i.e. oral tolerance to ovalbumin and intestinal IgA anti-rotavirus response, in gnotobiotic mice. In: Proceedings of the 10th international congress of immunology, New Delhi, India; 1–6 November 1998.

[76] Mullié C, Yazourh A, Thibault H, Odou MF, Singer E, Kalach N, Kremp O, Romond MB. Increased poliovirus-specific intestinal antibody response coincides with promotion of Bifidobacterium longum-infantis and Bifidobacterium breve in infants: a randomized, double-blind, placebo-controlled trial. Pediatr Res 2004;56:791–5. https://doi.org/10.1203/01.pdr.0000141955.47550.a0.

[77] Pabst O, Hornef M. Gut microbiota: a natural adjuvant for vaccination. Immunity 2014;41:349–51. https://doi.org/10.1016/j.immuni.2014.09.002.

[78] Gilmartin AA, Petri Jr WA. Exploring the role of environmental enteropathy in malnutrition, infant development and oral vaccine response. Philos Trans R Soc Lond Ser B Biol Sci 2015;370. https://doi.org/10.1098/rstb.2014.0143.

[79] Harper KM, Mutasa M, Prendergast AJ, Humphrey J, Manges AR. Environmental enteric dysfunction pathways and child stunting: a systematic review. PLoS Negl Trop Dis 2018;12:e0006205. https://doi.org/10.1371/journal.pntd.0006205.

[80] McKay S, Gaudier E, Campbell DI, Prentice AM, Albers R. Environmental enteropathy: new targets for nutritional interventions. Int Health 2010;2:172–80. https://doi.org/10.1016/j.inhe.2010.07.006.

[81] Xavier RJ. Growth and the microbiome—integrating global health with basic science. N Engl J Med 2020; 383:391–3. https://doi.org/10.1056/NEJMe2017496.

[82] Dahan S, Roth-Walter F, Arnaboldi P, Agarwal S, Mayer L. Epithelia: lymphocyte interactions in the gut. Immunol Rev 2007;215:243–53. https://doi.org/10.1111/j.1600-065X.2006.00484.x.

[83] Smith PM, Garrett WS. The gut microbiota and mucosal T cells. Front Microbiol 2011;2:111. https://doi.org/10.3389/fmicb.2011.00111.

[84] Becattini S, Taur Y, Pamer EG. Antibiotic-induced changes in the intestinal microbiota and disease. Trends Mol Med 2016;22:458–78. https://doi.org/10.1016/j.molmed.2016.04.003.

[85] Bach JF. The effect of infections on susceptibility to autoimmune and allergic diseases. N Engl J Med 2002;347: 911–20. https://doi.org/10.1056/NEJMra020100.

[86] West CE. Gut microbiota and allergic disease: new findings. Curr Opin Clin Nutr Metab Care 2014;17:261–6. https://doi.org/10.1097/mco.0000000000000044.

[87] Fagundes-Neto U, Viaro T, Wehba J, Patrício FR, Machado NL. Tropical enteropathy (environmental enteropathy) in early childhood: a syndrome caused by contaminated environment. J Trop Pediatr 1984;30: 204–9. https://doi.org/10.1093/tropej/30.4.204.

[88] Taniuchi M, Sobuz SU, Begum S, Platts-Mills JA, Liu J, Yang Z, Wang XQ, Petri Jr WA, Haque R, Houpt ER. Etiology of diarrhea in Bangladeshi infants in the first year of life analyzed using molecular methods. J Infect Dis 2013;208:1794–802. https://doi.org/10.1093/infdis/jit507.

[89] Lagos R, Fasano A, Wasserman SS, Prado V, San Martin O, Abrego P, Losonsky GA, Alegria S, Levine MM. Effect of small bowel bacterial overgrowth on the immunogenicity of single-dose live oral cholera vaccine CVD 103-HgR. J Infect Dis 1999;180: 1709–12. https://doi.org/10.1086/315051.

[90] Clem AS. Fundamentals of vaccine immunology. J Global Infect Dis 2011;3:73–8. https://doi.org/10.4103/0974-777x.77299.

[91] Coster TS, Hoge CW, VanDeVerg LL, Hartman AB, Oaks EV, Venkatesan MM, Cohen D, Robin G, Fontaine-Thompson A, Sansonetti PJ, et al. Vaccination against shigellosis with attenuated Shigella flexneri 2a strain SC602. Infect Immun 1999;67:3437–43. https://doi.org/10.1128/iai.67.7.3437-3443.1999.

[92] World Health Organization. Future needs and directions for Shigella vaccines. Relev Epidemiol Hebd 2006;81: 51–8.

[93] Kirkpatrick BD, Colgate ER, Mychaleckyj JC, Haque R, Dickson DM, Carmolli MP, Nayak U, Taniuchi M, Naylor C, Qadri F, et al. The "Performance of Rotavirus and Oral Polio Vaccines in Developing Countries" (PROVIDE) study: description of methods of an interventional study designed to explore complex biologic problems. Am J Trop Med Hyg 2015;92:744–51. https://doi.org/10.4269/ajtmh.14-0518.

[94] Naylor C, Lu M, Haque R, Mondal D, Buonomo E, Nayak U, Mychaleckyj JC, Kirkpatrick B, Colgate R, Carmolli M, et al. Environmental enteropathy, oral vaccine failure and growth faltering in infants in Bangladesh. EBioMedicine 2015;2:1759–66. https://doi.org/10.1016/j.ebiom.2015.09.036.

[95] Taniuchi M, Platts-Mills JA, Begum S, Uddin MJ, Sobuz SU, Liu J, Kirkpatrick BD, Colgate ER, Carmolli MP, Dickson DM, et al. Impact of enterovirus and other enteric pathogens on oral polio and rotavirus vaccine performance in Bangladeshi infants. Vaccine 2016;34: 3068–75. https://doi.org/10.1016/j.vaccine.2016.04.080.

[96] Nakaya HI, Wrammert J, Lee EK, Racioppi L, Marie-Kunze S, Haining WN, Means AR, Kasturi SP, Khan N, Li GM, et al. Systems biology of vaccination for seasonal influenza in humans. Nat Immunol 2011;12:786–95. https://doi.org/10.1038/ni.2067.

[97] Oh JZ, Ravindran R, Chassaing B, Carvalho FA, Maddur MS, Bower M, Hakimpour P, Gill KP,

Nakaya HI, Yarovinsky F, et al. TLR5-mediated sensing of gut microbiota is necessary for antibody responses to seasonal influenza vaccination. Immunity 2014;41:478–92. https://doi.org/10.1016/j.immuni.2014.08.009.

[98] Benyacoub J, Rochat F, Saudan KY, Rochat I, Antille N, Cherbut C, von der Weid T, Schiffrin EJ, Blum S. Feeding a diet containing *a* fructooligosaccharide mix can enhance salmonella vaccine efficacy in mice. J Nutr 2008;138: 123–9. https://doi.org/10.1093/jn/138.1.123.

[99] Ciabattini A, Olivieri R, Lazzeri E, Medaglini D. Role of the microbiota in the modulation of vaccine immune responses. Front Microbiol 2019;10:1305. https://doi.org/10.3389/fmicb.2019.01305.

[100] Turnbaugh PJ, Ley RE, Mahowald MA, Magrini V, Mardis ER, Gordon JI. An obesity-associated gut microbiome with increased capacity for energy harvest. Nature 2006;444:1027–31. https://doi.org/10.1038/nature05414.

[101] Russell SL, Gold MJ, Hartmann M, Willing BP, Thorson L, Wlodarska M, Gill N, Blanchet MR, Mohn WW, McNagny KM, et al. Early life antibiotic-driven changes in microbiota enhance susceptibility to allergic asthma. EMBO Rep 2012;13:440–7. https://doi.org/10.1038/embor.2012.32.

[102] Vijay-Kumar M, Aitken JD, Carvalho FA, Cullender TC, Mwangi S, Srinivasan S, Sitaraman SV, Knight R, Ley RE, Gewirtz AT. Metabolic syndrome and altered gut microbiota in mice lacking Toll-like receptor 5. Science 2010; 328:228–31. https://doi.org/10.1126/science.1179721.

[103] Ishikawa H, Barber GN. The STING pathway and regulation of innate immune signaling in response to DNA pathogens. Cell Mol Life Sci 2011;68:1157–65. https://doi.org/10.1007/s00018-010-0605-2.

[104] Thompson MR, Kaminski JJ, Kurt-Jones EA, Fitzgerald KA. Pattern recognition receptors and the innate immune response to viral infection. Viruses 2011;3:920–40. https://doi.org/10.3390/v3060920.

[105] Kato H, Takeuchi O, Sato S, Yoneyama M, Yamamoto M, Matsui K, Uematsu S, Jung A, Kawai T, Ishii KJ, et al. Differential roles of MDA5 and RIG-I helicases in the recognition of RNA viruses. Nature 2006; 441:101–5. https://doi.org/10.1038/nature04734.

[106] Radiation-Killed NIH. In: News N, editor. Bacteria vaccine induces broad immune response in mice. MD, USA: National Institute of Allergy and Infectious Diseases; 2006.

[107] NIAID. Types of vaccines. MD, USA: National Institute of Allergy and Infectious Diseases; 2009.

[108] Palm NW, Medzhitov R. Pattern recognition receptors and control of adaptive immunity. Immunol Rev 2009;227:221–33. https://doi.org/10.1111/j.1600-065X.2008.00731.x.

[109] Hayashi F, Smith KD, Ozinsky A, Hawn TR, Yi EC, Goodlett DR, Eng JK, Akira S, Underhill DM, Aderem A. The innate immune response to bacterial flagellin is mediated by Toll-like receptor 5. Nature 2001;410: 1099–103. https://doi.org/10.1038/35074106.

[110] Ichinohe T, Pang IK, Kumamoto Y, Peaper DR, Ho JH, Murray TS, Iwasaki A. Microbiota regulates immune defense against respiratory tract influenza *A* virus infection. Proc Natl Acad Sci USA 2011;108:5354–9. https://doi.org/10.1073/pnas.1019378108.

[111] Welsh RM, Che JW, Brehm MA, Selin LK. Heterologous immunity between viruses. Immunol Rev 2010;235:244–66. https://doi.org/10.1111/j.0105-2896.2010.00897.x.

[112] Spaulding AC, Kurane I, Ennis FA, Rothman AL. Analysis of murine CD8(+) T-cell clones specific for the dengue virus NS3 protein: flavivirus cross-reactivity and influence of infecting serotype. J Virol 1999;73:398. https://doi.org/10.1128/JVI.73.1.398-403.1999.

[113] Nilges K, Höhn H, Pilch H, Neukirch C, Freitag K, Talbot PJ, Maeurer MJ. Human papillomavirus type 16 E7 peptide-directed CD8+ T cells from patients with cervical cancer are cross-reactive with the coronavirus NS2 protein. J Virol 2003;77:5464–74. https://doi.org/10.1128/jvi.77.9.5464-5474.2003.

[114] Wedemeyer H, Mizukoshi E, Davis AR, Bennink JR, Rehermann B. Cross-reactivity between hepatitis C virus and influenza A virus determinant-specific cytotoxic T cells. J Virol 2001;75:11392. https://doi.org/10.1128/JVI.75.23.11392-11400.2001.

[115] Acierno PM, Newton DA, Brown EA, Maes LA, Baatz JE, Gattoni-Celli S. Cross-reactivity between HLA-A2-restricted FLU-M1:58-66 and HIV p17 GAG:77-85 epitopes in HIV-infected and uninfected individuals. J Transl Med 2003;1:3. https://doi.org/10.1186/1479-5876-1-3.

[116] Cornberg M, Sheridan BS, Saccoccio FM, Brehm MA, Selin LK. Protection against vaccinia virus challenge by CD8 memory T cells resolved by molecular mimicry. J Virol 2007;81:934. https://doi.org/10.1128/JVI.01280-06.

[117] Kim SK, Cornberg M, Wang XZ, Chen HD, Selin LK, Welsh RM. Private specificities of CD8 T cell responses control patterns of heterologous immunity. J Exp Med 2005; 201:523–33. https://doi.org/10.1084/jem.20041337.

[118] Clute SC, Watkin LB, Cornberg M, Naumov YN, Sullivan JL, Luzuriaga K, Welsh RM, Selin LK. Cross-reactive influenza virus-specific CD8+ T cells contribute to lymphoproliferation in Epstein-Barr virus-associated infectious mononucleosis. J Clin Invest 2005;115: 3602–12. https://doi.org/10.1172/jci25078.

[119] van Stipdonk MJ, Lemmens EE, Schoenberger SP. Naïve CTLs require a single brief period of antigenic stimulation for clonal expansion and differentiation. Nat Immunol 2001;2:423–9. https://doi.org/10.1038/87730.

[120] Veiga-Fernandes H, Walter U, Bourgeois C, McLean A, Rocha B. Response of naïve and memory CD8+ T cells to antigen stimulation in vivo. Nat Immunol 2000;1: 47–53. https://doi.org/10.1038/76907.

[121] Abbas KM, Kang GJ, Chen D, Werre SR, Marathe A. Demographics, perceptions, and socioeconomic factors affecting influenza vaccination among adults in the United States. PeerJ 2018;6:e5171. https://doi.org/10.7717/peerj.5171.

[122] Randolph HE, Barreiro LB. Herd immunity: understanding COVID-19. Immunity 2020;52:737–41. https://doi.org/10.1016/j.immuni.2020.04.012.

[123] Dowdy D, D'Souza G. Early herd immunity against COVID-19: a dangerous misconception. In: Center JHCR, editor. Baltimore, MD, USA: Johns Hopkins University of Medicine; 2020.

[124] Black FL. The role of herd immunity in control of measles. Yale J Biol Med 1982;55:351–60.

[125] Ye Q, Wang B, Mao J. The pathogenesis and treatment of the "Cytokine Storm" in COVID-19. J Infect 2020;80: 607–13. https://doi.org/10.1016/j.jinf.2020.03.037.

[126] Zeng MY, Cisalpino D, Varadarajan S, Hellman J, Warren HS, Cascalho M, Inohara N, Núñez G. Gut microbiota-induced immunoglobulin G controls systemic infection by symbiotic bacteria and pathogens. Immunity 2016;44:647–58. https://doi.org/10.1016/j.immuni.2016.02.006.

[127] Gou W, Fu Y, Yue L, Chen G-D, Cai X, Shuai M, Xu F, Yi X, Chen H, Zhu YJ, et al. Gut microbiota may underlie the predisposition of healthy individuals to COVID-19. medRxiv 2020. https://doi.org/10.1101/2020.04.22.20076091.

[128] Raftis EJ, Delday MI, Cowie P, McCluskey SM, Singh MD, Ettorre A, Mulder IE. Bifidobacterium breve MRx0004 protects against airway inflammation in *a* severe asthma model by suppressing both neutrophil and eosinophil lung infiltration. Sci Rep 2018;8:12024. https://doi.org/10.1038/s41598-018-30448-z.

[129] 4dpharmaplc. Clinical update- phase II COVID-19 study. 2020. Leeds UK.

[130] Xiao F, Tang M, Zheng X, Liu Y, Li X, Shan H. Evidence for gastrointestinal infection of SARS-CoV-2. Gastroenterology 2020;158:1831–1833.e1833. https://doi.org/10.1053/j.gastro.2020.02.055.

[131] Wu F, Zhao S, Yu B, Chen YM, Wang W, Song ZG, Hu Y. A new coronavirus associated with human respiratory disease in China. Nature 2020;579:265–9. https://doi.org/10.1038/s41586-020-2008-3.

[132] Zuo T, Zhang F, Lui GCY, Yeoh YK, Li AYL, Zhan H, Wan Y, Chung A, Cheung CP, Chen N, et al. Alterations in gut microbiota of patients with COVID-19 during time of hospitalization. Gastroenterology 2020;159:944–955.e948. https://doi.org/10.1053/j.gastro.2020.05.048.

[133] Moore SC, Penrice-Randal R, Alruwaili M, Dong X, Pullan ST, Carter D, Bewley K, Zhao Q, Sun Y, Hartley C, et al. Amplicon based MinION sequencing of SARS-CoV-2 and metagenomic characterisation of nasopharyngeal swabs from patients with COVID-19. medRxiv 2020. https://doi.org/10.1101/2020.03.05.20032011.

[134] Miquel S, Martín R, Rossi O, Bermúdez-Humarán LG, Chatel JM, Sokol H, Thomas M, Wells JM, Langella P. Faecalibacterium prausnitzii and human intestinal health. Curr Opin Microbiol 2013;16:255–61. https://doi.org/10.1016/j.mib.2013.06.003.

[135] Verdu EF, Hayes CL, O' Mahony SM. Chapter 9—Importance of the microbiota in early life and influence on future health. In: Hyland N, Stanton C, editors. The gut-brain Axis. Cambridge, MA, USA: Academic Press; 2016. p. 159–84. https://doi.org/10.1016/B978-0-12-802304-4.00009-8.

[136] Bao L, Zhang C, Dong J, Zhao L, Li Y, Sun J. Oral microbiome and SARS-CoV-2: beware of lung co-infection. Front Microbiol 2020;11:1840. https://doi.org/10.3389/fmicb.2020.01840.

[137] Vatti A, Monsalve DM, Pacheco Y, Chang C, Anaya JM, Gershwin ME. Original antigenic sin: *a* comprehensive review. J Autoimmun 2017;83:12–21. https://doi.org/10.1016/j.jaut.2017.04.008.

[138] Zhang A, Stacey HD, Mullarkey CE, Miller MS. Original antigenic sin: how first exposure shapes lifelong anti-influenza virus immune responses. J Immunol 2019;202: 335–40. https://doi.org/10.4049/jimmunol.1801149.

[139] Gursel M, Gursel I. Is global BCG vaccination coverage relevant to the progression of SARS-CoV-2 pandemic? Med Hypotheses 2020. https://doi.org/10.1016/j.mehy.2020.109707.

[140] Johny R. A review on possibility of polio vaccine to combat Covid-19. Int J Innov Sci Res Technol 2020;5:2020.

[141] Arts RJW, Moorlag S, Novakovic B, Li Y, Wang SY, Oosting M, Kumar V, Xavier RJ, Wijmenga C, Joosten LAB, et al. BCG vaccination protects against experimental viral infection in humans through the induction of cytokines associated with trained immunity. Cell Host Microbe 2018;23:89–100 e105. https://doi.org/10.1016/j.chom.2017.12.010.

[142] Roth A, Jensen H, Garly ML, Djana Q, Martins CL, Sodemann M, Rodrigues A, Aaby P. Low birth weight infants and Calmette-Guerin bacillus vaccination at birth: community study from Guinea-Bissau. Pediatr Infect Dis J 2004;23:544–50. https://doi.org/10.1097/01.inf.0000129693.81082.a0.

[143] Jensen KJ, Larsen N, Biering-Sorensen S, Andersen A, Eriksen HB, Monteiro I, Hougaard D, Aaby P, Netea MG, Flanagan KL, et al. Heterologous immunological effects of early BCG vaccination in low-birth-weight infants in Guinea-Bissau: a randomized-controlled trial. J Infect Dis 2015;211:956–67. https://doi.org/10.1093/infdis/jiu508.

[144] Chen YW, Yiu CB, Wong KY. Prediction of the SARS-CoV-2 (2019-nCoV) 3C-like protease (3CL (pro)) structure: virtual screening reveals velpatasvir, ledipasvir, and other drug repurposing candidates. F1000Research 2020;9:129. https://doi.org/10.12688/f1000research.22457.2.

[145] Netea MG, Quintin J, van der Meer JW. Trained immunity: a memory for innate host defense. Cell Host Microbe 2011;9:355–61. https://doi.org/10.1016/j.chom.2011.04.006.

[146] Vita R, Mahajan S, Overton JA, Dhanda SK, Martini S, Cantrell JR, Wheeler DK, Sette A, Peters B. The immune epitope database (IEDB): 2018 update. Nucleic Acids Res 2019;47:D339–43. https://doi.org/10.1093/nar/gky1006.

[147] Altschul SF, Gish W, Miller W, Myers EW, Lipman DJ. Basic local alignment search tool. J Mol Biol 1990;215:403–10. https://doi.org/10.1016/S0022-2836(05)80360-2.

CHAPTER 5

Tackling COVID-19 Using Small-Molecule Drugs

SUBHENDU KARMAKAR

After the two epidemics of severe acute respiratory syndrome (SERS) and Middle East respiratory syndrome (MERS) in 2003 and 2012, respectively, in relatively small geographic locations, human coronaviruses have now emerged as a pandemic of COVID-19, creating major ongoing public health and social and economic problems globally [1,2]. Despite controversies in the routes of transmission to humans [3–5], it can be stated that the human 'novel' coronaviruses, responsible for COVID-19, are of the same class as the past viruses severe acute respiratory syndrome coronavirus (SARS-CoV) and Middle East respiratory syndrome coronavirus (MERS-CoV) and hence the World Health Organization (WHO) has termed the virus severe acute respiratory syndrome coronavirus 2 (SARS-CoV-2) [6,7]. The virus is an enveloped single-stranded RNA beta-coronavirus that enters the host cell by binding with the receptor angiotensin-converting enzyme 2 (ACE2) of the host cell using the viral structural spike (S) protein and then the viral RNA gets replicated following multiple cellular processes [8].

At the beginning of the pandemic outbreak, clinicians had to rely on various supportive treatments such as oxygen therapy and the use of antibiotics and steroids because of the unavailability of established treatment options for COVID-19 [9]. Over time, the invention of several antiviral agents (drugs) from the vault of already approved drugs gives clinicians freedom from the initial treatment limitations. This is acceptable because the drug development process is too long, and therefore to fulfill the demands of urgent pandemic situation, approved drugs or marketed drugs are being used for treatment of patients suffering from COVID-19. The various clinical research efforts can be classified in three main categories [10]. The first and the major category involves the clinical trials of already approved antiviral drugs (drug repurposing or drug repositioning). The benefit of this method is the prior knowledge of the drug's mechanism of action (MOA). This consists of dose estimation, drug metabolism, and side effects. The majority of these drugs are the outcome of extensive research from the previous virus outbreaks (e.g., SARS-CoV, MERS-CoV, Ebola virus). Yet they have disadvantages of poor selectivity owing to the difference in virus families and also the issues that made them unsuccessful in the past outbreaks in the beginning of this century. The second category depends on the theoretic screening to generate drugs that can possess specific targets against the mode of action of viruses (e.g., RNA polymerase inhibition) or other off-target effects such as inhibition of kinase. The third category contains the generation of completely new class of antiviral drugs by studying the molecular mechanism of the disease characteristics.

Small-molecule drugs are a molecularly defined chemical entity having low molecular weight (usually <1000 Da) that can pass the cell membrane to reach the intracellular target of action [11]. Owing to their small structure, drug development and manufacture processes are simpler than those of large molecules. Small-molecule drugs also have the advantages of oral bioavailability and their MOAs are normally more predictable [12]. They have remained as the backbone of the pharmaceutical industry for nearly a century and shown predominance even in the latest WHO Model List of Essential Medicines published in 2019 [13]. During the emergency period of COVID-19 also, numerous small-molecule drugs have been tested clinically and partially serve to eradicate the disease in hospitalized patients. In this chapter, I will discuss about some widespread small-molecule drugs and their mode of action.

COVID-19: Tackling Global Pandemics through Scientific and Social Tools. https://doi.org/10.1016/B978-0-323-85844-1.00013-1

1 REMDESIVIR

Remdesivir (Veklury®, GS-5734, developed by Gilead Sciences) is a prodrug of its active triphosphate nucleoside, GS-443902 [14]. It has been used for the treatment of Ebola [14,15] and one of its derivatives (GS-6620) also showed hepatitis C virus polymerase inhibition activity [16]. Remdesivir acts in a similar way like other anticancer DNA antimetabolite drugs such as gemcitabine and methotrexate. But due to the low bioavailability and low solubility of the active antimetabolite of remdesivir (not an oral drug, given as an intravenous injection), a mixture of prodrug strategies is inserted to gain maximum pharmacologic results. After entering the cell, several enzymatic and nonenzymatic chemical conversions (Fig. 5.1) resulted in the active drug molecule [16,17] (and hence may be termed as a prodrug of GS-443902). The pyrrolotriazine moiety of remdesivir has similarity with the adenine and guanine moieties of the nucleobase and thus behaves as the bioisostere of the nucleobases necessary for the synthesis of viral RNA. The active triphosphate inhibits the RNA-dependent RNA polymerase (RdRp), and thereby replication of the viral RNA is inhibited [15]. This indiscrimination ability of viral RdRp forces it to incorporate nucleotide analogues (drugs) and thereby is a key point for exploitation during many antiviral drug designs.

Some clinical trials of hospitalized COVID-19 patients using remdesivir have shown faster improvement from the disease, and many trials are still underway [18,19]. On May 1, 2020, the initial successful clinical studies using remdesivir forced the FDA to approve the use of remdesivir for the treatment of COVID-19 on an emergency basis [20]. Some trials also reported termination of treatment with remdesivir because of the statistically nonsignificant clinical benefits [21]. Yet the FDA has extended the Emergency Use Authorization (EUA) on October 1, 2020, for the use of remdesivir to treat hospitalized COVID-19 patients [20].

2 HYDROXYCHLOROQUINE AND CHLOROQUINE

Hydroxychloroquine (HCQ) and chloroquine (CQ) (9-aminoquinoline derivative) had a rich history of use in the treatment of malaria, rheumatoid arthritis, and autoimmune diseases such as systemic lupus erythematosus [22]. HCQ is the relatively less toxic derivative of CQ, and only for this reason, HCQ is preferred more in COVID-19 clinical trials than CQ to get better pharmacotoxicologic effects [23]. CQ already had a proven history of in vitro antiviral activity against a broad range of viruses including SARS-CoV and MERS-CoV [22,24,25]. The drugs (HCQ and CQ) are likely to act in different ways against SARS-CoV-2 than they do against the malarial parasite. The antimalarial MOA of CQ is mainly the inhibition of the conversion of heme to hemozoin and thus prevention of the action of heme polymerase in trophozoites of the parasite [26]. On the contrary the antiviral mechanism of CQ derivatives is primarily related to the inhibition of cellular entry of the virus. Being weakly basic and lipophilic, CQ passes the cell membrane by passive diffusion following the invasion in the endosomal organelles where they become protonated. This results in the rise of endosomal pH needed for the virus/cell fusion and thus causes hindrance to the SARS-CoV particle from entering the host cell [24,27]. In addition, CQ is known to inhibit the terminal glycosylation of ACE2 receptor chains [24]. SARS-CoV-2 targets the ACE2 receptor for cellular entry using the SARS-CoV-2 spike protein [8]. The virus spike protein interacts less efficiently with the ACE2 receptor when the receptor remains unglycosylated. CQ also possesses immune-modulating activity [22], which is shown (by in vivo studies) to increase the overall antiviral activity synergistically. The active metabolite of CQ derivatives is desethylhydroxychloroquine formed after the dealkylation of *N*-dialkyl moiety by the cytochrome P450 3A4 (CYP3A4) enzyme of the liver, along with the

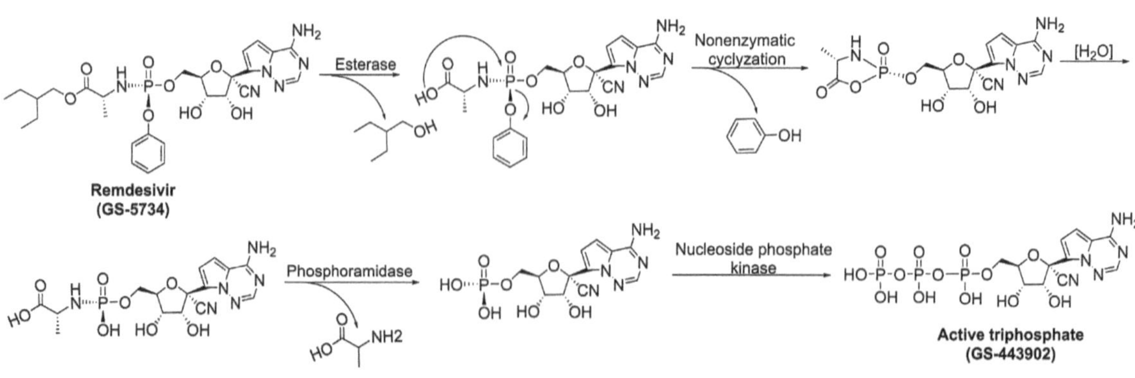

FIG. 5.1 Metabolism of remdesivir.

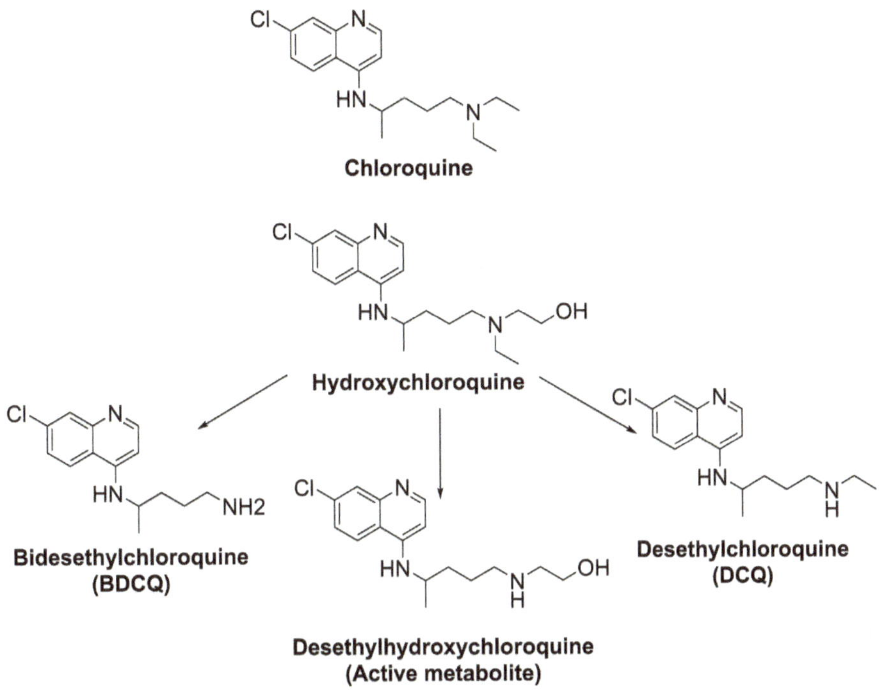

FIG. 5.2 Metabolism of hydroxychloroquine.

formation of two inactive metabolites desethylchloroquine (DCQ) and bidesethylchloroquine (BDCQ) (Fig. 5.2) [28].

3 FAVIPIRAVIR

Favipiravir (T-705, Avigan®, developed by the FUJIFILM Toyama Chemical Co., Ltd) is a chemically modified anti-influenza pyrazine derivative 3-hydroxy-2-pyrazinecarboxamide (T-1105) [29,30]. It also has a broad range of antiviral activity, other than against the influenza virus, and has been approved for clinical trials for COVID-19 in many countries [31]. Favipiravir is also a prodrug and needs intracellular activation. It actually inhibits the RNA polymerase of the influenza virus by interacting with RdRp [30]. After entering the virus cell, it undergoes phosphoribosylation to produce its active form (Fig. 5.3). This phosphoribosylated favipiravir is then recognized by the viral enzyme RdRp as a purine nucleotide analogue and is incorporated to the nascent RNA strand during the extension of viral RNA. If only one phosphoribosylated favipiravir is incorporated in the RNA strand, the extension of the RNA strand is partially blocked, and the attachment of a second molecule of phosphoribosylated favipiravir completely prevents the extension. The active drug actually competes with real purine nucleotides, adenosine and guanosine triphosphate, but not with pyrimidine nucleotides. The carboxamide group ($-CONH_2$) helps to form hydrogen bonds with cytosine (Fig. 5.4).

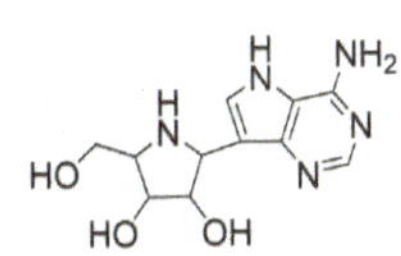

Galidesivir

4 GALIDESIVIR

Galidesivir (BCX4430) is also an RNA-targeting drug having structural similarity to remdesivir, except the sugar ring where the oxygen atom is released by a nitrogen atom. Owing to its antiviral activity against SARS-CoV, galidesivir is also clinically tested against SARS-CoV-2 , as both viruses have large similarities in their nonstructural proteins [32]. Inside the cell, galidesivir is phosphorylated to produce triphosphate nucleotide. Then it is transformed to galidesivir monophosphate following pyrophosphate cleavage. This

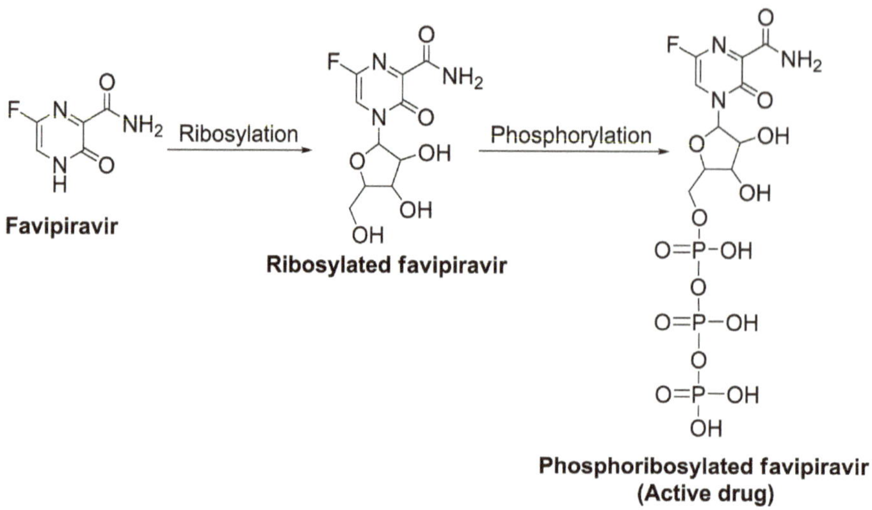

FIG. 5.3 Metabolism of favipiravir.

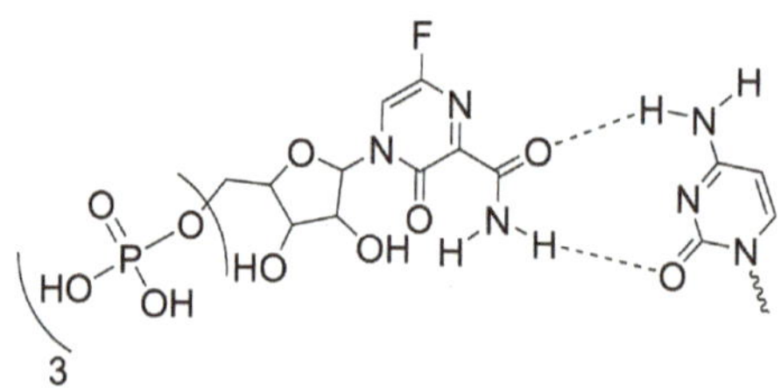

FIG. 5.4 Hydrogen bonding interaction with cytosine and carboxamide of favipiravir.

active monophosphate blocks RdRp chain extension soon after incorporation into the nascent RNA strand of the virus [33].

5 RIBAVIRIN

Ribavirin is also mimic of the guanine nucleotide substrate for RdRp after cellular transformation to the triphosphate analogue (ribavirin triphosphate) and added to the nascent RNA of virus [34]. Alike favipiravir, the carboxamide group ($-CONH_2$) of ribavirin behaves similarly to from hydrogen bonding with cytosine [35]. The monophosphate analogue of ribavirin can also affect the RNA polymerase activity by inhibition of the cellular enzyme inosine monophosphate dehydrogenase (IMPDH) causing a reduction in the cytoplasmic concentration of guanosine triphosphate [36]. The clinical antiviral activity of ribavirin was majorly evaluated against hepatitis C, but the activity against SARS-CoV-2 is found not so much remarkable because of the requirement of high doses and resistance.

6 PIRFENIDONE

Pirfenidone (Esbriet®, marketed by Roche) does not have any antiviral activity, even though it is planned to be used as supportive treatment for COVID-19 patients with severe infection [37,38]. Pirfenidone is an approved drug for the treatment of idiopathic pulmonary fibrosis (IPF), unclassifiable interstitial lung disease (uILD), and invasive pneumococcal disease (IPD) [39,40]. SARS-CoV-2 infection causes uncontrolled hyperactivation of the human immune system (cytokine storm, hypercytokinemia) leading to fatality of the patients by multiorgan failure and particularly acute respiratory distress syndrome (ARDS) becomes severe to patients with comorbidity [41–43]. Cytokine storm generates inflammation, fibrosis, oxidative stress, and apoptosis of healthy cells. Pirfenidone is an immunosuppressant [39] and has potent anti-inflammatory [44], antifibrotic [45], antioxidant [46], and antiapoptotic [47] activities, which are useful drug properties for the treatment of the aforesaid hyperimmune response [48]. The antifibrotic activity of pirfenidone involves downregulation of profibrotic gene expression, collagen secretion, and reduction of overexpression of transforming growth factor β during inflammation. Pirfenidone also inhibits the production of tumor necrosis factor α and other inflammatory cytokines, correlating the anti-inflammatory behavior [47].

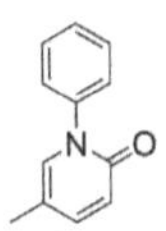

Pirfenidone

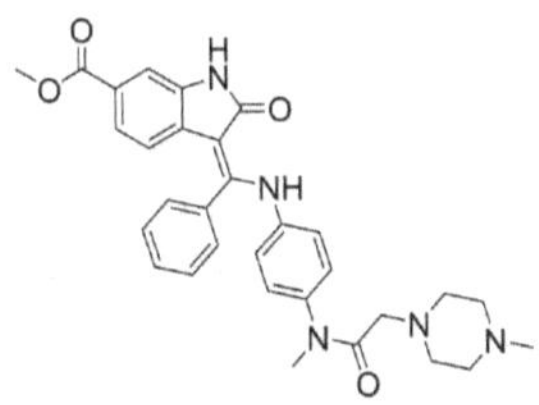

Nintedanib

7 NINTEDANIB

Nintedanib also possesses the antifibrotic property and is expected to have a similar clinical effect as pirfenidone [37], which has been proved by several clinical trials either as standalone [49,50] or in combination with pirfenidone [51].

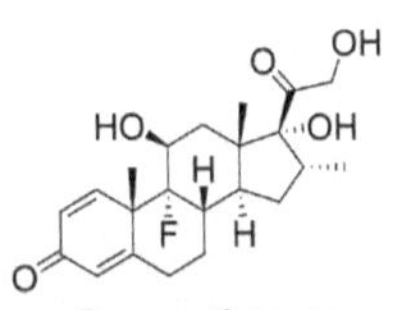

Dexamethasone

8 DEXAMETHASONE

The anti-inflammatory and immunosuppressant corticosteroid drug dexamethasone (an essential medicine in WHO's list [13]) is also approved by the WHO for treatment in patients with severe and critical COVID-19 to tackle ARDS [52].

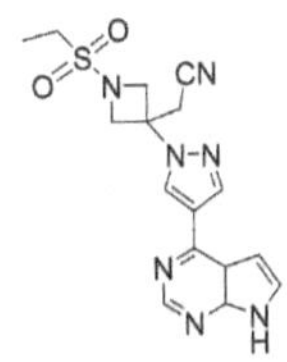

Baricitinib

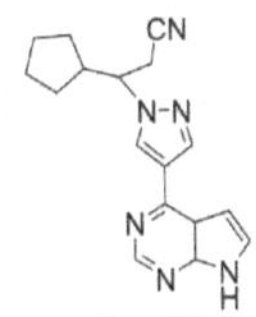

Ruxolitinib

9 BARICITINIB AND RUXOLITINIB

Like many other viruses, SARS-CoV-2 also enters the host cell by receptor-mediated endocytosis. SARS-CoV follows the clathrin-mediated endocytosis [53], and it is one of the predicted ways of entry for SARS-CoV-2 also [54]. Adaptor protein 2 (AP2)-associated protein kinase 1 (AAK1) modulates the clathrin-mediated endocytosis, and hence inhibition of AAK1 can distress the entry of the SARS-CoV-2 [55]. Baricitinib (Olumiant, developed by Lilly) is the inhibitor of AAK1 [56]. It is also the clinically approved inhibitor of Janus-associated kinase (JAK). JAK has crucial roles in proinflammatory signaling pathways in various autoimmune disorders [57]. Altogether, use of baricitinib in COVID-19 hampers the SARS-CoV-2 entry and inflammation-related complications by the inhibition of AAK1 and JAK (numb-associated kinase, NAK family), respectively, which is proved by numerous clinical trials [58]. Ruxolitinib also shows similar activities as baricitinib.

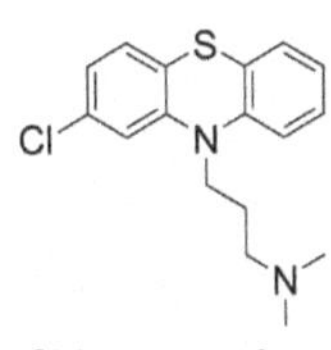

Chlorpromazine

10 CHLORPROMAZINE

Chlorpromazine is an antipsychotic agent and its encouraging antiviral activity was invented in vitro when tested against SARS-CoV and MERS-CoV [59]. The most accepted antiviral activity of chlorpromazine comes from the inhibition of relocation of clathrin lattices and AP2 on the cell surface and endosomes [60,61]. This directly affects the virus entry utilizing the endocytic pathway. Chlorpromazine was also shown to have immunomodulatory effects by increasing blood levels of IgM [59]. These effects have driven chlorpromazine to be tested clinically against SARS-CoV-2 [62,63].

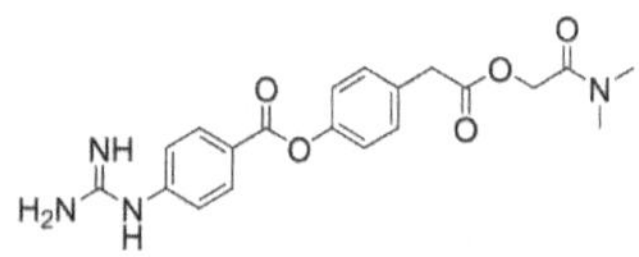

$MeSO_3H$ **Camostat mesylate**

11 CAMOSTAT MESYLATE

Mesylate salt of camostat is a serine protease inhibitor of trypsin and several inflammatory proteinases, and it is used in the treatment of chronic pancreatitis and postoperative reflux esophagitis in Japan [64]. It also has been found to have inhibitory activity against the transmembrane protease, serine 2 (TMPRSS2) of SARS-CoV [65]. TMPRSS2 also serves as an important drug target against SARS-CoV-2 because it mediates the entry and replication of SARS-CoV-2 in the host cell. This protease helps in the fusion with the host cell by cleaving the viral S protein followed by the replication inside the host cell [8].

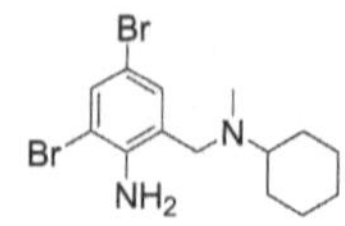

Bromhexine

12 BROMHEXINE

Bromhexine is also employed to inhibit the TMPRSS2 of virus [66]. It is a mucolytic cough medicine, approved in many countries for all age groups. Negligible side effects and low cost have made this drug an ideal candidate for the treatment of COVID-19 clinical trial [67].

13 LOPINAVIR/RITONAVIR

Lopinavir is an approved anti-HIV drug for the treatment of AIDS in combination with ritonavir [68]. They are protease inhibitors. The HIV protease-inhibiting activity of lopinavir is utilized to inhibit the enzyme 3-chymotrypsin-like protease ($3CL^{pro}$) and thus interferes with the replication of the RNA of SARS-CoV-2 [69]. Ritonavir (CYP3A4 inhibitor) is used as a supportive drug to increase the plasma concentration and efficacy [70]. A phase 2 trial was also carried out with the triple combination of lopinavir/ritonavir, ribavirin, and interferon beta-1b (natural immunomodulator) and was found more superior than only lopinavir/ritonavir [71].

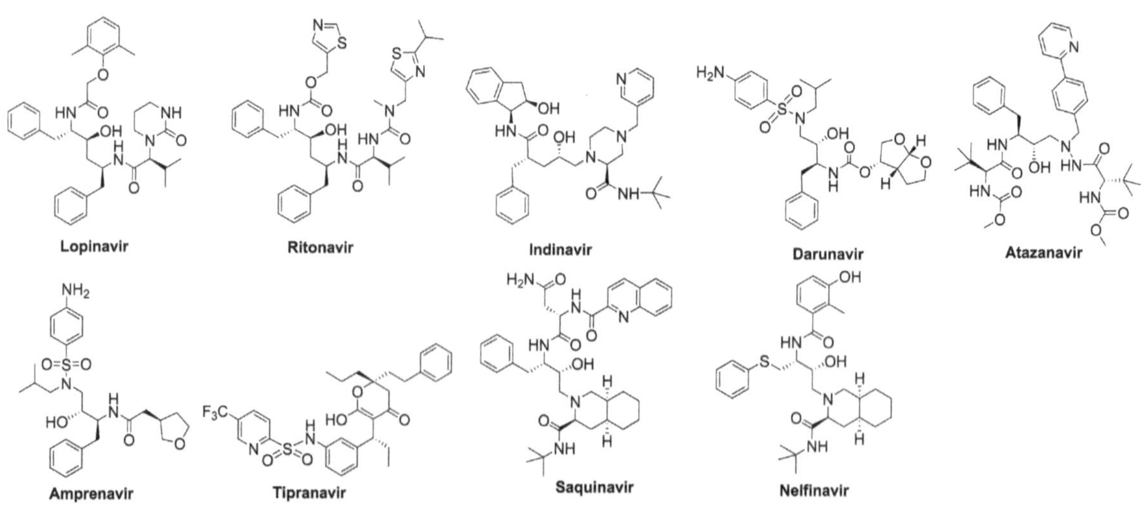

There are many other small-molecule HIV protease inhibitors to be mentioned here, which are tested either in vitro or clinically against SARS-CoV-2. These are indinavir, darunavir, amprenavir, tipranavir, saquinavir, nelfinavir, and atazanavir.

14 FAMOTIDINE

Famotidine (Pepcid®, developed by Yamanouchi Pharmaceutical Co.) is an approved drug for the treatment of gastroesophageal reflux disease and gastric ulcers. It is a histamine-2 receptor (H2R) antagonist that decreases the acid secretion in stomach by blocking the action of histamine [72]. The probable reason of antiviral activity of famotidine is demonstrated by computational studies revealing its targets of two important enzymes of SARS-CoV-2, namely, 3C-like protease ($3CL^{pro}$) and papain-like protease (PL^{pro}) [73,74]. In spite of having a large number of optimistic clinical trials, recent in vitro studies ruled out the possibility of the antiviral MOA based on the two aforementioned proteases (as predicted by computational studies) [75].

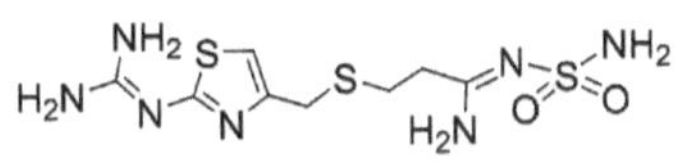

Famotidine

15 OSELTAMIVIR

Oseltamivir is a neuraminidase inhibitor and is used for the treatment of influenza. Although computational studies have shown that the $3CL^{pro}$ of SARS-CoV-2 is similar to that of neuraminidase of influenza A, oseltamivir is found ineffective in vitro against SARS-CoV-2 [76]. The clinical studies are also not encouraging.

16 NICLOSAMIDE

Niclosamide is used as an antihelminthic drug for the treatment of various parasitic infections (and hence registered in the WHO list of essential medicines) [77]. It is found effective in inhibiting the S-phase kinase-associated protein 2 (SPK2), which is involved in the viral nsp6-dependent blockage of autophagy in favor of virus replication. Thus the inhibition of SPK2 increases the autophagy of the host cell, which in turn decreases the virus replication [78,79]. Niclosamide also assists in increasing the intracellular pH, leading to an unfavorable condition to replicate virus [80].

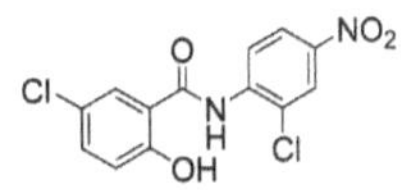

Niclosamide

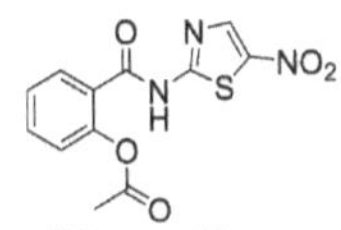

Nitazoxanide

17 NITAZOXANIDE

Owing to have structural similarity with niclosamide, nitazoxanide (an antiprotozoal and antihelminthic agent) also has the same MOA against SARS-CoV-2 and has been tested clinically [81,82].

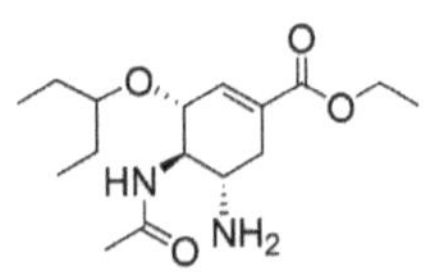

Oseltamivir

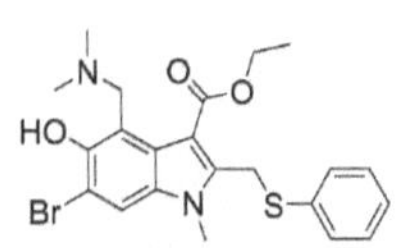

Umifenovir

18 UMIFENOVIR

Umifenovir (Arbidol®, manufactured by Pharmstandard) is an approved drug for use in the treatment of influenza and prophylaxis in some countries [83]. Although it is known to have activity against various ranges of RNA/DNA and enveloped/nonenveloped viruses, its exact mode of action is not well established. A few studies have shown that umifenovir interacts with the membrane lipids causing fusion between plasma membranes and virus particles and between virus particles and the membranes of endosomes; it can also serve as an immunomodulator and hence has interferon induction and/or macrophage activation ability [84]. The treatment outcomes of umifenovir treatment do not reveal the expectation [85].

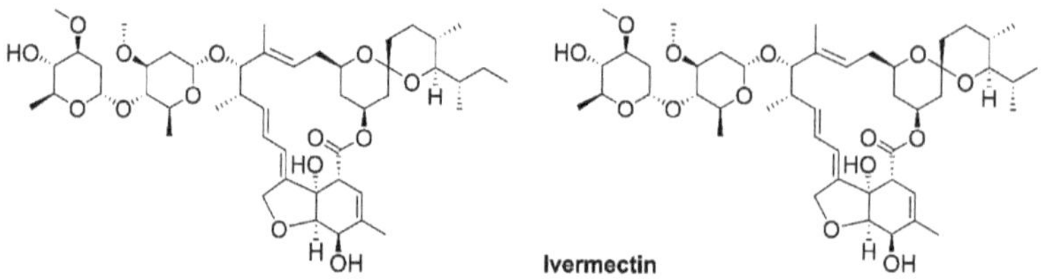

Ivermectin

19 IVERMECTIN

Ivermectin is an antiparasitic agent and a mixture of two 16-membered macrocyclic lactones of avermectin family and has antiviral activity against various viruses such as HIV, dengue virus, West Nile virus, Venezuelan equine encephalitis virus, and influenza virus, as well as against a few DNA viruses [86]. The antiviral activity of ivermectin involves hampering the transport of viral

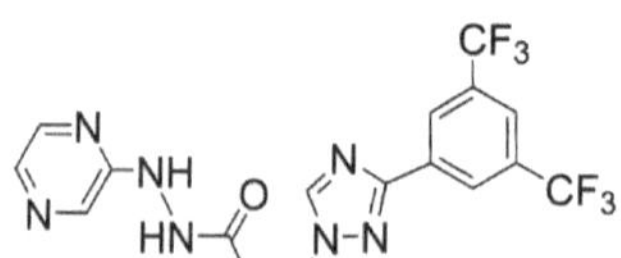

Selinexor

proteins in the nucleus during the replication process by dissociation of the importin (IMP) α/β1 heterodimer of host cells [87]. Various clinical trials of ivermectin are registered as standalone [88] or in combination with other drugs [82,89–91].

20 SELINEXOR

Selinexor (Xpovio®, KPT-330, manufactured by Karyopharm Therapeutics) is an approved drug for the treatment of refractory multiple myeloma and diffuse large B-cell lymphoma. It is a selective inhibitor of nuclear export that blocks the cellular protein exportin 1 (XPO1) and thus induces apoptosis of cancer cells following the reactivation of tumor suppressor protein [92]. The virus takes over the control on nuclear transport machinery of the host cell and localizes its own protein but mislocalizes host cell proteins. Therefore replication of the virus can be damaged by the specific inhibition of XPO1 [93]. Selinexor also has anti-inflammatory property that is useful to tackle the post-virus-attack cytokine storm. One placebo-controlled phase 2 clinical trial on 230 COVID-19 patients has been reported [94].

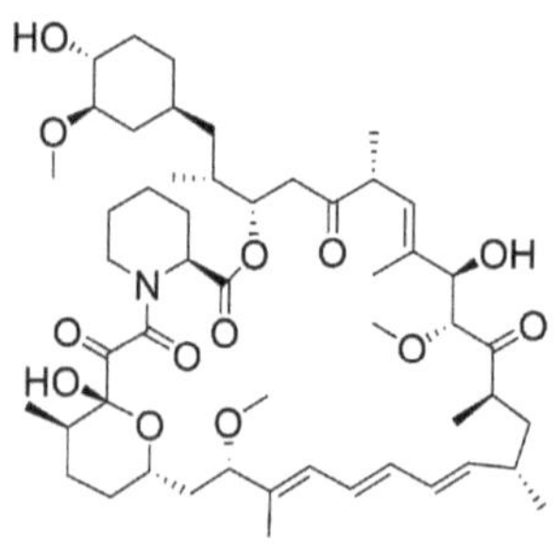

Sirolimus

21 SIROLIMUS (RAPAMYCIN)

Sirolimus is a macrolide class of compound and an immunosuppressant. It is an inhibitor of the serine/threonine kinase mTOR (mechanistic Target Of Rapamycin). This results in the restoration of T-cell function and reduction of cytokine storm [95].

22 VALSARTAN

Valsartan (Diovan®, by Novartis) is used as a vasodilator in patients with high blood pressure and heart diseases [96]. Valsartan is used as a supportive drug in COVID-19 patients without having any direct antiviral MOA. It is used to reduce the high burden of morbidity and mortality arising from ARDS. Valsartan blocks the angiotensin II type 1a receptor (AT1R) and also upregulates ACE2 expression. These two complementary processes result in decreases in the local concentration of angiotensin II (Ang II, which is responsible for increased lung pathology in COVID-19 patients) and at the same time increase the production of the other protective vasodilator angiotensin, which prevents ARDS [97].

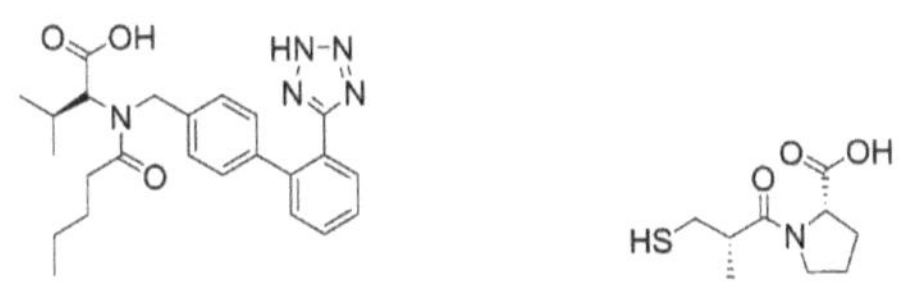

23 CAPTOPRIL

Captopril is a drug that inhibits ACE but at the same time upregulates ACE2 [98]. Apparently, it seems that upregulation of ACE2 opens up the gateway for viruses, but oppositely, ACE2 also has anti-inflammatory, antifibrotic, antioxidant, and vasodilator properties that are the essential to diminish lung damage (from ARDS) in hospitalized COVID-19 patients with comorbidity [99].

In summary, it can be stated that the small-molecule drugs used in the clinical trials against COVID-19 have assisted a larger group of patients to decrease the mortality or to recover early in the absence of any standard drug or treatment during the pandemic; at the same time, it can also be said that the contribution is not enough when human health comes into consideration. They have to be prior tested by in vitro and/or in vivo studies to prove their efficiency against SARS-CoV-2 strains, even the drugs that are repurposed because anything cannot be compromised with human health. The COVID-19 pandemic was an eye-opener for scientists around the world to tackle the numerous diseases instigated by various microorganisms.

REFERENCES

[1] Global Preparedness Monitoring Board, World Health Organisation. A World in Disorder. https://apps.who.int/gpmb/assets/annual_report/GPMB_AR_2020_EN.pdf (Accessed 13 October 2020).

[2] United Nations Development Programme. COVID-19 Socio-economic impact. 2020. https://www.undp.org/content/undp/en/home/coronavirus/socio-economic-impact-of-covid-19.html. [Accessed 13 October 2020].

[3] Zhu N, Zhang D, Wang W, Li X, Yang B, Song J, et al. A novel coronavirus from patients with pneumonia in China, 2019. N Engl J Med 2020;382(8):727–33.

[4] Zhou P, Yang X-L, Wang X-G, Hu B, Zhang L, Zhang W, et al. A pneumonia outbreak associated with a new coronavirus of probable bat origin. Nature 2020;579(7798):270–3.

[5] Calisher C, Carroll D, Colwell R, Corley RB, Daszak P, Drosten C, et al. Statement in support of the scientists, public health professionals, and medical professionals of China combatting COVID-19. Lancet 2020;395(10226):e42–3.

[6] The species severe acute respiratory syndrome-related coronavirus: classifying 2019-nCoV and naming it SARS-CoV-2. Nat Microbiol 2020;5(4):536–44.

[7] World Health Organisation. Naming the coronavirus disease (COVID-19) and the virus that causes it. 2020. https://www.who.int/emergencies/diseases/novel-coronavirus-2019/technical-guidance/naming-the-coronavirus-disease-(covid-2019)-and-the-virus-that-causes-it. [Accessed 13 October 2020].

[8] Gil C, Ginex T, Maestro I, Nozal V, Barrado-Gil L, Cuesta-Geijo MA, et al. COVID-19: drug targets and potential treatments. J Med Chem 2020. https://doi.org/10.1021/acs.jmedchem.0c00606.

[9] Huang C, Wang Y, Li X, Ren L, Zhao J, Hu Y, et al. Clinical features of patients infected with 2019 novel coronavirus in Wuhan, China. Lancet 2020;395(10223):497–506.

[10] Nitulescu GM, Paunescu H, Moschos SA, Petrakis D, Nitulescu G, Ion GND, et al. Comprehensive analysis of drugs to treat SARS-CoV-2 infection: mechanistic insights into current COVID-19 therapies (Review). Int J Mol Med 2020;46(2):467–88.

[11] Thierauch K-H. Small molecule drugs. In: Schwab M, editor. Encyclopedia of cancer. Berlin, Heidelberg: Springer Berlin Heidelberg; 2011. p. 3448–51.

[12] Ngo HX, Garneau-Tsodikova S. What are the drugs of the future? MedChemComm 2018;9(5):757–8.

[13] World Health Organisation. WHO model list of essential medicines, 21st list. 2019. https://apps.who.int/iris/bitstream/handle/10665/325771/WHO-MVP-EMP-IAU-2019.06-eng.pdf?ua=1. [Accessed 13 October 2020].

[14] Siegel D, Hui HC, Doerffler E, Clarke MO, Chun K, Zhang L, et al. Discovery and synthesis of a phosphoramidate prodrug of a pyrrolo[2,1-f][triazin-4-amino] adenine C-nucleoside (GS-5734) for the treatment of ebola and emerging viruses. J Med Chem 2017;60(5):1648–61.

[15] European Medicines Agency. Summary on compassionate use. Remdesivir Gilead; 2020. 03 April 2020, https://www.ema.europa.eu/en/documents/other/summary-compassionate-use-remdesivir-gilead_en.pdf. [Accessed 13 October 2020].

[16] Cho A, Zhang L, Xu J, Lee R, Butler T, Metobo S, et al. Discovery of the first C-nucleoside HCV polymerase inhibitor

(GS-6620) with demonstrated antiviral response in HCV infected patients. J Med Chem 2014;57(5):1812–25.
[17] Murakami E, Tolstykh T, Bao H, Niu C, Steuer HMM, Bao D, et al. Mechanism of activation of PSI-7851 and its diastereoisomer PSI-7977. J Biol Chem 2010; 285(45):34337–47.
[18] Grein J, Ohmagari N, Shin D, Diaz G, Asperges E, Castagna A, et al. Compassionate use of remdesivir for patients with severe Covid-19. N Engl J Med 2020; 382(24):2327–36.
[19] https://clinicaltrials.gov/ct2/results?cond=&term=remdesivir&cntry=&state=&city=&dist=. (Accessed 13 October 2020).
[20] https://www.fda.gov/media/137564/download. (Accessed 13 October 2020).
[21] Wang Y, Zhang D, Du G, Du R, Zhao J, Jin Y, et al. Remdesivir in adults with severe COVID-19: a randomised, double-blind, placebo-controlled, multicentre trial. Lancet 2020;395(10236):1569–78.
[22] Savarino A, Boelaert JR, Cassone A, Majori G, Cauda R. Effects of chloroquine on viral infections: an old drug against today's diseases? Lancet Infect Dis 2003;3(11): 722–7.
[23] Liu J, Cao R, Xu M, Wang X, Zhang H, Hu H, et al. Hydroxychloroquine, a less toxic derivative of chloroquine, is effective in inhibiting SARS-CoV-2 infection in vitro. Cell Discovery 2020;6(1):16.
[24] Vincent MJ, Bergeron E, Benjannet S, Erickson BR, Rollin PE, Ksiazek TG, et al. Chloroquine is a potent inhibitor of SARS coronavirus infection and spread. Virol J 2005;2 [No pp given].
[25] de Wilde AH, Jochmans D, Posthuma CC, Zevenhoven-Dobbe JC, van Nieuwkoop S, Bestebroer TM, et al. Screening of an FDA-approved compound library identifies four small-molecule inhibitors of Middle East respiratory syndrome coronavirus replication in cell culture. Antimicrob Agents Chemother 2014;58(8):4875–84. 11 pp.
[26] Slater AFG, Cerami A. Inhibition by chloroquine of a novel heme polymerase enzyme activity in malaria trophozoites. Nature (London) 1992;355(6356):167–9.
[27] Devaux CA, Rolain J-M, Colson P, Raoult D. New insights on the antiviral effects of chloroquine against coronavirus: what to expect for COVID-19? Int J Antimicrob Agents 2020;55(5):105938.
[28] Browning DJ. Pharmacology of chloroquine and hydroxychloroquine. Hydroxychloroquine and chloroquine retinopathy. 2014. p. 35–63.
[29] Sangawa H, Komeno T, Nishikawa H, Yoshida A, Takahashi K, Nomura N, et al. Mechanism of action of T-705 ribosyl triphosphate against influenza virus RNA polymerase. Antimicrob Agents Chemother 2013; 57(11):5202–8.
[30] Furuta Y, Komeno T, Nakamura T. Favipiravir (T-705), a broad spectrum inhibitor of viral RNA polymerase. Proc Jpn Acad Ser B Phys Biol Sci 2017;93(7):449–63.
[31] https://clinicaltrials.gov/ct2/results?cond=favipiravir&term=&cntry=&state=&city=&dist=. (Accessed 14 October 2020).
[32] Wu A, Peng Y, Huang B, Ding X, Wang X, Niu P, et al. Genome composition and divergence of the novel coronavirus (2019-nCoV) originating in China. Cell Host Microbe 2020;27(3):325–8.
[33] Warren TK, Wells J, Panchal RG, Stuthman KS, Garza NL, Van Tongeren SA, et al. Protection against filovirus diseases by a novel broad-spectrum nucleoside analogue BCX4430. Nature 2014;508(7496):402–5.
[34] Vanderlinden E, Vrancken B, Van Houdt J, Rajwanshi VK, Gillemot S, Andrei G, et al. Distinct effects of T-705 (favipiravir) and ribavirin on influenza virus replication and viral RNA synthesis. Antimicrob Agents Chemother 2016;60(11):6679–91.
[35] De Clercq E. New nucleoside analogues for the treatment of hemorrhagic fever virus infections. Chem Asian J 2019; 14(22):3962–8.
[36] Crotty S, Maag D, Arnold JJ, Zhong W, Lau JYN, Hong Z, et al. The broad-spectrum antiviral ribonucleoside ribavirin is an RNA virus mutagen. Nat Med 2000;6(12): 1375–9.
[37] George PM, Wells AU, Jenkins RG. Pulmonary fibrosis and COVID-19: the potential role for antifibrotic therapy. Lancet Respir Med 2020;8(8):807–15.
[38] https://clinicaltrials.gov/ct2/show/study/NCT04282902. (Accessed 14 October 2020).
[39] Margaritopoulos GA, Vasarmidi E, Antoniou KM. Pirfenidone in the treatment of idiopathic pulmonary fibrosis: an evidence-based review of its place in therapy. Core Evid 2016;11:11–22.
[40] Maher TM, Corte TJ, Fischer A, Kreuter M, Lederer DJ, Molina-Molina M, et al. Pirfenidone in patients with unclassifiable progressive fibrosing interstitial lung disease: a double-blind, randomised, placebo-controlled, phase 2 trial. Lancet Respir Med 2020;8(2):147–57.
[41] Chau Vinh Q, Oliveros E, Mahmood K, Singhvi A, Lala A, Moss N, et al. The imperfect cytokine storm: severe COVID-19 with ARDS in a patient on durable LVAD support. Netherlands FIELD Citation:: Division of Cardiology, Department of Medicine, Zena and Michael A. Wiener Cardiovascular Institute, Icahn School of Medicine at Mount Sinai, New York, New York. 2020. Report No.: 2666-0849.
[42] Qin C, Zhou L, Hu Z, Zhang S, Yang S, Tao Y, et al. Dysregulation of immune response in patients with coronavirus 2019 (COVID-19) in Wuhan, China. Clin Infect Dis 2020;71(15):762–8.
[43] Lescure F-X, Bouadma L, Nguyen D, Parisey M, Wicky P-H, Behillil S, et al. Clinical and virological data of the first cases of COVID-19 in Europe: a case series. Lancet Infect Dis 2020;20(6):697–706.
[44] Iyer SN, Hyde DM, Giri SN. Anti-inflammatory effect of pirfenidone in the bleomycin-hamster model of lung inflammation. Inflammation 2000;24(5):477–91.

[45] Schaefer CJ, Ruhrmund DW, Pan L, Seiwert SD, Kossen K. Antifibrotic activities of pirfenidone in animal models. Eur Respir Rev 2011;20(120):85–97.
[46] Giri SN, Leonard S, Shi X, Margolin SB, Vallyathan V. Effects of pirfenidone on the generation of reactive oxygen species in vitro. J Environ Pathol Toxicol Oncol 1999;18(3):169–77.
[47] Pirfenidone SS. A novel hypothetical treatment for COVID-19. Med Hypotheses 2020;144:110005.
[48] Mehta P, McAuley DF, Brown M, Sanchez E, Tattersall RS, Manson JJ. COVID-19: consider cytokine storm syndromes and immunosuppression. Lancet 2020; 395(10229):1033–4.
[49] https://clinicaltrials.gov/ct2/show/NCT04541680.Ahead. (Accessed 16 October 2020).
[50] https://clinicaltrials.gov/ct2/show/NCT04338802.Ahead. (Accessed 16 October 2020).
[51] https://clinicaltrials.gov/ct2/show/NCT02598193.Ahead. (Accessed 16 October 2020).
[52] https://www.who.int/publications/i/item/WHO-2019-nCoV-Corticosteroids-2020.1. (Accessed 31 October 2020).
[53] Inoue Y, Tanaka N, Tanaka Y, Inoue S, Morita K, Zhuang M, et al. Clathrin-dependent entry of severe acute respiratory syndrome coronavirus into target cells expressing ACE2 with the cytoplasmic tail deleted. J Virol 2007;81(16):8722–9.
[54] Bayati A, Kumar R, Francis V, McPherson PS. SARS-CoV-2 uses clathrin-mediated endocytosis to gain access into cells. bioRxiv 2020:1–31.
[55] Stebbing J, Phelan A, Griffin I, Tucker C, Oechsle O, Smith D, et al. COVID-19: combining antiviral and anti-inflammatory treatments. Lancet Infect Dis 2020; 20(4):400–2.
[56] Richardson P, Griffin I, Tucker C, Smith D, Oechsle O, Phelan A, et al. Baricitinib as potential treatment for 2019-nCoV acute respiratory disease. Lancet 2020; 395(10223):e30–1.
[57] Genovese MC, Kremer J, Zamani O, Ludivico C, Krogulec M, Xie L, et al. Baricitinib in patients with refractory rheumatoid arthritis. N Engl J Med 2016;374(13): 1243–52.
[58] https://clinicaltrials.gov/ct2/results?cond=baricitinib&term=&cntry=&state= &city=&dist=. (Accessed 14 October 2020).
[59] Plaze M, Attali D, Petit AC, Blatzer M, Simon-Loriere E, Vinckier F, et al. Repurposing chlorpromazine to treat COVID-19: the reCoVery study. L'Encephale 2020; 46(3):169–72.
[60] Stip E, Stip E. Psychiatry and COVID-19: the role of chlorpromazine. Can J Psychiatry Rev canadienne de psychiatrie 2020;65(10):739–40.
[61] Sathyamoorthy NK, Chintamaneni PK, Chinni S. Plausible role of combination of chlorpromazine hydrochloride and teicoplanin against COVID-19. Med Hypotheses 2020;144:110011.
[62] https://clinicaltrials.gov/ct2/show/NcT04354805. (Accessed 17 October 2020).
[63] https://clinicaltrials.gov/ct2/show/NCT04366739. (Accessed 17 October 2020).
[64] http://www.shijiebiaopin.net/upload/product/201272318373223.PDF. (Accessed 14 October 2020).
[65] Kawase M, Shirato K, van der Hoek L, Taguchi F, Matsuyama S. Simultaneous treatment of human bronchial epithelial cells with serine and cysteine protease inhibitors prevents severe acute respiratory syndrome coronavirus entry. J Virol 2012;86(12):6537–45.
[66] Depfenhart M, Depfenhart M, de Villiers D, Lemperle G, Meyer M, Di Somma S, et al. Potential new treatment strategies for COVID-19: is there a role for bromhexine as add-on therapy? Int Emergency Med 2020;15(5): 801–12.
[67] https://clinicaltrials.gov/ct2/show/NCT04355026. (Accessed 17 October 2020).
[68] Croxtall JD, Perry CM. Lopinavir/ritonavir: a review of its use in the management of HIV-1 infection. Drugs 2010; 70(14):1885–915.
[69] Chen Y, Liu Q, Guo D. Emerging coronaviruses: genome structure, replication, and pathogenesis. J Med Virol 2020;92(4):418–23.
[70] https://clinicaltrials.gov/ct2/results?cond=lopinavir&term=covid&cntry=&state=&city=&dist=&Search=Search. (Accessed 14 October 2020).
[71] Hung IF-N, Lung K-C, Tso EY-K, Liu R, Chung TW-H, Chu M-Y, et al. Triple combination of interferon beta-1b, lopinavir-ritonavir, and ribavirin in the treatment of patients admitted to hospital with COVID-19: an open-label, randomised, phase 2 trial. Lancet 2020; 395(10238):1695–704.
[72] Keithley JK. Histamine H2-receptor antagonists. Nurs Clin 1991;26(2):361–73.
[73] Wu C, Zheng M, Li H, Liu Y, Yang Y, Zhang P, et al. Analysis of therapeutic targets for SARS-CoV-2 and discovery of potential drugs by computational methods. Acta Pharmaceutica Sinica B 2020;10(5):766–88.
[74] Shaffer L. 15 drugs being tested to treat COVID-19 and how they would work. New York, NY, United States: Nature Medicine; 2020 [Ahead of Print].
[75] Loffredo M, Lucero H, Chen D-Y, O'Connell A, Bergqvist S, Munawar A, et al. The effect of famotidine on SARS-COV-2 proteases and virus replication. bioRxiv 2020:1–31.
[76] Tan Q, Duan L, Ma Y, Wu F, Huang Q, Mao K, et al. Is oseltamivir suitable for fighting against COVID-19: in silico assessment, in vitro and retrospective study. Bioorg Chem 2020;104:104257.
[77] Chen W, Mook Jr RA, Premont RT, Niclosamide WJ. Beyond an antihelminthic drug. Cell Signal 2018;41: 89–96.
[78] Wu C-J, Jan J-T, Chen C-M, Hsieh H-P, Hwang D-R, Liu H-W, et al. Inhibition of severe acute respiratory syndrome coronavirus replication by niclosamide. Antimicrob Agents Chemother 2004;48(7):2693–6.
[79] Gassen NC, Papies J, Bajaj T, Dethloff F, Emanuel J, Weckmann K, et al. Analysis of SARS-CoV-2-controlled autophagy reveals spermidine, MK-2206, and

niclosamide as putative antiviral therapeutics. bioRxiv 2020:1−13.
[80] Jurgeit A, McDowell R, Moese S, Meldrum E, Schwendener R, Greber UF. Niclosamide is a proton carrier and targets acidic endosomes with broad antiviral effects. PLoS Pathog 2012;8(10):e1002976.
[81] https://clinicaltrials.gov/ct2/show/NcT04359680. (Accessed 14 October 2020).
[82] https://clinicaltrials.gov/ct2/show/NcT04351347. (Accessed 14 October 2020).
[83] Blaising J, Polyak SJ, Pecheur E-I. Arbidol as a broad-spectrum antiviral: an update. Antivir Res 2014;107: 84−94.
[84] Haviernik J, Stefanik M, Fojtikova M, Kali S, Tordo N, Rudolf I, et al. Arbidol (umifenovir): a broad-spectrum antiviral drug that inhibits medically important arthropod-borne flaviviruses. Viruses 2018;10(4). 184/1-/8.
[85] Huang D, Yu H, Wang T, Yang H, Yao R, Liang Z. Efficacy and safety of umifenovir for coronavirus disease 2019 (COVID-19): A systematic review and meta-analysis. J Med Virol 2021;93(1):481−90.
[86] Mudatsir M, Yufika A, Nainu F, Frediansyah A, Megawati D, Pranata A, et al. Antiviral activity of ivermectin against SARS-CoV-2: an old-fashioned dog with a new trick—a literature review. Sci Pharm 2020;88(3):36.
[87] Yang SNY, Atkinson SC, Wang C, Lee A, Bogoyevitch MA, Borg NA, et al. The broad spectrum antiviral ivermectin targets the host nuclear transport importin α/β1 heterodimer. Antivir Res 2020;177:104760.
[88] https://clinicaltrials.gov/ct2/show/NCT04445311. (Accessed 14 October 2020).
[89] https://clinicaltrials.gov/ct2/show/NCT04523831. (Accessed 14 October 2020).
[90] https://clinicaltrials.gov/ct2/show/NcT04343092. (Accessed 14 October 2020).
[91] https://clinicaltrials.gov/ct2/show/NcT04360356 (Accessed 14 October 2020).
[92] Crochiere Marsha L, Baloglu E, Lee M, Kauffman M, Shacham S, Landesman Y, et al. XPO1 target occupancy measurements confirm the selinexor recommended phase 2 dose. Oncotarget 2017;8(66):110503−16.
[93] Uddin MH, Zonder JA, Azmi AS. Exportin 1 inhibition as antiviral therapy. Drug Discov Today 2020;25(10): 1775−81.
[94] https://clinicaltrials.gov/ct2/show/NCT04349098. [Ahead of Print].
[95] Omarjee L, Janin A, Perrot F, Laviolle B, Meilhac O, Mahe G. Targeting T-cell senescence and cytokine storm with rapamycin to prevent severe progression in COVID-19. Clin Immunol 2020;216:108464 (Amsterdam, Netherlands).
[96] https://www.drugs.com/mtm/valsartan.html. (Accessed 14 October 2020).
[97] https://clinicaltrials.gov/ct2/show/NCT04335786. (Accessed 14 October 2020).
[98] Ghafouri-Fard S, Noroozi R, Omrani MD, Branicki W, Pospiech E, Sayad A, et al. Angiotensin converting enzyme: a review on expression profile and its association with human disorders with special focus on SARS-CoV-2 infection. Vasc Pharmacol 2020;130:106680.
[99] https://clinicaltrials.gov/ct2/show/NCT04355429. (Accessed 14 October 2020).

CHAPTER 6

Modeling of COVID-19 Outbreak in Reference to Physical Parameters

VARUN AGIWAL • JITENDRA KUMAR • PRABHAT TRIVEDI • SAURABH KUMAR • ASHOK KUMAR

1 INTRODUCTION

There has been a history of the outbreak of viruses from time to time, like the emergence of Ebola virus, H1N1, Astrovirus, etc. Coronaviruses (CoVs) having ribonucleic acid (RNA) as their genetic material came into limelight during the outbreak of severe acute respiratory syndrome CoV (SARS-CoV) in 2002 and further during the outbreak of Middle East respiratory syndrome coronavirus (MERS-CoV) in 2012. The recent outbreak of 2019 novel coronavirus (SARS-CoV-2) in Wuhan, China, and its further spread throughout the world has become a global threat to the existence of humankind [1]. Several reports have suggested the relatively rapid spread of SARS-CoV-2 and the disease caused by it, coronavirus disease 2019 (COVID-19). This leads to the official declaration of COVID-19 by the World Health Organization (WHO) as a pandemic [2].

COVID-19 has developed an unrivaled challenge before the humankind. It becomes essential for scientists and researchers to analyze the effect of the spread of the virus. This chapter is focused on measuring the impact of this pandemic and tries to analyze the spread using numerous physical parameters. Epidemiologic studies have always evidenced that there is a strong association between the physical parameters and virus spread. Various studies had confirmed that temperature plays a very significant role in the community transmission of viral infections, see Refs. [3–6]. This advocates to study the progression of COVID-19 in different geographic locations concerning various physical parameters such as the maximum and minimum temperatures and the relative humidity of those locations. Qi et al. [7] and Wu et al. [8] examined the associations of daily average temperature (AT) and average relative humidity (ARH) with the daily count of COVID-19 cases based on a generalized additive model. In both the studies, COVID-19 variables are negatively correlated with AT and ARH. Conversely, Ma et al. [9] reported that there was a positive association between daily deaths and AT range but a negative association between daily death and ARH. Ujiie et al. [10] recorded an association among monthly inbound visitors from China, old-age dependency ratio, and low temperature with the cumulative number of COVID-19 cases. So information on the spread of COVID-19 in a geographic area, that is, about the demographic and climatic conditions, is needed to devise better strategies to combat the novel CoV infection. In retrospective studies, the outbreak of severe acute respiratory syndrome (SARS) in Guangdong in 2003 gradually faded with the warming weather coming and ended in July [11]. Tan et al. [12] documented that the temperature and its variations might have affected the SARS outbreak. A study in Korea found that the risk of influenza incidence was significantly increased with low daily temperature and low relative humidity. A significantly positive association was observed for the diurnal temperature range (DTR) [13]. Moreover, DTR [14] and temperature [15] were linked to death from respiratory diseases. Metz and Finn [16] demonstrated that absolute humidity had significant correlations with influenza virus survival and transmission rates. A few studies reported that COVID-19 was related to meteorologic factors, which decreased with increasing temperature see Refs. [17] and [18], but their effects on mortality had not been reported. Therefore weather conditions might also contribute to the spread of COVID-19.

Thus this chapter aims to evaluate the global spread in reference to physical parameters such as maximum temperature, minimum temperature, and relative

COVID-19: Tackling Global Pandemics through Scientific and Social Tools. https://doi.org/10.1016/B978-0-323-85844-1.00012-X

humidity. We use the cumulative number of confirmed cases and cumulative deaths to study COVID-19, along with other numerous variables such as physical parameters and demographic parameters, which are also used for extracting information about COVID-19. Important distributions are fitted and descriptive measures are used for comparing the differences between theoretic and descriptive parameters.

2 MATERIALS AND METHODS

The causal agent for COVID-19 was identified as SARS-CoV-2. The fast spread of COVID-19 crossed the boundaries of different geographic locations and infected people in almost all countries of the world within a few months. This study is conducted for the period May 1, 2020, to September 3, 2020. The global COVID-19 dataset is derived from the online dataset maintained by Our World in Data. It is updated daily and includes data on confirmed cases, deaths, and testing. The demographic data is obtained from the online site (Worldometer). In order to correlate the number of COVID-19 cases in different countries based on physical parameters, the distribution of confirmed cases in reference to maximum and minimum temperatures and relative humidity of those countries is derived from the National Oceanic and Atmospheric Administration (NOAA), Global Summary of The Day, made available for the research purposes.

This study has been targeted to explore the COVID-19 cases of different countries. We have entered the data in Microsoft Excel, and data analysis was carried out using the R software with 3.6.1 version. This provides a wide variety of statistical analyses and graphical techniques and is highly extensible. Various important distributions such as exponential, Weibull, gamma, and lognormal are used for fitting using fitdistr R package. The descriptive statistics are used for comparing the difference between theoretic and descriptive parameters. Imputation of missing observations are done using various approaches, and several evaluation metrics such as Akaike information criterion (AIC) and Bayesian information criterion (BIC) are used to come up with the best results.

3 RESULTS AND DISCUSSION

The main aim of this study is to explore the effect of temperature on the outbreak and spread of COVID-19. Under this motive, the study analyzes the temporal effect of climatic factors such as maximum and minimum temperatures and relative humidity on confirmed COVID-19 cases. From May 1, 2020, we recorded the climatic factor data at 2-week intervals. We chose May 14 as the initial date because most of the countries were affected by COVID-19 and approximately more than 90% of the total world population was under its threat. This work was carried out up to September 3, 2020, so that we obtain nine study intervals. To deduce whether survivability of SARS-CoV-2 is dependent on environmental temperature or not, we extract the number of COVID-19 cases reported on those dates.

3.1 Modeling on the Confirmed Cases About the Maximum Temperature

We recorded the maximum temperature of all those countries that reported COVID-19. For every 2 weeks, we recorded the total number of confirmed cases for each country, along with its maximum temperature, and then distributed the number of infected countries based on maximum temperature interval. Fig. 6.1 shows the allocation of the number of countries based on maximum temperature intervals in each study period.

For the abovementioned period, all relevant data are available on the online platform where climatic data are daily updated. But we came across some of the missing observations in other dates and excluded those countries from the analysis for which the data on climatic factors is not available. Fig. 6.1 gives the idea of the proportion of the number of infected countries with different intervals of maximum temperature. This could be helpful to see the growth of COVID-19 spread worldwide, along with maximum temperature intervals. We observed that the COVID-19 outbreak mostly affected those countries that have a standard maximum temperature ranging from 20 to 35°C as compared to the high or low level of maximum temperature. This interval covered around 70% of the total number of infected countries, and its coverage increased, as all countries are infected by COVID-19. So we conclude that the spread of the virus causing COVID-19 is more at the normal temperature, as there is a high chance of survival of the virus in a country with such temperature.

Once the range of maximum temperature at which most of the countries are infected by COVID-19 is determined, the study further analyzes the growth pattern of CoV based on distribution methodology. In distribution theory, one may often use well-known distributions such as exponential and Weibull that contain constant, increasing, and decreasing hazard function.

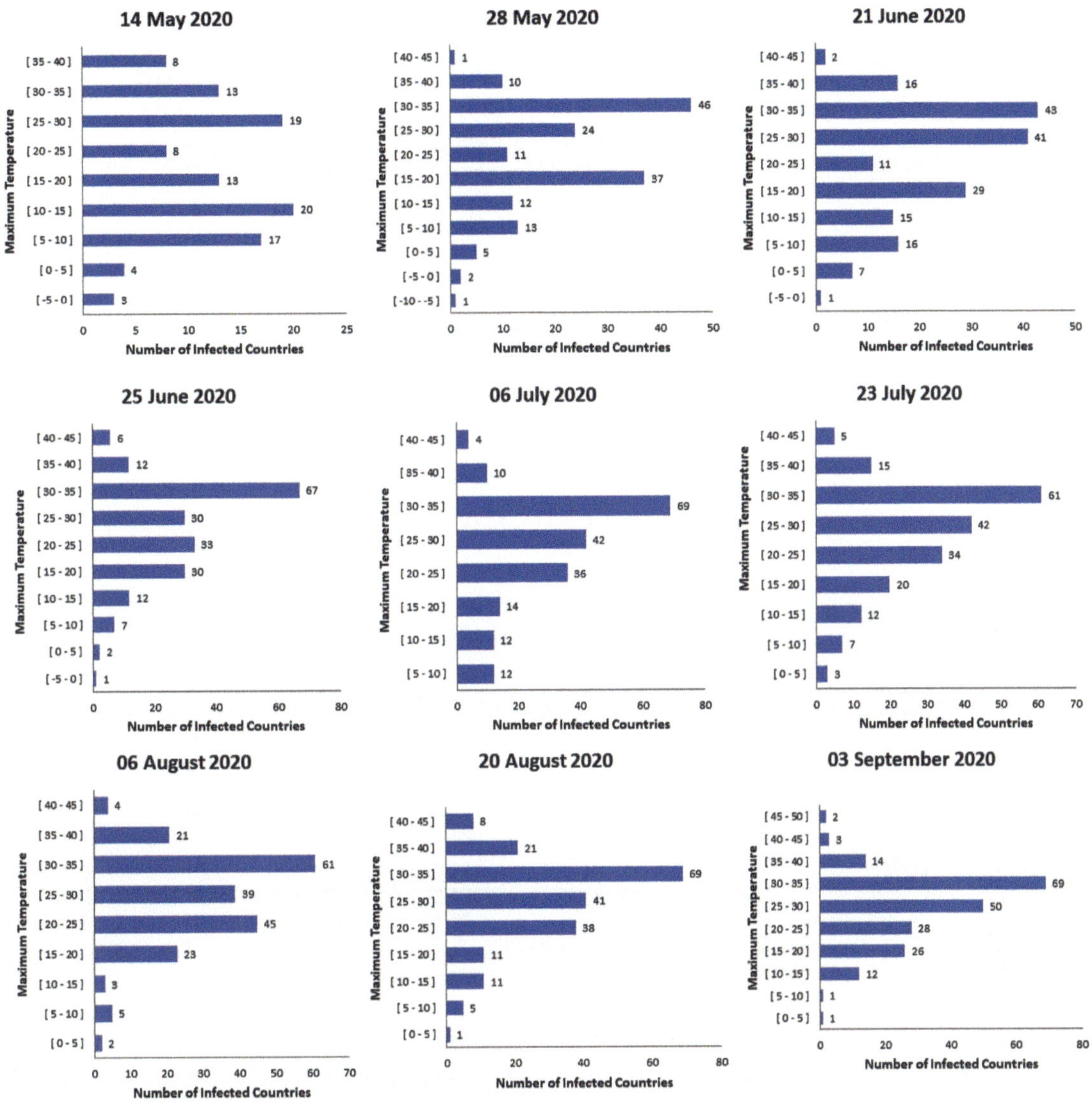

FIG. 6.1 The total number of infected countries in different maximum temperature intervals. X-axis denotes the number of infected countries and Y-axis denotes the various intervals of maximum temperature. The length of the bar is equal to the number of countries with coronavirus disease 2019 (COVID-19) in a particular maximum temperature interval.

But this may not be the situation in various real-life applications. So other hazard shapes are also considered, such as hump shape, bathtub, unimodal shape, etc. For each study period, the top three intervals of maximum temperature that cover most of the infected countries are considered to find out the best distribution in each interval. Based on the best fitted distribution, the growth pattern of CoV for a particular country is easily estimated as per the country's maximum temperature. The selection of the best distribution and its parameters' estimated values are described in Table 6.1.

TABLE 6.1
Best-Fitted Distribution Using Model Selection for Maximum Temperature Interval.

Date	Class Interval	NOC	Distribution	AIC	BIC	Estimate 1	Estimate 2
14-May	[5–10]	17	Gamma	−53.09	−51.67	0.53	8.01
	[10–15]	20	Lognormal	−22.38	−20.72	−2.92	2.06
	[25–30]	19	Exponential	−90.98	−90.04	31.40	-
28-May	[15–20]	37	Lognormal	−44.84	−42.04	−2.68	1.56
	[25–30]	24	Lognormal	−24.60	−24.00	−3.81	1.89
	[30–35]	46	Lognormal	−240.63	−237.30	−5.01	1.58
11-June	[15–20]	29	Exponential	6.85	7.94	2.43	-
	[25–30]	43	Lognormal	−121.27	−118.28	−3.91	1.82
	[30–35]	41	Lognormal	−167.34	−163.87	−3.79	1.39
25-June	[15–20]	30	Lognormal	108.91	111.35	0.13	1.72
	[20–25]	33	Weibull	81.62	84.28	0.61	1.29
	[25–30]	30	Lognormal	−61.46	−59.10	−3.21	1.53
	[30–35]	67	Lognormal	−181.05	−177.04	−3.48	1.46
9-July	[20–25]	36	Gamma	102.15	104.95	0.41	0.15
	[25–30]	42	Gamma	42.38	45.65	0.51	0.68
	[30–35]	69	Lognormal	−113.83	−109.74	−2.83	1.45
23-July	[20–25]	34	Gamma	130.18	133.05	0.61	0.20
	[25–30]	42	Lognormal	−27.71	−24.66	−2.39	1.65
	[30–35]	61	Lognormal	15.41	19.28	−1.87	1.76
6-August	[20–25]	45	Lognormal	160.13	163.35	0.01	1.97
	[25–30]	39	Weibull	−1.44	1.49	0.66	0.30
	[30–35]	61	Lognormal	9.21	13.00	−2.05	1.98
20-August	[20–25]	38	Gamma	167.10	170.09	0.51	0.10
	[25–30]	41	Lognormal	70.27	73.38	−1.16	1.99
	[30–35]	69	Lognormal	21.36	25.51	−1.94	1.95
3-September	[20–25]	28	Lognormal	122.99	125.34	0.32	2.09
	[25–30]	50	Lognormal	121.30	124.73	−0.68	2.01
	[30–35]	69	Lognormal	52.14	56.26	−1.71	2.02

AIC, Akaike information criterion; *BIC*, Bayesian information criterion; *NOC*, number of countries.

The results recorded in Table 6.1 show that the lognormal distribution is well fitted in the normal maximum temperature interval ranging from 25 to 35°C after a regular study period, followed by gamma and Weibull distribution. This selection is obtained based on the minimum values of AIC and BIC. The occurrence of lognormal distribution shows that the hazard function is hump-shaped, i.e., it initially increases, reaches a maximum, and then decreases toward no cases as time passes on. Therefore this distribution clearly explains the growth pattern of COVID-19 cases. As most of the times, any new disease pattern observes an increasing pattern, achieves a higher peak, and then comes down slowly. So lognormal distribution is well fitted to explain this as seen in COVID-19 pandemic also. The estimated values of distribution parameters are also recorded in Table 6.1. Overall, lognormal distribution is a better suited distribution to study the trend pattern of COVID-19 at any level of maximum temperature.

3.2 Modeling on the Confirmed Cases About the Minimum Temperature

Along with maximum temperature, the study wants to analyze the significance of minimum temperature in the global outbreak of COVID-19. Hence, the minimum temperatures of all those countries that reported COVID-19 infections at different study time points were recorded. To deduce whether the survivability of SARS-CoV-2 is dependent on environmental temperature or not, the study extrapolated the registered number of COVID-19 cases in reference to the minimum temperature of the respective country at different time points. Fig. 6.2 shows the allocation of the number of countries based on minimum temperature intervals in each study period.

The minimum temperature of most of the countries lies between 20 and 30°C. In this interval, CoV growth is rapidly increased and covers most of the region worldwide. This means the spread of COVID-19 occurred in all those countries where the minimum temperature has a range [20–30°C]. For the high or low level of minimum temperatures, a significantly less number of infected countries are included in these intervals. So the spread of CoV is moving up as the minimum temperature is increased. Here, the most favored

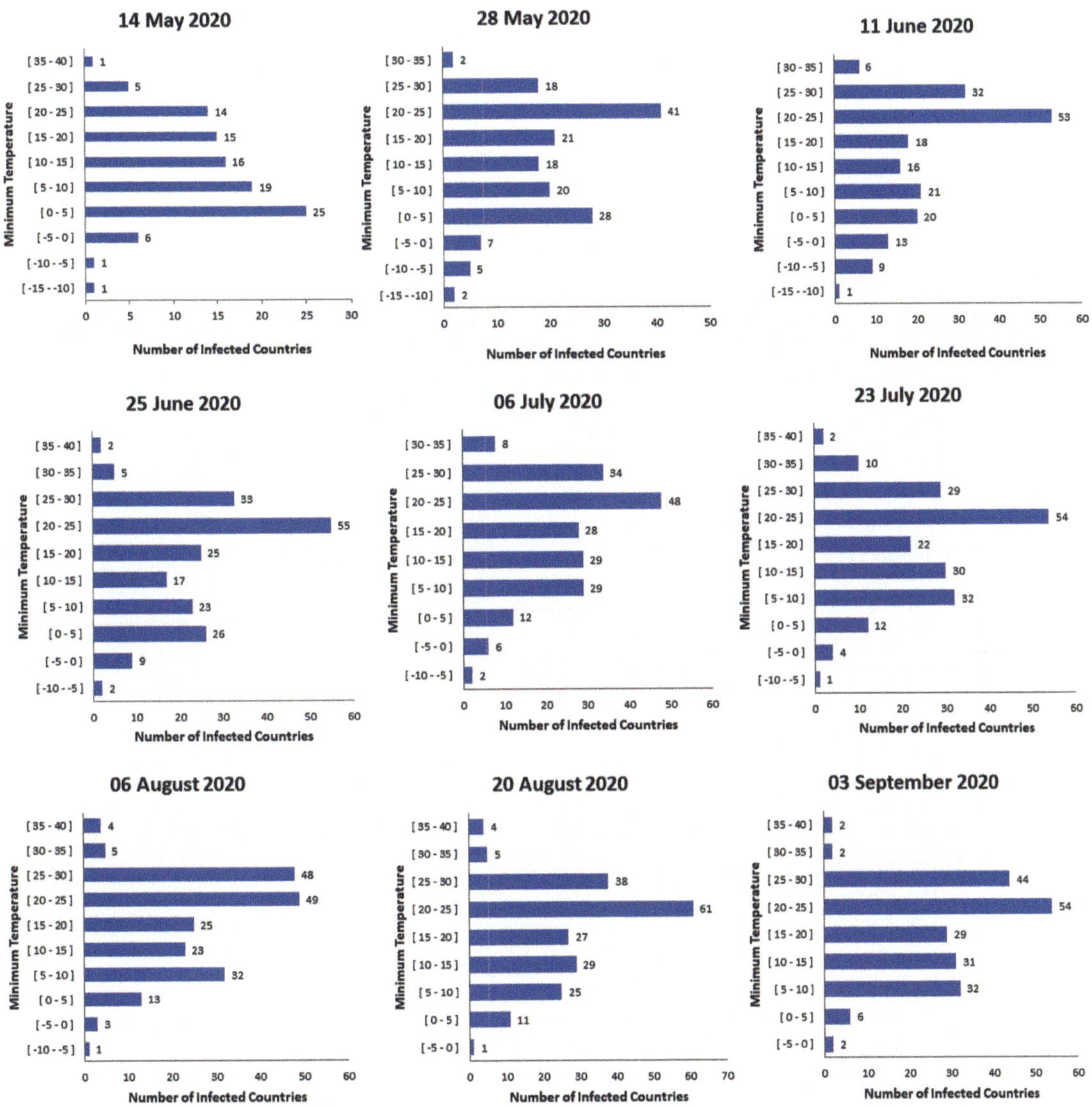

FIG. 6.2 The total number of infected countries in different minimum temperature intervals. X-axis denotes the number of infected countries and Y-axis denotes the various intervals of minimum temperature. The length of the bar is equal to the number of countries with coronavirus disease 2019 (COVID-19) in a particular minimum temperature interval.

minimum temperature for the spread of COVID-19 was found to be between the normal temperatures. As the minimum constructive interval for all infected countries worldwide was recorded, the next step is to study the development structure of COVID-19 in all countries using the distribution approach. We have used a similar distribution technique to determine the best suitable distribution corresponding to the top three minimum temperature intervals that include most of the infected countries. Table 6.2 records the best-fitted distribution using information criterion with its estimated parameter values. This selection in each minimum interval is based on the minimum values of AIC and BIC.

From Table 6.2, the best-fitted distribution for the interval [25−30°C] is lognormal distribution, whereas Weibull and lognormal distributions are fitted for the interval [20−25°C]. This is fitted well because in this minimum temperature interval, the increment of COVID-19 cases follows a hump shape. For a low level of minimum temperature, Weibull or gamma distribution is well fitted to define the trend of COVID-19 cases because at this temperature, most of the countries have recorded a higher number of cases, so an increasing pattern is observed in this interval. This is also seen based on the estimated values of the distribution parameters because the shape parameter (Estimate 1) of Weibull distribution is less than 1 as displayed in Table 6.2. This shows that an increasing hazard rate is suggested to determine the trend of COVID-19. These results conclude that not a single distribution has well explained the growth of COVID-19 cases at each interval, but on an average, the lognormal distribution is chosen as the best distribution to demonstrate the predictive values of infected patients in each country.

TABLE 6.2
Best-Fitted Distribution Using Model Selection for Minimum Temperature Interval.

Date	Class Interval	NOC	Distribution	AIC	BIC	Estimate 1	Estimate 2
14-May	[0–5]	25	Lognormal	−68.76	−66.67	−3.82	1.95
	[5–10]	19	Gamma	−41.86	−40.32	0.39	2.84
	[10–15]	16	Exponential	−81.51	−80.73	36.95	-
28-May	[0–5]	28	Lognormal	3.69	6.05	−1.92	1.64
	[15–20]	21	Lognormal	−87.41	−85.42	−4.42	2.04
	[20–25]	41	Lognormal	−193.15	−189.99	−4.71	1.75
11-June	[5–10]	21	Log normal	33.55	35.10	−0.69	1.21
	[20–25]	53	Lognormal	−144.19	−140.62	−3.50	1.49
	[25–30]	32	Lognormal	−155.10	−152.44	−4.55	1.33
25-June	[5–10]	26	Weibull	149.01	151.10	0.49	10.33
	[20–25]	55	Lognormal	−71.44	−67.82	−2.63	1.45
	[25–30]	33	Exponential	−136.96	−135.74	43.78	–
9-July	[5–10]	29	Weibull	139.31	141.75	0.69	4.92
	[10–15]	29	Exponential	55.98	57.12	0.84	–
	[20–25]	48	Lognormal	−19.40	−16.03	−2.13	1.51
	[25–30]	34	Exponential	−136.96	−135.74	43.78	–
23-July	[5–10]	32	Gamma	163.94	166.45	0.56	0.06
	[10–15]	30	Lognormal	−0.96	2.56	−1.97	1.63
	[20–25]	54	Lognormal	−45.77	−43.41	−2.95	1.64
6-August	[5–10]	32	Weibull	161.09	163.18	0.51	13.65
	[20–25]	49	Weibull	15.70	19.08	0.65	0.38
	[25–30]	48	Lognormal	16.22	19.69	−1.97	2.01
20-August	[10–15]	29	Gamma	128.82	131.17	0.49	0.08
	[20–25]	61	Lognormal	60.00	63.78	−1.51	1.94
	[25–30]	38	Lognormal	−19.87	−17.00	−2.38	1.79
3-September	[5–10]	32	Weibull	182.88	185.54	0.65	7.97
	[20–25]	45	Weibull	97.28	100.89	0.59	0.90
	[25–30]	44	Lognormal	−22.45	−19.34	−2.41	1.84

AIC, Akaike information criterion; *BIC*, Bayesian information criterion; *NOC*, number of countries.

3.3 Modeling on the Confirmed Cases About the Relative Humidity

Modeling on the confirmed cases about maximum and minimum temperatures gives fruitful results. So we further extend our modeling on the confirmed cases about relative humidity. Relative humidity is the ratio of the current absolute humidity to the highest possible absolute humidity, which depends on the current air temperature. In this section, the study will find out the transmission intensity of COVID-19 with an increase in relative humidity. Fig. 6.3 shows the allocation of the number of countries based on relative humidity for the considered humidity period.

As we know, the ARH level is from 40% to 50%. So there is a high chance that the country whose relative humidity is within the normal range has less COVID-19

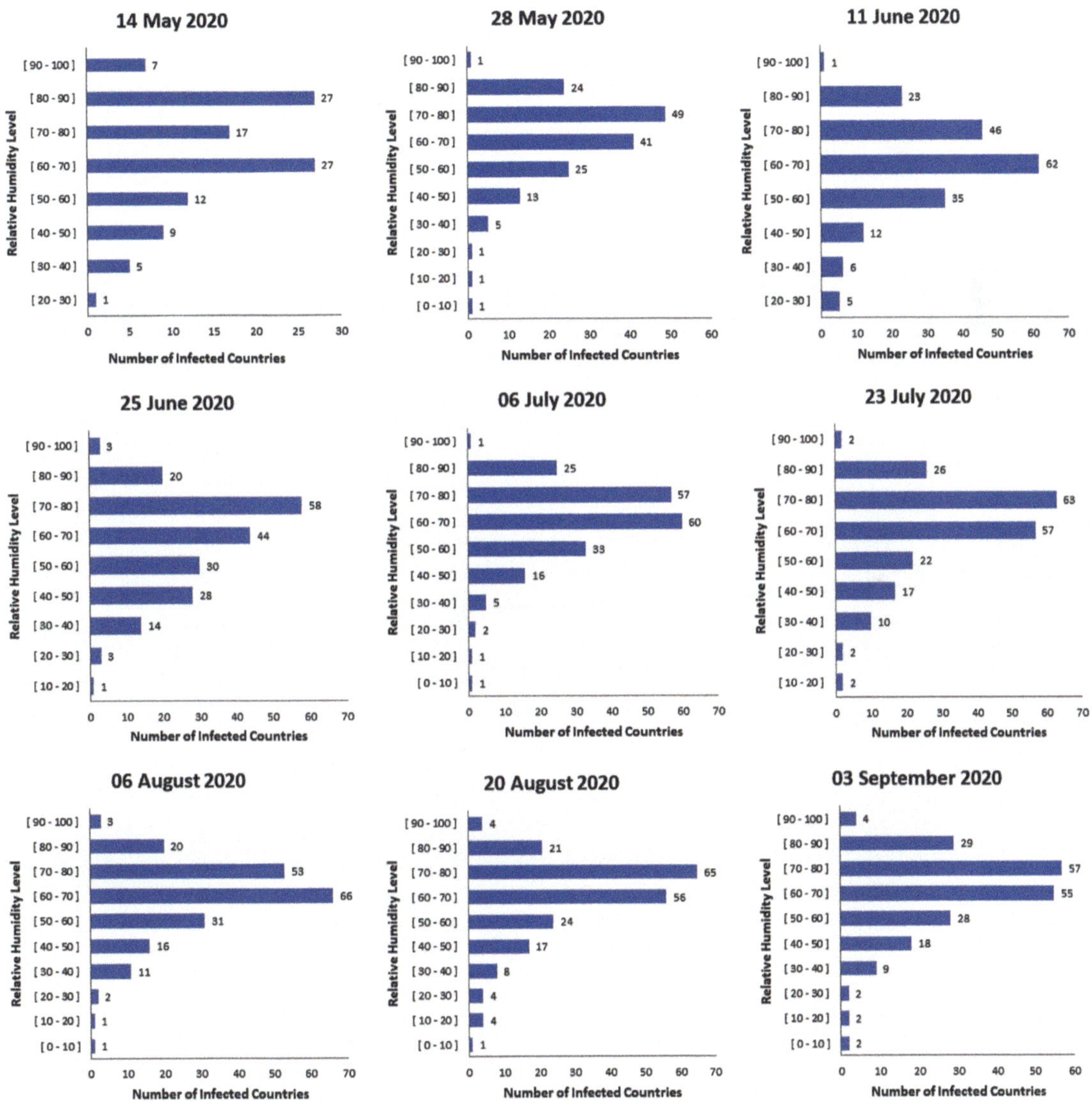

FIG. 6.3 The total number of infected countries in different ranges of relative humidity. X-axis denotes the number of infected countries and Y-axis denotes the various intervals of relative humidity. The length of the bar is equal to the number of countries with coronavirus disease 2019 (COVID-19) in a particular range of relative humidity.

transmission as compared with countries with high humidity level. This can be easily seen in Fig. 6.3 because the maximum number of infected countries has humidity levels more than the normal range (>50%). A smaller number of countries have relative humidity below the normal range. There is no such change observed in the humidity level during the study period every 2 weeks. We conclude that the COVID-19 pandemic will be automatically controlled under the normal humidity level and will spread over countries having a high relative humidity range.

Similarly, the study determines the best distribution that appropriately explains the spread of COVID-19 based on relative humidity. Table 6.3 records the best-fitted distribution of each humidity level with its estimated parameter values. During the study period, the spread of CoV rapidly increased with increase in the number of COVID-19 cases because a maximum number of countries come under the high humidity level. The growth pattern of COVID-19 cases based on relative humidity is somewhere lognormal and Weibull distributions for high humidity [60%–80%], whereas Weibull/gamma distribution for [40%–50%] normal range. In particular, Weibull distribution is chosen as the best distribution to explain relative humidity because increased level of humidity led to an increased pattern in COVID-19 cases. So the estimated value of the shape parameter of Weibull and gamma distributions comes to be less than 1 as recorded in Table 6.3.

3.4 Comparison Based on Empirical and Theoretic Data

The comparison has been carried out between empirical and theoretic results based on various study variables

TABLE 6.3
Best-Fitted Distribution Using Model Selection for Relative Humidity Interval.

Date	Class Interval	NOC	Distribution	AIC	BIC	Estimate 1	Estimate 2
14-May	[60–70]	27	Lognormal	−115.25	−112.89	−4.47	1.76
	[70–80]	17	Gamma	−65.49	−63.94	0.64	14.19
	[80–90]	27	Gamma	−67.70	−65.35	0.50	4.85
28-May	[50–60]	25	Exponential	−53.57	−52.57	10.91	–
	[60–70]	41	Gamma	−118.06	−114.90	0.46	5.17
	[70–80]	49	Weibull	−82.61	−79.13	0.59	0.11
11-June	[50–60]	35	Gamma	10.64	13.50	0.60	1.31
	[60–70]	62	Lognormal	−27.94	−23.96	−2.55	2.31
	[70–80]	46	Lognormal	−40.48	−37.20	−2.66	1.92
25-June	[50–60]	30	Weibull	57.95	60.30	0.50	1.01
	[60–70]	44	Weibull	16.94	20.16	0.56	0.38
	[70–80]	58	Lognormal	−53.52	−49.78	−2.67	1.93
9-July	[50–60]	33	Gamma	48.75	51.34	0.59	0.62
	[60–70]	60	Lognormal	77.45	81.32	−1.57	2.38
	[70–80]	57	Lognormal	31.93	35.71	−1.86	2.08
23-July	[60–70]	57	Lognormal	40.66	44.36	−1.83	2.23
	[70–80]	63	Lognormal	127.98	131.99	−0.91	1.85
	[80–90]	26	Lognormal	28.43	30.51	−1.51	1.96
6-August	[50–60]	31	Weibull	119.86	122.46	0.59	2.79
	[60–70]	66	Lognormal	121.48	125.46	−1.12	2.20
	[70–80]	53	Lognormal	49.22	52.75	−1.63	2.09
20-August	[50–60]	24	Gamma	69.81	71.81	0.44	0.16
	[60–70]	56	Weibull	131.80	135.51	0.50	1.25
	[70–80]	65	Weibull	84.53	88.43	0.63	0.70
3-September	[60–70]	57	Weibull	125.92	129.45	0.56	1.35
	[70–80]	55	Lognormal	153.62	157.44	−0.73	2.24
	[80–90]	29	Gamma	22.04	24.22	0.52	0.78

AIC, Akaike information criterion; *BIC*, Bayesian information criterion; *NOC*, number of countries.

on COVID-19 data. This is done to see the closeness of the empirical and theoretic distribution and to check the size and spread of the fitted distribution. Table 6.4 represents the comparison of empirical and theoretic parameters.

Based on different intervals of maximum temperature at different periods, we see that the theoretic mean and variability of the lognormal distribution is smaller than the empirical mean and variability. This shows that the mean growth of new confirmed cases in normal temperatures is more significant than then mean observed by fitting the lognormal distribution. It can be further observed that the empirical mean and the theoretic mean of gamma, Weibull, and exponential distributions are almost the same, but the observed variance and theoretic variance for these distributions are slightly different, thus indicating that these distributions are well fitted on the confirmed cases of countries lying in respective maximum temperature interval. Similarly, results are also seen in Table 6.5 for the minimum temperature variable.

Hence, a moderate temperature level contains most of the affected countries, as well as most cases, and fits lognormal distribution, with the theoretic mean less than the empirical mean and the theoretic variance less than the empirical variance with a significant difference. Table 6.6 shows that in all the distributions,

TABLE 6.4
Comparison Based on Empirical and Theoretic Parameters for Maximum Temperature.

Date	Class Interval	Distribution	Theoretic Mean	Empirical Mean	Theoretic Variance	Empirical Variance
14-May	[5–10]	Gamma	0.0657	0.0657	0.0082	0.0053
	[10–15]	Lognormal	0.1514	0.2432	0.1566	0.1400
	[25–30]	Exponential	0.0318	0.0318	0.0010	0.0011
28-May	[15–20]	Lognormal	0.1499	0.1866	0.0845	0.0692
	[25–30]	Lognormal	0.0570	0.0873	0.0183	0.0147
	[30–35]	Lognormal	0.0147	0.0232	0.0008	0.0016
11-June	[15–20]	Exponential	0.4108	0.4108	0.1687	0.1735
	[25–30]	Lognormal	0.0495	0.0918	0.0126	0.0281
	[30–35]	Lognormal	0.0454	0.0514	0.0062	0.0042
25-June	[15–20]	Lognormal	2.7088	3.7596	33.8388	37.5245
	[20–25]	Weibull	1.8979	1.9101	10.6675	9.5925
	[25–30]	Lognormal	0.0870	0.1099	0.0272	0.0231
9-July	[20–25]	Gamma	2.8094	2.8091	19.2658	17.8597
	[25–30]	Gamma	0.7477	0.7477	1.0997	0.9389
	[30–35]	Lognormal	0.1225	0.1653	0.0492	0.0679
23-July	[20–25]	Gamma	3.1024	3.1029	15.6951	13.4308
	[25–30]	Lognormal	0.2098	0.2909	0.1859	0.2205
	[30–35]	Lognormal	0.3708	0.5318	0.6592	0.6532
6-August	[20–25]	Lognormal	2.7125	4.3728	45.4824	54.3207
	[25–30]	Weibull	0.4012	0.4027	0.3973	0.3423
	[30–35]	Lognormal	0.3468	0.6061	0.7537	1.0857
20-August	[20–25]	Gamma	5.3154	5.3186	55.2579	49.0776
	[25–30]	Lognormal	0.8475	1.1941	4.5416	4.0311
	[30–35]	Lognormal	0.3811	0.5952	0.8794	0.7573
3-September	[20–25]	Lognormal	3.9266	8.1845	109.3914	221.4312
	[25–30]	Lognormal	1.3750	2.2544	12.1505	15.2167
	[30–35]	Lognormal	0.4981	0.8132	1.6214	1.5302

TABLE 6.5
Comparison Based on Empirical and Theoretic Parameters for Minimum Temperature.

Date	Class Interval	Distribution	Theoretic Mean	Empirical Mean	Theoretic Variance	Empirical Variance
14-May	[0–5]	Lognormal	0.0582	0.0968	0.0204	0.0321
	[5–10]	Gamma	0.1377	0.1378	0.0485	0.0366
	[10–15]	Exponential	0.0271	0.0271	0.0007	0.0007
28-May	[0–5]	Lognormal	0.3327	0.5610	0.4621	1.1013
	[15–20]	Lognormal	0.0335	0.0627	0.0075	0.0083
	[20–25]	Lognormal	0.0215	0.0335	0.0022	0.0027
11-June	[5–10]	Lognormal	0.4490	0.4949	0.4485	0.4050
	[20–25]	Lognormal	0.1729	0.2851	0.1716	0.1692
	[25–30]	Lognormal	0.0635	0.0823	0.0139	0.0141
25-June	[5–10]	Weibull	21.4274	20.3430	2426.0681	1008.3645
	[20–25]	Lognormal	0.1491	0.1696	0.0723	0.0428
	[25–30]	Exponential	0.0228	0.0228	0.0005	0.0005
9-July	[5–10]	Weibull	6.3234	6.2902	88.4868	70.2318
	[10–15]	Exponential	1.1895	1.1895	1.4150	2.2334
	[20–25]	Lognormal	0.2543	0.3324	0.2293	0.2491
	[25–30]	Lognormal	0.0456	0.0520	0.0053	0.0050
23-July	[5–10]	Gamma	9.1650	9.1638	149.0675	133.4391
	[10–15]	Weibull	2.2071	2.2444	10.9495	12.2077
	[20–25]	Lognormal	0.3166	0.4183	0.4114	0.3954
6-August	[5–10]	Weibull	6.8410	6.8421	84.3368	69.9701
	[20–25]	Weibull	0.5161	0.5135	0.6868	0.4872
	[25–30]	Lognormal	0.3804	0.7070	0.9395	1.4649
20-August	[10–15]	Gamma	6.2117	6.2110	78.1922	64.4683
	[20–25]	Lognormal	0.5826	0.9619	2.0285	2.6219
	[25–30]	Lognormal	0.2251	0.3203	0.2521	0.1671
3-September	[5–10]	Weibull	10.8985	10.7564	302.7341	213.8874
	[20–25]	Weibull	1.3772	1.3765	6.0083	4.2695
	[25–30]	Lognormal	0.2262	0.3641	0.2705	0.3343

except lognormal, the empirical mean and theoretic mean are almost the same and there is a slight difference between empirical variance and theoretic variance, which is not much significant under relative humidity. Thus Weibull, gamma, and exponential distributions are well fitted on the cases of countries lying in respective intervals. Still, the lognormal distribution shows a significant difference between the theoretic and empirical parameters. Overall, the study concludes that the current trend of COVID-19 cases may be seen if the best distribution in any interval follows exponential, Weibull, and gamma distributions because the empirical and theoretic parameters are almost similar.

4 CONCLUSION

This chapter analyzes the effect of physical parameters on the spread of COVID-19 worldwide. This study concludes that the spread of COVID-19 covered almost all the countries and populations all over the globe. Most of the countries getting affected have a normal temperature, neither high nor low, and a relative humidity range between 60% and 80%. The confirmed cases about maximum [20–35°C] and minimum [20–30°C] temperature intervals follow lognormal distribution in most of the intervals and about relative humidity follow gamma and Weibull distributions. The comparison of the theoretic and descriptive parameters showed that gamma and Weibull distributions are well

TABLE 6.6
Comparison Based on Empirical and Theoretic Parameters Concerning Relative Humidity.

Date	Class Interval	Distribution	Theoretic Mean	Empirical Mean	Theoretic Variance	Empirical Variance
14-May	[60–70]	Lognormal	0.0276	0.0359	0.0037	0.0018
	[70–80]	Gamma	0.0453	0.0453	0.0032	0.0027
	[80–90]	Gamma	0.1028	0.1028	0.0212	0.0154
28-May	[50–60]	Exponential	0.0917	0.0917	0.0084	0.0053
	[60–70]	Gamma	0.0893	0.0893	0.0173	0.0127
	[70–80]	Weibull	0.1764	0.1729	0.0998	0.0557
11-June	[50–60]	Gamma	0.4564	0.4565	0.3483	0.2698
	[60–70]	Lognormal	0.2474	0.4848	0.5578	0.5769
	[70–80]	Lognormal	0.1833	0.2940	0.1956	0.2125
25-June	[50–60]	Weibull	2.0500	1.9048	21.4818	8.4749
	[60–70]	Weibull	0.6277	0.6182	1.4205	0.9272
	[70–80]	Lognormal	0.1808	0.2949	0.1921	0.2346
9-July	[50–60]	Gamma	0.9459	0.9459	1.5179	1.2606
	[60–70]	Lognormal	0.6867	1.4249	4.6252	5.5408
	[70–80]	Lognormal	0.4376	0.9689	1.3367	3.1457
23-July	[60–70]	Lognormal	0.4888	1.0331	1.9900	3.5231
	[70–80]	Lognormal	1.0181	1.4681	5.5629	4.7984
	[80–90]	Lognormal	0.5884	1.2811	2.1125	6.7979
6-August	[50–60]	Weibull	4.2603	4.1424	57.8741	35.4994
	[60–70]	Lognormal	0.9807	1.9831	7.7351	14.4050
	[70–80]	Lognormal	0.5570	1.1157	2.1964	4.4901
20-August	[60–70]	Weibull	2.4690	2.4591	29.7334	19.9576
	[70–80]	Weibull	0.9949	1.0017	2.7454	2.3538
	[80–90]	Gamma	5.3241	5.3239	81.9666	55.0927
3-September	[50–60]	Weibull	5.2773	5.3288	93.1727	75.7065
	[60–70]	Weibull	2.2160	2.1383	17.8809	11.4247
	[70–80]	Lognormal	1.4780	2.9887	18.2988	24.9784

fitted on the cases rather than lognormal distribution, as the values of empirical mean and variance are almost the same as descriptive mean and variance for gamma and Weibull distributions, but a significant difference could be observed in the case of the lognormal distribution. The future scope is to determine the best prediction model to know the upcoming trend of COVID-19 cases that includes the effects of physical parameters and other associated factors.

REFERENCES

[1] Cui J, Li F, Shi ZL. Origin and evolution of pathogenic coronaviruses. Nat Rev Microbiol 2019;17(3):181–92.

[2] Zhao S, Lin Q, Ran J, Musa SS, Yang G, Wang W, Lou Y, Gao D, Yang L, He D, Wang MH. Preliminary estimation of the basic reproduction number of novel coronavirus (2019-nCoV) in China, from 2019 to 2020: a data-driven analysis in the early phase of the outbreak. Int J Infect Dis 2020;92:214–7.

[3] Casanova LM, Jeon S, Rutala WA, Weber DJ, Sobsey MD. Effects of air temperature and relative humidity on coronavirus survival on surfaces. Appl Environ Microbiol 2010;76(9):2712–7.

[4] Lowen AC, Steel J. Roles of humidity and temperature in shaping influenza seasonality. J Virol 2014;88(14):7692–5.

[5] Banerjee A, Kulcsar K, Misra V, Frieman M, Mossman K. Bats and coronaviruses. Viruses 2019;11(1):41.

[6] Marr LC, Tang JW, Van Mullekom J, Lakdawala SS. Mechanistic insights into the effect of humidity on airborne influenza virus survival, transmission and incidence. J R Soc Interface 2019;16(150):20180298.

[7] Qi H, Xiao S, Shi R, Ward MP, Chen Y, Tu W, Su Q, Wang W, Wang X, Zhang Z. COVID-19 transmission in

Mainland China is associated with temperature and humidity: a time-series analysis. Sci Total Environ 2020: 138778.

[8] Wu Y, Jing W, Liu J, Ma Q, Yuan J, Wang Y, Du M, Liu M. Effects of temperature and humidity on the daily new cases and new deaths of COVID-19 in 166 countries. Sci Total Environ 2020:139051.

[9] Ma Y, Zhao Y, Liu J, He X, Wang B, Fu S, Yan J, Niu J, Zhou J, Luo B. Effects of temperature variation and humidity on the death of COVID-19 in Wuhan, China. Sci Total Environ 2020:138226.

[10] Ujiie M, Tsuzuki S, Ohmagari N. Effect of temperature on the infectivity of COVID-19. Int J Infect Dis 2020;95: 301–3.

[11] Wallis P, Nerlich B. Disease metaphors in new epidemics: the UK media framing of the 2003 SARS epidemic. Soc Sci Med 2005;60(11):2629–39.

[12] Tan J, Mu L, Huang J, Yu S, Chen B, Yin J. An initial investigation of the association between the SARS outbreak and weather: with the view of the environmental temperature and its variation. J Epidemiol Community 2005; 59(3):186–92.

[13] Park JE, Son WS, Ryu Y, Choi SB, Kwon O, Ahn I. Effects of temperature, humidity, and diurnal temperature range on influenza incidence in a temperate region. Influenza Other Respir Viruses 2020;14(1):11–8.

[14] Luo Y, Zhang Y, Liu T, Rutherford S, Xu Y, Xu X, u W, Xiao J, Zeng W, Chu C, Ma W. Lagged effect of diurnal temperature range on mortality in a subtropical megacity of China. PloS One 2013;8(2):e55280.

[15] Pinheiro SDLLDA, Saldiva PHN, Schwartz J, Zanobetti A. Isolated and synergistic effects of PM10 and average temperature on cardiovascular and respiratory mortality. Rev Saude Publica 2014;48:881–8.

[16] Metz JA, Finn A. Influenza and humidity—why a bit more damp may be good for you! J Infect 2015;71:S54–8.

[17] Oliveiros B, Caramelo L, Ferreira NC, Caramelo F. Role of temperature and humidity in the modulation of the doubling time of COVID-19 cases. MedRxiv; 2020.

[18] Wang M, Jiang A, Gong L, Luo L, Guo W, Li C, Zheng J, Li C, Yang B, Zeng J, Chen Y. Temperature significant change COVID-19 transmission in 429 cities. MedRxiv; 2020.

CHAPTER 7

Tackling Coronavirus Disease 2019 by Nonpharmaceutical Interventions

SANMITRA GHOSH • SAPTARSHI CHATTERJEE • SHATARUPA BISWAS • MEGHA DUTTA

1 INTRODUCTION

The severe acute respiratory syndrome coronavirus 2 (SARS-CoV-2) is a novel strain of coronavirus causing the contagious coronavirus disease 2019 (COVID-19) that was identified as a global health emergency by the World Health Organization (WHO) on January 30, 2020, and as a pandemic on March 11, 2020 [1]. Since the first emergence of the disease in Wuhan (Hubei province, China) on December 31, 2019, this zoonotic disease has spread to almost all the countries worldwide resulting in over 44,888,869 confirmed cases and 1,178,475 deaths globally as on October 30, 2020 [2]. In close unprotected direct or indirect contact with an infected person or fomite the disease spreads by droplet transmission with an incubation period of 2−14 days [3]. Spread of COVID-19 is also significantly contributed by healthcare setting transmissions [4]. SARS-CoV-2 appears to have greater transmissibility and lower pathogenicity than genetically similar coronaviruses such as SARS-CoV and Middle East respiratory syndrome coronavirus (MERS-CoV). The basic reproduction number (R0) for SARS-CoV and MERS-CoV was reported to be 3.0 and <1, respectively, and that for SARS-CoV-2 as an estimate of transmissibility ranges from 2.8 to 5.5, in the absence of intense quarantine and social distancing measures [5]. Around 80% of COVID-19 patients either are asymptomatic or have mild symptoms such as dry cough, fever, shortness of breath, or general symptoms such as headache, myalgias, or fatigue and usually recover from the disease within 2 weeks. People with the severe form suffer from dyspnea and pneumonia, and ~5%−6% of COVID-19 patients are critically ill with respiratory failure, sepsis, or multiorgan failure [3]. The disease severity is variable from mild to severe symptoms in different people based on their age and health status [6]. However, high mortality is recorded among the aged people and people with underlying chronic diseases. About 2% of COVID-19 patients are under 18 years of age, out of which, fewer than 3% developed severe conditions [2]. Mortality rate for COVID-19 ranges from 2% to 3%, which is significantly less than 10% of SARS in 2003 and 35% of MERS in 2012 [7].

At present, no effective pharmacologic interventions or vaccines are available to treat or prevent COVID-19 and the available treatments are mainly supportive. Therefore governments across the world have resorted to nonpharmacologic public health measures such as isolation, social distancing, and quarantine as the only available effective means to curtail the spread of the disease from one person to another, as well as transborder spread from country to country. Isolation denotes the separation of symptomatic patients, while quarantine is the restriction of asymptomatic healthy people who have come in contact with confirmed or suspected cases. Implementation of quarantine may be voluntary or legally enforced at an individual, group, or community level [8]. With the objective of disease control with minimal risk as well as assessment of the consequences of interventional measures in the present situation with no antiviral treatment or vaccines being available, it is crucial to investigate the optimal impact of the available control measures. So this chapter aims at evaluating the effect of short-term government-enforced strategies and self-imposed prevention measures on mitigating and delaying the pandemic of COVID-19 among common people and health workers. This chapter is segmented in two parts. The first part concerns with the nonpharmaceutical interventions adapted by common people to cease the spread of the pandemic in the global population (Fig. 7.1), whereas the second part elaborates the role of nonpharmaceutical interventions in determining the safety of health workers in the COVID-19 era.

COVID-19: Tackling Global Pandemics through Scientific and Social Tools. https://doi.org/10.1016/B978-0-323-85844-1.00010-6

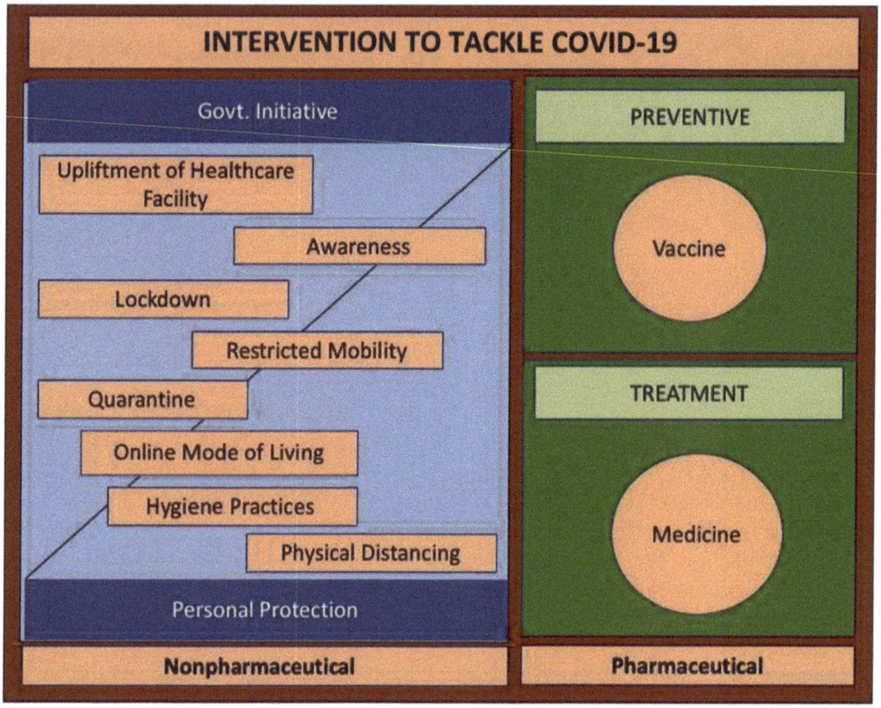

FIG. 7.1 Tackling coronavirus disease 2019 (COVID-19) by nonpharmaceutical intervention.

2 PART 1: THE ROLE OF NONPHARMACEUTICAL INTERVENTIONS IN THE MANAGEMENT OF COVID-19 PANDEMIC IN THE WORLD POPULATION

Initially strategies to prevent the exponential spread of COVID-19 were limited to reporting cases, strict isolation of patients with severe symptoms, home isolation for mild cases, and contact tracing [9]. But very soon, identification of the significance of asymptomatic and presymptomatic spread of the illness revealed the insufficiency of the case-based interventions in containment of the COVID-19 pandemic [10].With the rapid increase in incidence rate of the disease and the risk of exceeding critical care bed capacities, social distancing was implemented as a short-term measure in many countries intending at reducing the contact rate in the population and, successively, transmission in order to flatten the curve of the pandemic [11]. Several governments have enforced partial or complete lockdowns nationwide by closing schools, public places, and nonessential businesses; canceling mass events; and issuing stay-at-home orders [11]. Self-imposed prevention measures such as washing hands, use of masks, and social distancing could also contribute to slowing down the pace of the epidemic [12]. Broadcast of information regarding current situation of the pandemic and official recommendations about COVID-19 can create awareness and motivate individuals to adopt such measures and therefore alter the course of an epidemic reducing its rate of attack [13]. At the same time, it can also lead to undesirable outcomes such as the appearance of multiple epidemic peaks [13].

2.1 Awareness and Knowledge Dissemination

Within weeks of the emergence of novel coronavirus infection, COVID-19, in China, misleading rumors and conspiracy theories about the origin of the disease, extent of outbreak, outcome, and other aspects such as racism had rapid dissemination similar to the spread of the virus creating overwhelming panic among the public globally. To get rid of all these misconceptions, a proper awareness program needed to be initiated.

Many such commendable initiatives were partaken by several government and nongovernment organizations, people from academics and scientific communities, WHO, the Centers for Disease Control and Prevention (CDC), clinicians, and other health workers through television, social media, and newspaper.

- TheCDC has released several awareness programs that include videos, journals, and podcast. They published one journal related to stopping the spread of the virus in children [14]. The CDC is supporting multiple efforts to better understand the impact of COVID-19 during pregnancy on both the mother and the infant [15]. It also published information about considerations for wearing mask to provide sufficient knowledge among the common people [16]. The WHO also published several information for the general public as an awareness program (Mythbusters, How to report misinformation, general guidelines if someone is unwell, travel advices, and many more). Apart from information, they also arranged a Q&A session to clear all the misconceptions of people [17].

2.2 Lockdown

Prior to its implementation due to the outbreak of the COVID-19 pandemic, the term "lockdown" was unfamiliar to most of the people. Taking into account the incubation period of COVID-19 to be 14 days, lockdown for the tenure of 3 weeks was suggested to allow the emergence of symptomatic cases of COVID-19 and also to provide the government sufficient time to plan a treatment response. Lockdown allowed the governments of various countries not only to break the infection chain but also to isolate the infected from the general population and treat those who require it with varying degrees of hospitalization. During lockdown, except medical and other essential services, all sorts of public places, mass events, festivals, educational institutes, and nonessential businesses, along with all sorts of transport and communications, were stopped.

The lockdown started in China on January 23, 2020, and it extended for about 2 months. As all approaches aimed at the containment of SARS-CoV-2 in China were unsuccessful, the pandemic started to spread to other countries exponentially. A significant number of asymptomatic cases and the likelihood of presymptomatic transmission hampered the accuracy of airport screening of national and international travelers [18]. Quarantine of 14 days combined with fever surveillance happened to be inadequate in the virus containment because of the high variability of the incubation period. As the disease started to spread to other countries, lockdown was imposed in each country. For instance, in Italy, lockdown was announced on March 9, 2020, and in India, it was announced on March 25, 2020. Most of the complete lockdown extended for about 1–2 months. Based on the number of cases, the unlock procedure was started statewise.

The implementation of lockdown had both advantages and disadvantages. It prevented community spread of the virus. It is observed that the environment has started to remediate itself because the industrial activities, vehicle movement, and social activities of people were temporarily but significantly reduced. For example, one study showed that the quality of air due to the lockdown in Delhi had a positive effect. Another study showed that Dantas et al. have calculated the CO emission level as approximately 30.3%–48.5% due to the lockdown in Rio de Janeiro, Brazil. Yunus et al. have quantitatively determined with the remote sensing imaging method that the quality of water of the Vembanad Lake has increased approximately 15.6% in India [19]. COVID-19 outbreak caused not only global health crisis but also an unprecedented catastrophic economic disaster during the lockdown. Global industrial activity has been largely affected because of the dissolution of all sorts of transport and communication between different parts of the world. In Europe, the GDPs of France, Spain, and Italy fell by 21.3%, 19.2%, and 17.5%, respectively.

Due to the increase in unemployment after the lockdown, many people were affected by stress and severe depression that led to an increase in the number of suicide cases. The CDC has arranged an immediate helpline number for people undergoing stress during the COVID situation [20].

2.3 Physical Distancing

According to the CDC, social distancing, also known as "physical distancing," means maintenance of a safe space between yourself and other people who are not from your household. To practice social or physical distancing, it is suggested to maintain at least 6 ft (about two arms' length) from outsiders. Social distancing should be practiced in combination with other self-imposed everyday preventive measures to reduce the spread of COVID-19, including wearing masks, avoiding touching your face with unwashed hands, frequent handwashing with soap and water, use of hand and surface sanitizers, etc.

According to the WHO, aerosol transmission can occur in specific settings, particularly in indoor, crowded, and inadequately ventilated spaces, where infected person(s) spend long periods with others, such as restaurants, choir practices, fitness classes, nightclubs, offices, schools, colleges, community halls, places of worship, and other public places. So physical distancing is very important to prevent the spread of the virus. Recently in Chennai, India, corporation officials sealed Kumaran Silks because they were not able to maintain physical distancing and other safety regulations [21]. The emphasis on physical distancing over social distancing emerged as a critical aspect during this time [22].

2.4 Prohibition on Social Gathering

To prevent the spread of COVID-19, several countries prohibited social gathering, as this disease is aerosol transmitted. According to the CDC, more the number of people an individual interacts with and the longer that interaction lasts, the higher will be the potential risk of becoming infected with COVID-19 and COVID-19 spreading. The embargo is applicable to both public and private areas including conferences, symposium, outdoor meetings, academic areas (schools and colleges), cultural events, movie theaters, private events (wedding, birthday celebration), and religious activities.

In England, social gathering of more than six people was banned from September 14, 2020. In India, celebration of several events were restricted, such as Muharram. In October 2020, Durga Puja, the biggest festival of Bengalis, was celebrated with strict restraint on social gathering and the puja pandals were denoted as containment zone. A number of award ceremonies such as the 40th Golden Raspberry Awards and the 19th Lunas del Auditorio, Mexico, were canceled in various parts of the world. Military events like the 75th Vietnam National Day parade and the Bulgarian Armed Forces Day Parade and beauty pageants like Miss World 2020 were also annulled [23].

2.5 Restricting Mobility

Lockdown in cities across the globe has imposed severe travel restrictions to limit the spread of the Coronavirus disease. The travel behavior and operations will not be the same as before because of the requirements such as physical (social) distancing. To minimize the contact with each other or any common surface, extreme measures such as lockdown of cities with very limited mobility for essential services is imposed. Railways, airways (both national and international), road transportation, and water transportation were not in service during the lockdown. Use of public transport is reported to enhance the risk of an individual of acquiring a respiratory infection (the flu or common cold) because of the proximity of people to each other. Herd immunity to other infections may not hold true for the coronavirus because it is a new virus strain.

The IOM UN Migration has developed a COVID-19 global mobility database, DTM (Displacement Tracking Matrix), mapping the different restrictions to provide a global overview. For this, the DTM uses the International Air Transport Association (IATA) site as the primary source on restrictions with information reported per country, territory, and area to the country [24]. The WHO reports the incidences per country, territory, or area. This database is daily updated. The data collected includes

- date of restriction;
- country, territory, and area of restriction;
- countries, territories, or areas on which restrictions were imposed;
- type of restrictions—total restriction or conditional restriction, such as medical and visa restriction.

After the lockdown, flight services (both national and international) have been initiated but railways in India have not yet started. Train service in other countries have started but with certain restrictions. New innovative ideas have been invented by several countries to confront the spread of COVID-19. Some transit systems are using technology to monitor at-risk individuals. Taiwan's metro agency installed an infrared tool to measure people's body temperature before allowing them to enter a train, barring anyone with a temperature above 38°C (or 100°F) from riding. A Chinese tech firm recently developed an algorithm for Beijing subway officials to identify commuters not wearing a mask or wearing it improperly, and Hong Kong's railway operator rolled out a robot to disinfect trains, in addition to deploying cleaning staff [25]. One of the main purposes of cross-border mobility is international trading and investment, but due to the spread of COVID-19, international trading has been greatly affected, as the World Trade Organization (WTO) has imposed temporary mobility-related measures such as entry bans, exit guidelines, quarantines, and several travel restrictions. Overall, all these restrains caused economic activity to reach a near standstill leading to massive setback in global economy to an extent that the world has not confronted in several decades.

2.6 Improved Hygiene Practices: Mask and Sanitizers

According to the WHO, "Hygiene refers to conditions and practices that help to maintain health and prevent the spread of diseases." The coronavirus pandemic has impacted our daily habits. Social distancing is now a buzzword around the world, so as wearing mask and washing or sanitizing hands frequently. Certain guidelines have been made by the CDC about how to select, wear, and clean our masks.

DO choose masks that

- have two or more layers of washable, breathable fabric;
- completely cover your nose and mouth;

- snugly fit against the sides of your face and do not have gaps.

DO NOT choose masks that

- are made of fabric that makes it hard to breathe, for example, vinyl;
- have exhalation valves or vents, which allow virus particles to escape;
- are intended for healthcare workers (HCWs), including N95 respirators or surgical mask.

How to wear:

- Wear a mask correctly and consistently for the best protection.
- Be sure to wash your hands before putting on a mask.
- Do NOT touch the mask when wearing it.

How to take off a mask:

- Carefully untie the strings behind your head.
- Handle only by the ear loops or ties.
- Hold outside corners together.
- Do not touch your eyes, nose, or mouth when removing and wash hands immediately.

How to clean:

Masks should be washed regularly. Always remove masks correctly and wash your hands after handling or touching a used mask.

- Include your mask with your regular laundry.
- Use regular laundry detergent and the warmest appropriate water setting for the cloth used to make the mask.
- Use the highest heat setting and leave in the dryer until completely dry.

Washing of hands: Wash hands frequently with soap and water for at least 20–30 s.

The use of hand sanitizer: The WHO and the Australian Government Department of Health both recommend alcohol-based hand sanitizers to remove the novel coronavirus. The sanitizer must contain ethanol/isopropanol/n-propanol or a combination of these three alcohols. This sanitizer works mainly against the proteins of the microbes but sometimes the high alcohol content based sanitizer can also work against the lipid coating that surrounds the coronavirus. To be effective, hand sanitizers must contain at least 60% alcohol. A higher alcohol content sanitizer is not that effective, as it coagulates the protein instantly creating a protein layer that protects the other proteins from further coagulation, whereas a lower alcohol content sanitizer penetrates the cell at a slower rate and coagulates all the proteins of the cell wall and the microbes die.

Disinfection of all the surrounding and inanimate objects is necessary to prevent contamination of infection. Wearing head caps and gloves is also considered as a good hygiene practice.

2.7 Adapting Online Mode of Living: Telemedicine, Online Class, and Work From Home

It is hard to believe that several months before, people were going to work, sitting in class, spending their weekends at restaurants, and traveling around the world. Now, everyone has settled into a new normal of remote work, online classes, and virtual events. This online mode of telemedicine, online class, and work from home (WFH) is the only way to prevent the spread of COVID-19 to some extent.

Telemedicine is the use of technology that enables remote healthcare (telehealth). Basically it makes it possible for physicians to treat patients whenever needed and wherever the patient is by using a computer or Smartphone. In this situation, telemedicine is very much in use, as there is restriction on traveling.

The situation has pushed many educators to turn to remote teaching, which is rapidly becoming the "new normal" in the era of social distancing and community quarantine. Several schools and colleges have shifted to online mode of taking classes via Zoom, Webex, Google Meet, Microsoft Teams, and other such platforms. This online mode of education is helping the students to learn new technologies, and also they are protected from the pandemic situation by staying home. But this sudden shift from traditional classrooms to virtual learning has plunged many teachers into uncharted waters. Issues such as lack of access to the internet for some students and stress problem created by sitting in front of laptops for so many hours have been experienced by many students.

WFH is another new normal adopted by several countries during the pandemic situation. The main attempt of WFH is to keep the employees safe. WFH is a savior for many companies that would have otherwise shut down and cost billions of professionals their jobs. Even leading Public Sector Undertakings (PSUs) and the government sector that previously never accepted the idea of WFH have embraced it for the greater good. Therefore WFH complies with the guidelines issued by the WHO in terms of social distancing and kept isolated from the rest of the world. WFH has been instrumental in maintaining the social distancing norms that would have been impossible otherwise.

As the number of cases of COVID-19 increases throughout the world, the period of WFH also increases for example, the Department of Telecommunications (DoT) extended relaxation in WFH norms till December

31, 2020, given to other service providers (OSPs), including IT sectors. One study found that willingness to WFH is entirely dependent on the presence of one's children at home, comfortable space at home, quite environment at home, and good internet connectivity. On the contrary, many of the respondents are also of the opinion that even if they are ready to WFH, they do not like to WFH.

3 PART 2: INTERROGATING THE ROLE OF NONPHARMACEUTICAL INTERVENTIONS IN THE SAFETY OF HEALTH WORKERS

The COVID-19 pandemic has led to an outpour of patients surpassing the available human and physical resources in diverse settings, triggering the implementation of crisis standards of care. To mitigate the threat of COVID-19 contagion, public health measures have been adopted by all the countries that are either already afflicted by the disease or predisposed to it [26]. As the SARS CoV-2 viral load has been reported to be present in both the lower and upper respiratory tracts [27], and also because the virus can persist in the air for 3 h after aerosolization [28], the virus may spread through airborne transmission. The WHO has issued guidelines for contact and droplet precautions for HCWs caring for suspected COVID-19 patients, while the US CDC has recommended airborne precautions [29]. But unfortunately, basic protective equipment and safety protocols are not always available in many medical institutions dealing with COVID-19 patients.

During an outbreak, HCWs are expected to work long hours under significant stress with often inadequate resources, while accepting the risks intrinsic to close interaction with ill patients. It is known that the average person infected with COVID-19 is likely to infect 1.5–3.5 others [30]. COVID-19 infection of a single HCW can have a dramatic effect on the healthcare itself and might severely reduce the capacity of an entire hospital [31]. An escalation in the number of infections and the associated fatalities of HCWs intimidate the capacity of the global healthcare system to care for additional COVID-19 patients [32]. Such explosion downgrades the care delivery system to bottomless slumps, with unknown consequences, especially since illness unrelated to COVID-19 proceeds, often unattended. Besides, families bear the burden as HCWs fall ill [33]. So our nation's health, security, and economic prosperity will be saved if the health and safety of our HCWs are taken care of [33].

Healthcare systems globally could be operating at more than maximum capacity for many more months to come. But HCWs cannot be run at 100% tenancy for long periods in this emergency situation. It is vital that HCWs are not considered simply as pawns to be deployed but as human individuals. In the global response, the safety of HCWs must be safeguarded. Besides adequate provision of personal protective equipment (PPE), other practical measures must be considered, including canceling nonessential events to prioritize resources; provision of food, rest, and family support; and providing psychologic support. Presently, HCWs are the most valuable resource of each and every country.

3.1 Infection Prevention and Control

3.1.1 Well-resourced dedicated hospital facility

To activate the surge capacity of healthcare system in the present scenario of COVID-19-mediated devastation, the WHO recommends designation of a dedicated hospital facility (DHF) to receive COVID-19 patients. A well-resourced DHF to deal with these cases with adequate protections for HCWs and other patients should be the key to the standard of care. The DHF must be a self-contained institution with only essential but limited contact with the outside world. Basic amenities that need to be warranted include a continuous supply of clean water; proper cleaning practices; sufficient space for beds; appropriate handwashing facilities; suitable ventilation for isolation rooms and procedure rooms; adequate isolation facilities for airborne, droplet, contact isolation; regulated and a coherent traffic flow to enable patient and clinical material transport; precautions to control rodents, pests, and other vectors; and appropriate waste management facilities/practices. The DHF should be divided into two units. The first unit will be designated as an isolation space for laboratory-confirmed COVID-19-positive cases that can accommodate multiple patients in the same room. To prevent nosocomial infections, barrier nursing practices and protective isolation facility must be implemented. The second unit will be dedicated to suspected cases like family and hospital contacts who are supposed to have possible contact with confirmed cases but await laboratory confirmation. Medical workers appointed to this facility should undergo focused training in infection prevention and control, otherwise ignorance regarding the disease management may enhance infection susceptibility.

3.1.2 Preventing intrahospital infection and transmission of COVID-19 in healthcare workers

Daily monitoring of the temperature of all HCWs (twice a day) should be made compulsory to identify those who are unwell and to prevent intrahospital propagation of the disease. To minimize the risk of occupational exposure, care teams are to be segregated in two dimensions. Cross-institution coverage by doctors is suspended, and medical staff are limited to practice in single primary institution. Care teams must also be kept small to reduce the impact of any inadvertent occupational exposure on personnel downtime due to the need to quarantine contacts. Meal times for HCWs should be staggered. Didactic teaching and departmental meetings should be conducted using videoconferencing [34].

3.1.3 Strategies to improve respiratory protection

- **Standard hygiene**
  Standard precautions must always be followed by all HCWs, for all patients and also for themselves and their family [35]. This broadly encompasses hand hygiene practices; suitable donning, doffing, and disposal of the PPE; and maintaining respiratory hygiene and cough etiquette [36]. Hand hygiene with alcohol-based hand rubs (ABHRs) or soap and water is recommended. Adequate supply of ABHR (with 60%–80% ethanol) and antiseptic soap solution (chlorhexidine gluconate 2% and alcohol combination) has been seen to be synergistic [37].
- **Contact precaution and droplet precaution**
  A medical/surgical mask is recommended for routine care, while a respirator (airborne precautions) is recommended if HCWs are conducting an aerosol-generating procedure, such as endotracheal intubation, bronchoscopy, or airway suctioning, along with droplet precautions [9]. Droplet precautions include the recommendation to preserve spatial separation of 1 m with an infected patient, with the apprehension that large droplets can only spread horizontally to a maximum of 1 m. On the other hand, airborne precautions as implemented by the US CDC are much more stringent and include the use of a mask by the patient apart from following airborne precautions by HCWs.
- **PPE**
  All HCWs involved should be trained with the correct steps of donning and doffing, along with appropriate disposal, of PPE. Each room should have dedicated equipment such as sphygmomanometer, thermometer, and stethoscopes. Nonpowdered latex-free gloves should be used by all HCWs. Eye protection and face shield should be used [36,38]. Respiratory hygiene and cough etiquette must be followed. Gowns should be long-sleeved and made of nonabsorbent materials. The same gown should not be worn for all patients. In case of inaccessibility of gowns, waterproof aprons should be used. The WHO recommends the use of medical masks (surgical or procedure masks that may be flat or pleated) and particulate respirators (NIOSH-certified N95, EU standard FFP2 or equivalent) for contact and airborne precautions and for aerosol-generating procedures, respectively [39]. If the patient is under droplet precautions and requires moving or transporting, he/she should wear a mask. Hand hygiene should be ensured after removing the masks. However, it is noteworthy that there is no sufficient evidence to conclude the adequacy of surgical masks in providing protection against this pathogen. HCWs must wear at least N95 respirators and full-face shields when working in high-risk settings, where notably patients with critical ailments may be infected with SARS-CoV-2 but not yet diagnosed [40–42].
  Some healthcare settings are reported to lack respiratory protective equipment for frontline employees. The WHO discourages homemade facial coverings for HCWs because of concerns for moisture and pathogen retention. Institutions must relentlessly secure sufficient supplies of rigorously tested PPE [33].

3.1.4 Sample handling

HCWs at both clinical and diagnostic setups assigned with sample handling must have adequate knowledge/training regarding the disease prognosis and causal agent. Timely communication between the staff of health departments with that of public health laboratory regarding reviewed and updated protocols, working practice policies, and biosafety protocols is essential to minimize the risk incurred in handling specimens from patients with probable SARS-CoV-2 infection. Routine viral testing of specimens can be handled in a Biosafety Level (BSL)-2 laboratory, whereas viral isolation from specimens collected from people suspected to have COVID-19 should be performed in a BSL-3 laboratory. For procedures with a high possibility to generate aerosols or droplets, a certified Class II Biological Safety Cabinet (BSC) needs to be operated. Otherwise additional precautions such as the use of PPE (surgical mask/face shield/splash shield), centrifuge safety cups, and sealed centrifuge rotors should be implemented to provide an obstruction between the specimen and personnel.

For healthcare providers who are collecting specimens or are working within 6 ft of patients suspected to be infected with SARS-CoV-2, practice of proper infection control and use of recommended PPE (N95 or higher level respirator, face mask if a respirator is not available, eye protection, gloves, and a gown) are mandatory.

Providers who are handling specimens, but are not directly involved in collection and not working within 6 ft of the patient, should follow standard precautions and use gloves, face mask, or cloth face covering at all times while in the healthcare facility. Use of PPE can be minimized through patient self-collection while the healthcare provider maintains at least 6 ft of separation [43,44]. Specimens for transport must be sealed, properly labeled, and placed in leak-proof specimen bags, with a separate sealable pocket for the specimen (i.e., a plastic biohazard triple-layered specimen bag). All specimens are to be delivered by hand if possible. Any shipment of specimens should be done according to the prevailing International Air Transport Association and UN/WHO norms [44,45].

3.1.5 Use of appropriate disinfectants

Environmental Protection Agency (EPA)-registered disinfectants with validated activity against enveloped viruses include chlorine, alcohol, peroxygen, quaternary ammonium compounds, and phenolic compounds and should be adequate if used according to manufacturer's recommendations such as dilution, contact time, and safe handling. Clinical setups, work surfaces, and equipment should be decontaminated after specimens are collected from patients or processed [46]. We need to be cautious about some environment cleaning practices. For instance, dusting and carpeting of floor should be avoided and rooms should not have furnishings. In areas needing airborne precautions, high-touch surfaces should be cleaned daily and whenever required, while low-touch surfaces can be cleaned on a scheduled basis. Primary emphasis should be given to required PPE and additional entry/exit procedures. For undertaking droplet and/or contact precautions, cleaning may be done twice a day, or as needed, for high-touch surfaces, with a focus on all surfaces within the patient zone, noncritical patient care equipment, and any surface visibly soiled with blood or body fluids [47].

3.1.6 Food practices

Food should be served on disposable crockery and should be consumed with disposable cutlery.

Nondisposable crockery and cutlery should be washed using hot water (70°C) and detergent, rinsed, and dried. Wherever feasible, eating utensils should be cleaned in a dishwasher using a hot water cycle (reaching at least 70°C). A record of all patients or staff entering the room should be maintained.

3.1.7 Waste management

Improper treatment and disposal of healthcare waste derived from healthcare facilities, medical laboratories, and biomedical research facilities poses serious hazards of secondary disease transmission due to exposures to infectious agents among waste pickers, waste workers, health workers, patients, and the community in general where waste is improperly disposed. Open burning and incineration without adequate pollution control exposes waste workers and the surrounding community to toxic contaminants in air emissions and ash. It is estimated that about 0.5 kg of medical waste per bed per day is produced in hospitals globally. However, this amount, along with composition, is highly variable, with higher income countries generating far higher levels of waste and plastic. Because of this huge diversity, there is no single best solution to deal with medical waste. Generally 75%–90% of the waste produced by healthcare facilities is nonrisk general waste, comparable to domestic waste. The rest is infectious waste that is suspected to contain pathogens in sufficient concentration or quantity to cause disease in susceptible hosts. According to the WHO, management of waste associated with COVID-19 should be the same as any other infectious waste. Treatment of COVID-19 waste with disinfectant is not needed. For the collection and transport of waste, leak-proof, puncture-proof, double-layered containers labeled with the biohazard symbol should be used. Waste, segregated at source at the time of generation, should be deposited in an infectious waste bin with a suitably color-coded liner (yellow) and stored in a clean secured storage room for around 72 h separately after which it can be treated as regular municipal waste. In this period, it will pose minimal risk to the handlers [48]. Dedicated transport systems should be used that are regularly sanitized. Both the WHO and the United Nations Environment Programme (UNEP) have endorsed steam-based or other nonincineration methods of disinfection over incineration because of the persistent organic pollutants produced by incineration [49]. After disinfection, waste can be sent for disposal or recycling. Proper training, regular sanitization, and provision for adequate PPE should be ensured for workers involved in handling and collection of biomedical waste [50–52]. These rules are applicable for health centers, quarantines, and home-care facilities.

4 CONCLUSION

This chapter summarizes how practice of effective prevention measures, such as handwashing or mask usage, in combination with government-imposed social distancing intervention, may be crucial to attain control over the accelerated rate of spread of the COVID-19 epidemic. With interventions such as quarantine, isolation, and public health education, the number of exposed and infected individuals will reduce drastically within a short time but not to zero, leaving a residue of infected individuals with the potential to cause a further outbreak. But in the absence of specific antiviral medication or vaccine, these preventive measures appear to be the only available means, which are commendable too, to reduce the disease occurrence and case-fatality rate. The increased awareness and information on the escalating number of COVID-19 diagnoses and death rates provided by the media may trigger anxiety in the population. However the widespread thorough promotion of self-imposed measures with proven efficacy by governments or public health institutions together with strict implementation of social distancing policies adopted by governments may be a key weapon to tackle COVID-19 in the general population as well as among health workers.

REFERENCES

[1] World Health Organization (WHO). Rolling updates on coronavirus diseases (COVID-19). Available from: www.who.int/emergencies/diseases/novel-coronavirus-2019/events-as-they-happen.

[2] WHO coronavirus disease (COVID-19) dashboard. Available from: https://covid19.who.int/.

[3] World Health Organization (WHO). reportReport of the WHO-China joint mission on coronavirus disease 2019 (COVID-19). Available from: www.who.int/docs/default-source/coronaviruse/who-china-joint-mission-on-covid-19-final-report.pdf.

[4] Del Rio C, Malani PN. COVID-19-new insights on a rapidly changing epidemic. J Am Med Assoc 2020;323(14):1339–40. https://doi.org/10.1001/jama.2020.3072.

[5] Bauch CT, Lloyd-Smith JO, Coffee MP, Galvani AP. Dynamically modeling SARS and other newly emerging respiratory illnesses: past, present, and future. Epidemiology 2005;16(6):791–801. https://doi.org/10.1097/01.ede.0000181633.80269.4c.

[6] He S, Tang SY, Rong L. A discrete stochastic model of the COVID-19 outbreak: forecast and control. Math Biosci Eng 2020;17(4):2792–804. https://doi.org/10.3934/mbe.2020153.

[7] Tang B, Bragazzi NL, Li Q, Tang S, Xiao Y, Wu J. An updated estimation of the risk of transmission of the novel coronavirus (2019-nCov). Infect Dis Modell 2020;5:248–55. https://doi.org/10.1016/j.idm.2020.02.001.

[8] Cetron M, Landwirth J. Public health and ethical considerations in planning for quarantine. Yale J Biol Med 2005;78(5):329–34.

[9] European Centre for Disease Prevention and Control. Outbreak of acute respiratory syndrome associated with a novel coronavirus, China: first local transmission in the EU/EEA—third update. January 31, 2020. Available from: https://www.ecdc.europa.eu/sites/default/files/documents/novel-coronavirus-risk-assessment-china-31-january-2020_0.pdf.

[10] Ferretti L, Wymant C, Kendall M, Zhao L, Nurtay A, Abeler-Dörner L, Parker M, Bonsall D, Fraser C. Quantifying SARS-CoV-2 transmission suggests epidemic control with digital contact tracing. Science 2020;368(6491):eabb6936. https://doi.org/10.1126/science.abb6936.

[11] European Centre for Disease Prevention and Control. Coronavirus disease 2019 (COVID-19) pandemic: increased transmission in the EU/EEA and the UK—seventh update. March 25, 2020. Available from: https://www.ecdc.europa.eu/sites/default/files/documents/RRA-seventh-update-Outbreak-of-coronavirus-disease-COVID-19.pdf.

[12] European Centre for Disease Prevention and Control. Guidelines for the use of non-pharmaceutical measures to delay and mitigate the impact of 2019-nCoV. September 24, 2020. Available from: https://www.ecdc.europa.eu/en/publications-data/guidelines-use-non-pharmaceutical-measures-delay-and-mitigate-impact-2019-ncov.

[13] Funk S, Gilad E, Watkins C, Jansen VA. The spread of awareness and its impact on epidemic outbreaks. Proc Natl Acad Sci U S A 2009;106(16):6872–7. https://doi.org/10.1073/pnas.0810762106.

[14] Help stop the spread of COVID-19 in children. Available from: https://www.cdc.gov/coronavirus/2019-ncov/daily-life-coping/children/protect-children.html.

[15] Investigating the impact of COVID-19 during pregnancy. Available from: https://www.cdc.gov/coronavirus/2019-ncov/cases-updates/special-populations/pregnancy-data-on-COVID-19/what-cdc-is-doing.html.

[16] Considerations for wearing masks. Available from: https://www.cdc.gov/coronavirus/2019-ncov/prevent-getting-sick/cloth-face-cover-guidance.html.

[17] Q&As on COVID-19 and related health topics. Available from: https://www.who.int/emergencies/diseases/novel-coronavirus-2019/question-and-answers-hub.

[18] Gostic KM, Kucharski AJ, Lloyd-Smith JO. Effectiveness of traveller screening for emerging pathogens is shaped by epidemiology and natural history of infection. eLife 2015;4:e05564. https://doi.org/10.7554/eLife.05564.

[19] Atalan A. Is the lockdown important to prevent the COVID-9 pandemic? Effects on psychology, environment and economy-perspective. Ann Med Surg (2012) 2020;

56:38–42. https://doi.org/10.1016/j.amsu.2020.06.010. Available from: https://www.ncbi.nlm.nih.gov/pmc/articles/PMC7293850/.
[20] Coping with stress. Available from: https://www.cdc.gov/coronavirus/2019-ncov/daily-life-coping/managing-stress-anxiety.html.
[21] Textile showroom in T. Nagar sealed for failing to maintain physical distancing norms among customers. Available from: https://www.thehindu.com/news/cities/chennai/textile-showroom-in-t-nagar-sealed-for-failing-to-maintain-physical-distancing-norms-among-customers/article32898644.ece.
[22] Chatterjee S. Choosing physical distancing over social distancing in the era of technology: minimizing risk for older people. J Gerontol Soc Work 2020:1–2. https://doi.org/10.1080/01634372.2020.1774831. Advance online publication.
[23] List of events affected by the COVID-19 pandemic. Available from: https://en.wikipedia.org/wiki/List_of_events_affected_by_the_COVID-19_pandemic.
[24] Travel restrictions monitoring methodology. Available from: https://migration.iom.int/.
[25] We need to restrict use of crowded public transport to prevent transmission of corona viruses. Available from: https://indianexpress.com/article/opinion/columns/corona-virus-public-transport-COVID-19-trains-6320886/.
[26] Agarwal A, Nagi N, Chatterjee P, Sarkar S, Mourya D, Sahay RR, Bhatia R. Guidance for building a dedicated health facility to contain the spread of the 2019 novel coronavirus outbreak. Indian J Med Res 2020;151(2 & 3):177–83. https://doi.org/10.4103/ijmr.IJMR_518_20.
[27] Wölfel R, Corman VM, Guggemos W, Seilmaier M, Zange S, Müller MA, Niemeyer D, Jones TC, Vollmar P, Rothe C, Hoelscher M, Bleicker T, Brünink S, Schneider J, Ehmann R, Zwirglmaier K, Drosten C, Wendtner C. Virological assessment of hospitalized patients with COVID-2019. Nature 2020;581(7809):465–9. https://doi.org/10.1038/s41586-020-2196-x.
[28] van Doremalen N, Bushmaker T, Morris DH, Holbrook MG, Gamble A, Williamson BN, Tamin A, Harcourt JL, Thornburg NJ, Gerber SI, Lloyd-Smith JO, de Wit E, Munster VJ. Aerosol and surface stability of SARS-CoV-2 as compared with SARS-CoV-1. N Engl J Med 2020;382(16):1564–7. https://doi.org/10.1056/NEJMc2004973.
[29] reportNovel coronavirus (2019-nCoV) situation reports. Available from: https://www.who.int/emergencies/diseases/novel-coronavirus-2019/situation-reports.
[30] Italian doctors call for protecting healthcare workers and boosting community surveillance during covid-19 outbreak. March 26 , 2020. https://doi.org/10.1136/bmj.m1254. Available from:.
[31] FNOMCeO. List of doctors who died during the COVID-19 pandemic. Available from: https://portale.fnomceo.it/elenco-dei-medici-caduti-nel-corso-dellepidemia-di-covid-19/.
[32] Institute of medicine (US) committee on guidance for establishing standards of care for use in disaster situations. In: Altevogt BM, Stroud C, Hanson SL, Hanfling D, Gostin LO, editors. Guidance for establishing crisis standards of care for use in disaster situations: a letter report. National Academies Press (US); 2009.
[33] Friese CR, Veenema TG, Johnson JS, Jayaraman S, Chang JC, Clever LH. Respiratory protection considerations for healthcare workers during the COVID-19 pandemic. Health Secur 2020;18(3):237–40. https://doi.org/10.1089/hs.2020.0036.
[34] Gan WH, Lim JW, Koh D. Preventing intra-hospital infection and transmission of coronavirus disease 2019 in health-care workers. Saf Health Work 2020;11(2):241–3. https://doi.org/10.1016/j.shaw.2020.03.001.
[35] Juneja D, Nasa P, Singh O. Physician staffing pattern in intensive care units: have we cracked the code? World J Crit Care Med 2012;1(1):10–4. https://doi.org/10.5492/wjccm.v1.i1.10.
[36] World Health Organization. Infection prevention and control of epidemic- and pandemic-prone acute respiratory diseases in health care. Available from: https://www.who.int/csr/resources/publications/swineflu/WHO_CDS_EPR_2007_6/en/.
[37] Centers for Disease Control and Prevention. Guideline for hand hygiene in health-care settings: recommendations of the healthcare infection control practices advisory committee and the HICPAC/SHEA/APIC/IDSA hand hygiene task force. Available from: https://www.in.gov/isdh/files/Tab_4_Resource_CD.pdf.
[38] Lu CW, Liu XF, Jia ZF. 2019-nCoV transmission through the ocular surface must not be ignored. Lancet 2020;395(10224):e39. https://doi.org/10.1016/S0140-6736(20)30313-5.
[39] World Health Organization. Infection prevention and control of epidemic-and pandemic prone acute respiratory diseases in health care WHO/CDS/EPR/2007.6. Geneva: WHO; 2007. World Health Organization. Home care for patients with suspected novel coronavirus (nCOV) infection presenting with mild symptoms and management of contacts. Geneva: WHO; 2020. Available from: https://www.who.int/publications/i/item/home-care-for-patients-with-suspected-novel-coronavirus-(ncov)-infection-presenting-with-mild-symptoms-and-management-of-contacts.
[40] Infection prevention and control during health care when novel coronavirus (nCoV) infection is suspected. Interim guidance. 2020. Available from: https://www.who.int/publications-detail/infection-prevention-and-control.
[41] During-health-care-when-novel-coronavirus-(nCoV)-infection-is-suspected; World Health Organization. Management of ill travellers at points of entry - international airports, seaports and ground crossings - in the context of COVID-19 outbreak: interim guidance, 16 February 2020. World Health Organization; 2020. Available from: https://apps.who.int/iris/handle/10665/331003.
[41] . Interim laboratory biosafety guidelines for handling and processing specimens associated with coronavirus disease 2019 (COVID-19). Available from: https://www.cdc.

gov/coronavirus/2019-ncov/lab/lab-biosafety-guidelines.html.

[42] MacIntyre CR, Zhang Y, Chughtai AA, Seale H, Zhang D, Chu Y, Zhang H, Rahman B, Wang Q. Cluster randomised controlled trial to examine medical mask use as source control for people with respiratory illness. BMJ Open 2016;6(12):e012330. https://doi.org/10.1136/bmjopen-2016-012330.

[43] Interim guidelines for collecting, handling, and testing clinical specimens for COVID-19. Available from: https://www.cdc.gov/coronavirus/2019-nCoV/lab/guidelines-clinical-specimens.html.

[44] IATA. Infectious substances shipping guidelines. Available from: https://www.iata.org/en/publications/store/infectioussubstances-shipping-guidelines/. (Accessed 13 February 2020).

[45] UN3373 medical packaging. UN3373: biological substances category B. Available from: https://www.un3373.com/category-biological-substances/category-b/. (Accessed 13 February 2020).

[46] Laboratory biorisk management for laboratories handling human specimens suspected or confirmed to contain novel coronavirus: interim recommendations. Available from: https://www.who.int/csr/disease/coronavirus_infections/Biosafety_InterimRecommendations_NovelCoronavirus_19Feb13.pdf.

[47] Centers for Disease Control and Prevention. Best practices for environmental cleaning in healthcare facilities in resource-limited settings. 1st ed. Atlanta, GA: US Department of Health and Human Services; 2019 CDC; Cape Town, South Africa: Infection Control Africa Network; Available from: https://www.cdc.gov/hai/pdfs/resource-limited/environmental-cleaning-RLS-H.pdf.

[48] USEPA (19 March 2020) list N: disinfectants for use against SARS-CoV-2 | US EPA. Available from: https://www.epa.gov/pesticide-registration/list-n-disinfectants-coronavirus-covid-19.

[49] UNEP. Technical guidelines on the environmentally sound management of biomedical and healthcare wastes (Y1; Y3).BIOMEDICAL TG. 2003. Available from: http://archive.basel.int/pub/techguid/tech-biomedical.pdf.

[50] WHO training modules in health-care waste management. Available from: https://www.who.int/water_sanitation_health/facilities/waste/training_modules_waste_management/en/.

[51] Health care waste management: corona virus update. Available from: https://noharm-global.org/sites/default/files/documents-files/6339/HCWH%20Covid-19%20Waste%20Facts_0.pdf.

[52] WHO. Rational use of personal protective equipment for coronavirus disease 2019 (COVID-19). 2020. Available from: protective-equipment-for-coronavirus-disease-(covid-19)-and-considerations-during-severe-shortages.

CHAPTER 8

Belief System, Deity, and Pandemics: A Sociological Analysis

DEBLEENA BISWAS

1 INTRODUCTION

The world is experiencing an ongoing global medical crisis. The infection is caused by a virus and it seems to be defying all odds and is continuing to march across the globe. Coronavirus disease 2019 (COVID-19) is distributed by droplets; the simplest form of reproduction and the virus's transmissibility score ranges from 2.24 to 3.58 (cited in Ref. [1]). The novel coronavirus pandemic (severe acute respiratory syndrome coronavirus 2 [SARS-CoV-2]) has quickly spread around the globe. According to the World Health Organization (WHO), this infectious respiratory virus is thought to have arisen in China's Wuhan province, spreading over 140 countries on six continents as of mid-March 2020 [2]. Kumari and Shukla [3] put forth the view of how the novel corona virus has no boundaries, has no religion, and is beyond cast and creed. It is extremely infectious and volatile in nature. The world was not prepared for a pandemic of this magnitude, and we are racing to find an antidote to stop it from spreading. Kar et al. [4] provide a brief idea about people having different belief patterns and coping mechanisms. People are usually coping with depression during the COVID-19 pandemic by engaging in recreational activities, artistic hobbies, online learning, doing yoga, meditation, prayer, and indoor games.

Pirutinsky et al. [5] conducted a study that evaluated the religious coping strategy among the American orthodox Jewish community. The study found that positive religious coping was found to be associated with less stress, whereas negative religious coping was found to be associated with more stress (cited in Ref. [5]). Peter Berger used the phrase "sacred canopy" or "sheltering canopy" to explain about religion's role as a defensive cover to shield society from disorder and individuals from rootlessness [6]. Therefore religious belief not only tends to provide solutions to the issues and concerns of the people related to intent, fate and any kind of uncertainty but also provides protection and support in times of risk, grief, and demise.

2 STUDIES FROM LITERATURE REVIEW

2.1 Coalescing Pandemic and Deity: A Religious Discourse

The major pandemics that have occurred in history have been highlighted in the paper by Pokharel [7]. But in the past, the entire world was not affected, as it is now. Earlier the area was limited but because of the coronavirus pandemic, the entire globe has been affected. The paper begins with the prehistoric pandemic in China. The current pandemic first originated from the Wuhan province of China. The first pandemic is said to have occurred in China, which is ironic. It was around 3000 BCE when the pandemic took place, and it was in the archaeological excavation called the Hamin Manga. The skeletons that were derived from the spot suggested that the pandemic had not left any age group as skeletons of old as well as young were found. The researcher has also mentioned several pandemics from the pages of history, such as the plague in Greece, Rome, Justinin; Black Death of 1343; American Plague; Cocolitzli Epidemic; Great Plague of London; Great Plague of Marsellie; Russian Plague; Philadelphia yellow fever; Flu Epidemic; Spanish flu; Asian flu; AIDS pandemic and epidemic; H1N1 swine flu pandemic; West African Ebola epidemic; and Zika virus epidemic.

We pray in moments of crisis, and it is very usual for human beings to turn toward religion to get comfort and explanation. Bentzen [8] illustrates how the COVID-19 pandemic is no exception. Searches for prayer have increased to the highest point ever reported, as per the daily data from Google for 95 countries. The researcher describes that how some philosophers (Marx, Weber, Durkheim, and Freud) had earlier

COVID-19: Tackling Global Pandemics through Scientific and Social Tools. https://doi.org/10.1016/B978-0-323-85844-1.00011-8

depicted that religion might die out as the society get modernized, but this did not happen (described by Norris and Inglehart, Stark and Finke, Iannaccone). Religion is gaining a stronghold in some societies, with 83% of world's population having faith in God. There must be something in religion, which the modernization is unable to fulfill. It is important to see during the times of crisis who uses religion to understand the religion's function and its persevering and social and economic impacts. The WHO declared the 2020 pandemic on 11th March, with the amount of prayer searching in March 2020 being 50% higher than the average in February 2020. During these contemporary times, the book of prayer was also found on the internet. In March 2020, the "Coronavirus prayer" (p. 3) was the most readily searched prayer on the internet. According to the Pew Research Center survey conducted in March 2020, more than half of Americans prayed to eradicate the coronavirus. Google searches on topics like Allah, Quran, Bible, and Jesus increased and to a lesser extent, Buddha, Vishnu, and Shiva also increased. Prayer was also seen to increase in more poor, vulnerable, and unequal countries, but this is mainly due to the religious nature of these countries.

In their research, Alam and Rao [9] provide a number of studies on the worship of deities during the epidemic. Religion in ancient society depended on the worship of various gods and goddesses to reduce physical and emotional suffering. The villagers shared the belief that their village is surrounded by bad spirits that cause crop failure, fire, or indication of diseases, epidemics such as cholera, smallpox, fever, and illness of cattle. The villagers prayed to the guardian gods and satisfied them in order to avoid sickness and disaster. Traces of mother worship is found in the Valley of Indus, and this provided an idea that worship of goddesses was practiced in the treatment of diseases. The researchers have provided a plethora of studies from South India, like the temples of *Ankamma* and *Gangamma* goddesses for cholera and smallpox, respectively, in Andhra Pradesh. Temples of *Plagueamma* are built to prevent plague. In Tamil Nadu, *Mariamma* or *Mari* is venerated for curing smallpox. Performing certain *vazhipadus* in the Siva temple of Taliparamba cures illnesses of epilepsy. The *Kurumba Bhagavati* goddess temple at Kodungallur in Kerala is the *masuri devata* (Goddess of smallpox) and prayers are offered to keep the disease away. The temple in Guruvayur is also known for treating rheumatism and other illnesses. Muslims and Christians alike are led by superstitious beliefs, like the St. Thomas tradition, have beliefs in miracles. St. Sebastian possesses the power to get rid of infectious diseases and St. George provides relief from serpentine irritation. The Putupalli Church is visited for curing poisoning. Muslims in Kerala place their faith in their Saints, whether they are alive or dead. *Musaliars* are assigned with miraculous powers to cure diseases, exorcising the evil forces and providing blessing with goodness and welfare to families. The principal goal of worshipping the village gods and goddesses is to prevent any kind of anger. The aim is to avoid diseases such as smallpox, cholera, or any other calamities. This kind of worship takes place in many villages during an epidemic, while in some villages, it takes place annually.

According to Archibong et al. [10] the link between religion and medicine is long and complex. Religious healing seems to be found among the population with religious gifts offered in the matters of health. Two basic forms of belief exist in divine healing. One form of belief is that divine healing takes place by psychological mechanisms and it works for those with psychosomatic illness. The second form of belief is that divine healing works as a result of interference in the disease of the deity and such healing constitutes to miracle in the present day.

Sen [11] discusses about cholera, a "dehydrating diarrheal disease" (p. 345) that killed many people in the 19th century. The disease is thought to have begun in 1817 in Jessore (now Bangladesh) near Calcutta and spread to Europe by 1823. In the 19th century, five cholera pandemics occurred, the first of which started in 1817 in West Bengal and killed millions of people around the world. This pandemic was named for its magnitude rather than its length. The climatic condition, soil, and air, particularly of Bengal region, were deemed to be the potential cause of the disease. When the disease started spreading in Europe during the 1830s, cholera was termed as "Indian Cholera" (p. 345) by the Europeans and India earned the reputation of being the "Homeland of Cholera" (p. 345), as all the pandemics were thought to have originated in Bengal and spread throughout the country. The main intention of this paper is to negate the belief system surrounding "India/Bengal" being responsible for the cholera pandemic during the 19th century. Sen describes a study put forth by McNamara who had traced the genesis of cholera in Calcutta, before the year 1817. During the early 18th century, the root of the word "Ola-Ota" was documented, and the inhabitants used it to describe cholera. An old woman was walking in the jungle and she had found a stone that was revered as the Goddess of *Ola-ota* and was renamed as *Ola-bibi*. The desire to worship the *Ola-bibi* goddess, who drew worshippers from far away, suggested that the disease *Ola-ota*, may have caused a lot of health issues and mortality problems for the people of Calcutta and Bengal in the early 1800s. In 1720, the English merchant, Mr. Duncan gave Rs.4000/- for a temple to *Ola-bibi* goddess. In 1750, Mr. Duncan again donated Rs.6000/- for the construction of the second temple of the *Ola-bibi* goddess in Calcutta. An old rough stone was moved to the new residence and there was a very large idol that was constructed. This also in a way reveals that the mortality rate among the Europeans was large and the funds came from the Europeans to build two temples for the goddess *Ola-bibi*. Men, women, and children used to offer their prayers at a shrine by the roadside during the cholera epidemic in Poland, just like in Bengal where people worshipped the *Ola-bibi* goddess.

Bansal et al. [12] describe in their paper how even after the world entering the new millennium and touching new heights, the belief that epidemics are caused by evil spirits is still prevalent in the Northern part and Southern part of India. *Atharvaveda* contains a number of mantras for treating diseases that are thought to be caused by an evil force. In certain rural areas, *Bhai Moolchand* is revered for curing warts. A survey was conducted and it was found out that if the person who is affected offers common salt to God, the disease can get cured. In Northern parts of the country, the *Babas* are worshipped. Sometimes referred to as the 'Muslim peer'. There are several holy places called as the *Mazaar*. In Himachal Pradesh, there is a place named *Chintapurni*, where the goddess is said to eliminate all tensions. As the meaning clearly suggests, *Chinta* means tension and *Purni* means removal. Various planets are also revered to cure illnesses, such as *Surya* and *Mangal* who are worshipped for *Pitta* disorders such as fevers and skin-related problems. *Vata* disorders are cured by praying to *Rahu*, *Ketu*, and *Shanni*, while *Kapha* diseases are cured by worshipping *Shukra* and *Chandrama*. There is a temple in Sancharwada, Rajasthan, where *Bibayi Mata* is worshipped to heal cuts in the feet. In Jhalatala, Alwar district of Rajasthan, there is a man who is said to have supernatural powers and who recites mantras to cure dental diseases. There is a place called Abaneri in Rajasthan's Dosha district that is popular for curing typhoid. People take the patient's clothes and have the priests conduct some ceremonies on them, which causes the illness to vanish. In Gonda in Uttar Pradesh, there is a Muslim *Mazaar* known as *Deva Sharif*, which is known for curing infertility.

2.2 Beliefs Revolving Around COVID-19 Pandemic and Deity

A report [13] provided the details of how the superstitious beliefs ran deep in certain states. After Bihar, the coronavirus has been given a gender status and is worshipped in Uttar Pradesh, where the superstition is growing. Women in the villages of Tumkuhiraj, Kasia, Hata, Captanganj, and Khaddha Tehsil in Kushinagar district have begun to worship 'Corona *Mai*' so that she will protect people's lives. To please Corona *Mai*, the women have dug a small pit in the field and filled it with water. Each woman offers nine cloves and nine *laddoos* (round-shaped sweets) to her. However, some residents have appealed to the district administration to stop those activities that spread false information and superstitions.

According to the news reported in *The Hindu* [14], the world is coping with the COVID-19 pandemic, a small group of women from the state of West Bengal village has chosen to fight the disease in their own way. It all began in Asansol's Nichupara *Basti* (slum), where a small puja was conducted around the banks of Chinnamasta pond in the district, which they had done so in good conscience. Women of various ages assembled there at 8 a.m., and the rituals lasted until 9 a.m. As they walked to the venue, the women sang songs and chanted mantras. The residents of that area invited every woman to join them in the early hours of the day. They sat in a row on the grassy field, and flowers were set in a line in front of them. They lit the incense sticks and served fruit, vegetables, *ghee* (clarified butter), and jaggery to the *Devi*. Many of the women were daily wage earners, and they spent about Rs.500/- by buying the material for the rituals. They believed that the Coronavirus was a creation by *Shitala Devi*, as she is believed to be the goddess of relief in some culture. When asked how the Corona *Devi* looks like, a woman of Nichupara *Basti* replied that she is formless. She said that she can be the wind that creaks on the tile-roof house at night, may be the sun, or may be the water in the nearby stream.

Swamy [15], in *The Print*, reports about the Mysuru *Dasara* festival started with a special prayer to the *Chamundeshwari* goddess, the deity mainly revered during this festival, to guide the mind so that the vaccine for COVID-19 can be found out soon. The festival is normally marked by grand celebrations, but owing to the pandemic crisis, the 10-day Mysuru *Dasara* was observed in silence.

According to a report from the *India TV Trending Desk* [16]; the Corona *Devi puja* is being performed to fend off coronavirus. Videos of women performing the *puja* of "Corona *Devi*" or "Corona *Mai*" went viral on social media platform sending Twitter into an outburst. This image of worship was captured in rural parts of India, including several districts of Bihar like Nalanda, Gopalganj, Saran, Vaishali, Muzaffarpur, and also in the parts of Jharkhand, Uttar Pradesh, and West Bengal. The Twitterati finds it difficult to believe that people have developed such superstition. On the other side, scientists and doctors are struggling to develop the coronavirus vaccine, but the rural India have come up with this approach to please "Corona *Mai*".

The *Times of India* report by Khare [17] says that in Sindri Bokaro, more than the pandemic's spread, the superstitious belief in offering prayers to the Corona goddess to keep the virus at bay had gained pace in parts of Jharkhand. A group of women were found praying to "Corona *Mai*" and chanting her glory near the closed temple complex in Sindri. Nine hibiscus flowers and an equal number of *laddoos* were offered to the Corona goddess as part of the rituals and requests were made to her to push the virus out of the country. The flowers and sweets were eventually buried in a small ditch filled with soil. Nine vermillion streaks were laid on the surface, signaling the goddess's burial and the end of COVID-19. *"Corona Mata Ki Jai"* (hail to Goddess Corona) was also chanted by the women.

Ranjan [18], in *The New Indian Express*, reports the situation from Ranchi where transgenders in Dhanbad performed a *puja* on Friday to please "Corona *Mai*" (Mother Corona) who they believed appeared in their dreams and told them that if people started worshipping her then she would leave. To ward off the virus, a group of more than a dozen transgender people gathered in an open area at Jharia in Dhanbad to perform the special puja. They also urged people to worship "Corona *Mai*" in order to get rid of her. A transgender named *Nirmala Kinner* told that she saw a cow in her dream that then transformed into a lady and that if people in India worshipped her, she would leave the country, which is why they have gathered to perform the puja. *Bansi Kinner* said that there is no other choice and that people must worship Corona *Mai* in order to please her. More than 250 women assembled on the banks of River Baanki in Bodra village in Garhwa to carry out the *Chhath Puja* in order to please "Corona *Mai*" and escape the difficulties they face. Even interesting was the fact that women who gathered for worship had brought their bank passbooks and Aadhar cards, expecting money to be credited to their bank accounts if they succeed in making her happy. Similar incidents were recorded in Ranchi's Namkom and Tatisilwai, Jamshedpur's Bagbeda, Dompara, Jharia in Dhanbad, and parts of Giridih.

3 THEORETICAL ANALYSIS AND DISCUSSION

Ukuekpeyetan-Agbikimi [19] outline the essentialist philosophies that place an emphasis on the contents of religion, their purpose, and the implications of those contents for people. People have faith in belief because it makes a sense to them asserts the essentialist approach. From the abovementioned news reports, we can draw a picture that during the time of pandemic crisis, people are resorting to superstitious beliefs because those beliefs make sense to them and they are doing it so that they can get rid of the pandemic soon. Ukuekpeyetan-Agbikimi delineates Durkheim's theory of religion, which is based on the symbolic explanation of the social reality, without it the religious expressions find no meaning. Durkheim's approach to religion gives rise to the functionalist model that defines social institutions in collective terms. Therefore the collectivity that revolves around the social institution of religion among the people during the pandemic situation is seen when they perform the rituals together and they give meaning to those actions. Jha [20] describes the approach given by Malinowski and Brown. Malinowski has worked among the Trobriand Islanders, and discovered that religion is very closely connected with different emotional stages and those are the stages of tensions. Religion is used as a medium to deal with and get rid of tension so that the emotional states can be controlled. So from the abovementioned lines, we can see that during this pandemic situation, people are in an emotional state and by practicing those beliefs, they tend to control their emotions. On the other hand, Brown's theory as described by Jha depicts the idea that the role of religion is not to remove fear and other emotional tension from the human mind, but it establishes dependency in it. Hence, the beliefs practiced by people are giving them a sense of dependency so that they can get rid of the COVID-19 pandemic soon.

4 CONCLUSION

According to the findings of the studies presented, despite the best methods invented and discovered by science and medical science on the one hand, faith healing belief systems operate on the other. The cases are mostly reported from rural backgrounds. The belief systems play a very important role in the sociocultural lives of the people during times of pandemic. The myths about pandemics are important to those who believe them, and they are passed down from one generation to another. During the times of crisis, religion retains a strong hold and serves as one of the social tools available. The pandemic condition, which the major part of the globe is going through, is now no more confined to only a medical condition but has in a way restructured our ways of life. This pandemic situation is a kind of deadliest infection that has altered our existence. It has in a way engineered our existence to go from one mode to another. The chapter focuses on this condition from a sociological angle and has tried to bring forth the connection between the belief systems and the pandemic. The chapter in no way tries to glorify any kind of superstitious beliefs or practices. It is an attempt to understand people's tendency to respond to everything including the mechanisms to deal with the pandemic in a religious manner since time immemorial and even in the contemporary times.

REFERENCES

[1] Quadri SA. COVID-19 and religious congregations: implications for spread of novel pathogens. Int J Infect Dis 2020;96:219–21.

[2] Cron RQ, Chatham WW. The rheumatologist's role in COVID-19. 2020.

[3] Kumari T, Shukla V. Covid-19: towards confronting an unprecedented pandemic. Int J Biol Innovat 2020;2(1): 1–10.

[4] Kar SK, Arafat SY, Kabir R, Sharma P, Saxena SK. Coping with mental health challenges during COVID-19. In: Coronavirus disease 2019 (COVID-19). Singapore: Springer; 2020. p. 199–213.

[5] Pirutinsky S, Cherniak AD, Rosmarin DH. COVID-19, mental health, and religious coping among American Orthodox Jews. J Religion Health 2020;59(5): 2288–301.

[6] Johnstone R. Religion in society a sociology of religion. 8th ed. New Delhi: Pearson Education Inc; 2007.

[7] Pokharel JR. Past pandemics and their parallel situation in Nepal. Appl Sci Technol Ann 2020;1(1):136–41.

[8] Bentzen J. In crisis, we pray: religiosity and the COVID-19 pandemic. 2020.

[9] Alam MM, Rao BR. Religious practices in South India to cure diseases. Bull Indian Inst Hist Med 1998:1–6.

[10] Archibong EP, Enang EE, Bassey GE. Witchcraft beliefs in diseases causation and health-seeking behaviour in pregnancy of women in Calabar South-Nigeria. IOSR J Humanit Soc Sci 2017;22(6):24–8.

[11] Sen S. Indian cholera: a myth. Indian J Hist Sci 2012;47: 345–74.

[12] Bansal P, Acharya MV, Sharma KD, Veluchamy G. Religious practices in North India to cure diseases. Bull Indian Inst Hist Med 2001;XXXI:25–30.

[13] National Herald. Corona now gets 'Mai' status in Uttar Pradesh, too. 2020. Retrieved on 9 November 2020, from, https://www.nationalheraldindia.com/national/corona-now-gets-mai-status-in-uttar-pradesh-too.

[14] Samanta S. How a goddess called Corona Devi came to be worshipped in West Bengal. The Hindu; 2020. Retrieved on 9 November 2020, from, https://www.thehindu.com/society/a-goddess-called-corona-devi/article31795320.ece.

[15] Swamy R. Mysuru Dasara starts with a special prayer to deity Chamundeshwari for 'early Covid vaccine'. The Print; 2020. Retrieved on 9 November 2020, from, https://theprint.in/india/mysuru-dasara-starts-with-a-special-prayer-to-deity-chamundeshwari-for-early-covid-vaccine/526211/.

[16] India TV Trending Desk. Corona Devi pujas held to ward off coronavirus, superstition sends Twitter into frenzy. 2020. Retrieved 9 November 2020, from, Indiatvnews.com. https://www.indiatvnews.com/trending/news-corona-devi-pujas-corona-mai-held-to-ward-off-coronavirus-superstition-twitter-624091.

[17] Khare D. Women pray to 'corona Mai' to ward off virus in Sindri, Bokaro. The Times Of India; 2020. Retrieved on 9 November 2020, from, https://timesofindia.indiatimes.com/city/ranchi/women-pray-to-corona-mai-to-ward-off-virus-in-sindri-bokaro/articleshow/76224396.cms.

[18] Ranjan M. Jharkhand: transgenders, women perform puja to ward off coronavirus. The New Indian Express; 2020. Retrieved on 9 November 2020, from, https://www.newindianexpress.com/nation/2020/jun/05/jharkhand-transgenders-women-perform-puja-to-ward-off-coronavirus-2152749.html.

[19] Ukuekpeyetan–Agbikimi N. Current trends in theories of religious studies: a clue to proliferation of religions worldwide. Global J Arts Humanit Soc Sci 2014;2(7): 27–46.

[20] Jha M. An introduction to social anthropology. Uttar Pradesh: Vikas Publishing House Pvt Ltd; 2019.

CHAPTER 9

A Critical Study on Violation of a Decent Burial/Cremation Right of a Dead Person During COVID-19: Special Reference to India

PYALI CHATTERJEE

1 INTRODUCTION

After the terrorist attack in Bali, Indonesia, G. Tresna, a lifeguard at Kuta Beach rightly pointed out that 'We should treat the dead with respect. In death, money doesn't matter, material possessions don't matter; dignity is what we should care about.' [1].

According to Article 1 of the Universal Declaration of Human Rights, 'All human beings are born free and equal in dignity and rights. They are endowed with reason and conscience and should act towards one another in a spirit of brotherhood [2]'. This law rightly and equally protects every person who is born alive. However, there is an exception to this rule, which says that certain rights can be entrusted to an unborn child in relation to transfer of property. This means a person can transfer his/her property in the name of his unborn child. But the question that arises here is that whether a dead person has any legal right. In the case of *Pt. Parmanand Katara v Union of India 1995 (3) SCC*, the Supreme Court held that 'right to dignity and fair treatment under Article 21 of the Constitution of India is not only available to a living man but also to his body after his death' [3]. As per law, the right to life with dignity also consists of the right to a decent burial/cremation after death. The Division Bench of the Madras High Court comprising of Justice M. Sathyanarayanan and Justice M. Nirmal Kumar, on April 20, 2020, in a Suo Moto writ petition, was of the opinion that the scope and ambit of Article 21 includes the right to have a decent burial/cremation [4]. In another case *Common Cause (A Regd. Society) v Union of India (2018) 5 SCC 1*, the honorable Supreme Court of India held that 'In a certain context, it can be said, life sans dignity is an unacceptable defeat and life that meets death with dignity is a value to be aspired for and a moment for celebration' [5].

The right of a decent burial/cremation is not a newly established law. In legal jurisprudence, it is mentioned that after death, all rights cease to exist except three rights, which is conferred by law to a dead person, namely, a man's body, his reputation, and his estate [6]. A cadaver is the property of no one [6]. Even one cannot dispose it off through a will. Violation of a grave is a criminal offense under criminal law because it secures a decent burial of every dead person. The reputation of the dead person receives some degree of protection from the criminal law [6]. Libel upon a dead man is punishable, but only when its publication is an attack upon the interests of the living persons.

Section 297 of the Indian Penal Code states that 'Whoever, with the intention of wounding the feelings of any person, or of insulting the religion of any person, or with the knowledge that the feelings of any person are likely to be wounded, or that the religion of any person is likely to be insulted thereby, commits any trespass in any place of worship or on any place of sepulcher, or any place set apart from the performance of funeral rites or as a depository for the remains of the dead, or offers any indignity to any human corpse, or causes disturbance to any persons assembled for the performance of funeral ceremonies, shall be punished with imprisonment of either description for a term which may extend to one year, or with fine, or with both' [7]. According to section 297 of Indian Penal Code a person will be punished if he/she acts with an intention of wounding the feelings of any person;

COVID-19: Tackling Global Pandemics through Scientific and Social Tools. https://doi.org/10.1016/B978-0-323-85844-1.00007-6

insults the religion of any person ; commits trespass in a place that is used for paying last funeral rites, e.g., a religious or a cremation/burial place or a place that is used as a depository for the remains of the dead; does indignity to any human corpse; or causes any disturbance to any person assembled for the performance of funeral ceremonies. Thus the Indian Penal Code also gives protection to corpses relating to burial/cremation.

The recent pandemic situation gives rise to the genuine question whether a coronavirus disease 2019 (COVID-19) dead body has the right of decent burial/cremation. This is highlighted after several instances were recorded where dead bodies of people who died from coronavirus infection, popularly known as COVID-19, were denied a decent burial/cremation, despite there being a settled law that a dead body has the right of decent burial in every country.

2 RESEARCH METHODOLOGY

Both doctrinal and nondoctrinal research methods have been used for this research. For doctrinal research, various articles, case laws, journals, books, etc. have been referred and studied here. For the nondoctrinal method, an open interview method has been used for collecting opinions related to this topic.

3 RIGHT OF DECENT BURIAL/CREMATION UNDER INDIAN AND INTERNATIONAL LAWS

3.1 Indian Constitution

Article 21 of the Indian Constitution deals with the right to life with dignity, and with the span of time through judicial pronouncements the scope and ambit of Article 21 has increased. The right to life with dignity of a living person has also been extended to a person who is dead. It is the legal right of a person to have a decent burial/cremation after death.

In *Ashray Adhikar Abhiyan v Union of India, AIR 2002 SC 554*, the honorable Supreme Court has acknowledged that the government should take all the possible steps to give a decent burial to the unclaimed dead bodies found on road.

Even during the hearing of a public interest litigation filed by Vikas Chandra Guddu Baba related to the dumping of unidentified dead bodies in rivers, the bench headed by Justice B. S. Chauhan expressed that 'Dead person must be given some respect' [8].

But in this COVID-19 pandemic, many incidents have been reported in India wherein it has been found that there is a gross violation of decent burial/cremation right of a person dying from COVID-19. It is important to mention that the Indian government has already issued guidelines for the manner and procedure in which the burial/cremation of COVID-19 dead bodies needs to be carried out. The author will discuss the said guidelines in the later part of her study. Even according to the Samaj Sudhar Committee (SSC) [9], a decent burial/cremation is the fundamental right of every citizen of this country under Article 21 of the Constitution, so the denial of proper cremation of COVID-19 bodies is a heinous crime in the eyes of law.

Former union law minister and senior counsel Ashwani Kumar, in his letter to the Chief Justice of India (CJI), requested the CJI to take 'suo motu cognizance of various reports highlighting that corona virus patients are being ill-treated and dead bodies manhandled' [10] and he further stated that 'Right to die with dignity is a fundamental right and it includes the right to a decent burial/cremation' [10]. Even Justice Ashok Bhushan remarked that 'Covid-19 patients are treated worse than animals. In one of case, a dead body was found in the garbage' [11].

The Meghalaya High Court has strictly directed that burial or cremation of coronavirus-infected patients should be done in a decent way without any obstruction in the state [12]. Even Mr. K. Paul, Vice President of the Meghalaya High Court Bar Association, submitted that the actions of local bodies in preventing burial and cremation were in violation of the law and that they were liable to be punished accordingly [12].

What one needs to understand here is that a legal right does not end with the death of a person. Some rights exist even after the death of a person, such as the burial/cremation right, according to their religion.

3.2 International Law

For the decent burial of corpses, many international instruments have given strict guidelines relating to recovery, identification, and burial of human remains in accordance with their respective religious rituals and cultural beliefs. Different guidelines are given in the international law for the treatment of human remains in different situations.

- Interpol Guides—for those who die in disasters.
- Geneva Conventions and other protocols—for those who die in armed conflicts.
- Guiding Principles on Internal Displacement—deals with natural catastrophes.
- ICAO Standards—deals with aircraft accidents.
- Australia's Defense Force Manual, 1994 [13]: the remains of the dead, regardless of whether they are combatants, noncombatants, protected persons, or

civilians, are to be respected, in particular their honor, family rights, religious convictions and practices, and manners and customs at all times shall be humanely treated.

- The UK Military Manual, 1958 [13]: no maltreatment should be done with the dead.
- The United Nations Commission on Human Rights [13]: the 2005 resolution on human rights and forensic science underlined the importance of dignified handling of human remains, including their proper management and disposal, as well as respecting the needs of the families.

The common intention of this entire international instrument is to provide a decent burial right to all dead bodies according to the person's religion whether identified or unidentified.

4 GUIDELINES ISSUED BY THE INDIAN GOVERNMENT FOR THE MANAGEMENT OF DEAD BODIES AT THE BURIAL GROUND/CREMATORIUM DURING THE PANDEMIC [14]

Following are the guidelines given by the Indian government while handling COVID-19 dead bodies by the workers of burial/cremation sites and also by the family members for the decent burial of dead bodies:

i. The crematorium/burial ground staff should be sensitized that COVID-19 does not pose additional risk.
ii. The staff will practice standard precautions of hand hygiene and the use of masks and gloves.
iii. Viewing of the dead body by unzipping the face end of the body bag (by the staff using standard precautions) may be allowed for the relatives to see the body for one last time.
iv. Religious rituals such as reading from religious scripts, sprinkling holy water, and any other last rites that do not require touching of the body can be allowed.
v. Bathing, kissing, hugging, etc. of the dead body should not be allowed.
vi. The funeral/burial staff and family members should perform hand hygiene after cremation/burial. The ash does not pose any risk and can be collected to perform the last rites.
vii. Large gatherings at the crematorium/burial ground should be avoided as a social distancing measure because it is possible that close family contacts may be symptomatic and/or shedding the virus.

Another important point that is necessary to be mentioned here is regarding the transportation of a COVID-19 dead body, 'The body, secured in a body bag, exterior of which is decontaminated poses no additional risk to the staff transporting the dead body' [14]. Thus one should not be afraid of the infection if the body is properly packed in the body bag and the guidelines prescribed for handling of COVID-19 dead bodies are followed by the medical personnel and by the relatives.

The government has made it specifically clear in its guidelines that it should be the responsibility of the staff of crematorium/burial ground to show the face of the dead person for the last time to his/her family members by unzipping the face end of the body bag. However, recently a case was reported in Kolkata where the hospital officials at the crematorium blatantly asked the family members to pay Rs 51,000 in order to see the face of their relative. According to the statement given by the family members, 'it was only after the media pressure that the cremation authorities took only Rs 2,500/-, which is the deposit amount for showing the body' [15]. This incident clearly shows how the given guidelines are being violated by the public authorities who are supposed to follow the guidelines for the benefit of the public but they themselves violate the guidelines for mere pecuniary benefits. Death is the last stage of human life. Every relative has a right to see the face of their loved ones for the last time; otherwise it can create a mental agony to the person leaving behind, which will affect their mental health in some cases.

In yet another incident, a person cremated a COVID-19 dead body of another person named Balchandra Gaikwad thinking it to be the body of his father named Janardhan. Both Janardhan (67 years) and Gaikwad (72 years) were undergoing treatment for COVID-19 at the Global Hub COVID Hospital in Thane [16]. Such a grave mistake happened because the body was completely packed as per the guidelines of the government and the hospital authorities packed the body immediately after his death without showing it to the relative and even at the crematorium ground the face was not shown to the relative. It was after the fourth day of the incident that the hospital informed Janardhan's son that this time it was his father who actually passed away. Who is to be blamed for such a gross mistake, the hospital authorities or Janardhan's son? Due to the recklessness on the part of authorities, another son was not able to see the face of his father for the last time and also did not get the opportunity to perform his father's last rites. In this entire incident the Gaikwad family members were not even able to say a last good bye to their family member, and this has become a lifelong trauma and scar for the family members. On the other hand, Janardhan's family members performed two last rites: first one, by mistake and the second time, the actual ritual.

A similar incident was reported from Raipur, Chhattisgarh (Fig. 9.1).

This incident again shows that there is a lack of awareness regarding the guidelines among the public and also about their right to see the face of their relative for the last time at the cremation ground. Moreover, it is the duty of the authorities to show the face of the COVID-19 dead person to the relatives. This not only helps avoid the grave mistake of mix-up of dead bodies but also gives peace of mind to the relatives that for the last time before the cremation they have seen their beloved one.

The World Health Organization (WHO), with regard to Infection Prevention and Control for the safe management of a dead body in the context of COVID-19, issued the following guidelines:

i. If the family wishes only to view the body and not touch it, they may do so using standard precautions at all times, including hand hygiene. Give the family clear instructions not to touch or kiss the body.
ii. Family and friends may view the body after it has been prepared for burial, in accordance with customs. They should not touch or kiss the body and should wash hands thoroughly with soap and water after viewing.
iii. Those tasked with placing the body in the grave, on the funeral pyre, etc. should wear gloves and wash hands with soap and water after removal of the gloves once the burial is complete [17].

It is pertinent to mention here that this is not the first time that a virus or flu like SARS-COVID-19 has caused a pandemic. Earlier, the Spanish flu of 1918 was also declared as a pandemic.

अव्यवस्था • अस्पताल ने दिया था किसी दूसरे शख्स का शव

मुक्तिधाम में शव देखते ही बेटी बोली, ये मेरे पिता नहीं

दुर्ग (नईदुनिया प्रतिनिधि। भिलाई जुनवानी स्थित शंकराचार्य कोविड अस्पताल में कोरोना पॉजिटिव मृतक का शव बदलने का सनसनीखेज मामला सामने आया है। शव बदलने की भनक लगते ही परिजनों ने मुक्तिधाम में हंगामा मचाना शुरू कर दिया। इधर, शव सौंपने में लापरवाही उजागर होने के बाद विभागीय अधिकारियों के होश उड़ गए।

जानकारी के मुताबिक, तितुरडीह कर्मचारी नगर निवासी 52 वर्षीय दशरथ मारकंडे को कोरोना पॉजिटिव पाए जाने के बाद भिलाई जुनवानी स्थित शंकराचार्य कोविड अस्पताल में तीन दिन पहले भर्ती कराया गया था, जहां उसकी मौत हो गई। अस्पताल के कर्मचारियों ने मृतक दशरथ के परिजनों को इसकी जानकारी दी और शव को परिजनों को सौंप दिया। अस्पताल से शव मिलने के बाद परिजन अंतिम संस्कार के लिए सीधे शिवनाथ नदी मुक्तिधाम लेकर पहुंचे। इसी बीच मृतक दशरथ मारकंडे की बेटी ने अपने पिता के अंतिम दर्शन करने की इच्छा जताई। बेटी की इच्छा पर पीपीई किट के भीतर से ही चेहरा दिखाया गया, तो बेटी के होश उड़ गए। मुक्तिधाम में वह चिल्लाकर बोली ये मेरे पिता नहीं हैं। इसके बाद अंतिम संस्कार के लिए पहुंचे लोगों ने पास जाकर देखा तो जो शव अस्पताल ने दिया था वह दशरथ का नहीं था। इसके बाद मृतक के परिजनों ने मुक्तिधाम में हंगामा कर दिया।

शंकराचार्य कोविड अस्पताल के बाहर शुक्रवार को हंगामा करते मृतक के परिजन। • नईदुनिया

ऑक्सीजन की कमी से जूझ रहे पीड़ित: भिलाई जुनवानी स्थित शंकराचार्य कोविड अस्पताल में ऑक्सीजन की कमी बनी हुई है। अस्पताल में भर्ती मरीज ऑक्सीजन की कमी से जूझ रहे हैं। शुक्रवार को सोशल मीडिया में एक वीडियो वायरल हुआ। इसमें एक बुजुर्ग ऑक्सीजन की कमी से कराहता हुआ नजर आ रहा है। लेकिन चिकित्सक उन्हें ऑक्सीजन चढ़ाने के बजाय बिस्तर पर ही छोड़ दिए हैं।

शंकराचार्य कोविड अस्पताल में अव्यवस्था का आलम है। पार्ट-पार्ट में भर्ती मरीजों को ऑक्सीजन दिया जा रहा है। समय पर व्यवस्था नहीं सुधरी तो भर्ती मरीजों की स्थिति बिगड़ सकती है। मैं वहां मौजूद था। शव लेने के लिए परिजनों को दिक्कतों का सामना करना पड़ा। विभागीय जिम्मेदारी अधिकारी भी जवाब देने से कन्नी कटाते रहे।
-आयुश शर्मा, प्रत्यक्षदर्शी

जिला प्रशासन ने मामले को संज्ञान में लेते हुए शंकराचार्य अस्पताल प्रबंधन को दोबारा ऐसी लापरवाही नहीं बरतने की हिदायत दी है। इस पर अस्पताल प्रबंधन ने भविष्य में ऐसी घटना की पुनरावृत्ति नहीं करने का भरोसा दिलाया है।
-डॉ.सर्वेश्वर नरेंद्र भूरे, कलेक्टर दुर्ग

FIG. 9.1 Naidunia, Raipur, Saturday, September 5, 2020, 9 p.

4.1 Importance of Guidelines Related to the Management of COVID-19 Dead Bodies

During an interview the following questions were asked to Dr. Khan regarding the handling and management of COVID-19 dead bodies:

A. Why is it so dangerous to handle a COVID-19 dead body?
   Answer: It is a danger because it is a viral disease. Those who are handling the patients or dead bodies of COVID-19 may get infected.
B. How to handle it properly so that the dead person can get a decent burial/cremation?
   Answer: While dealing with COVID-19 dead bodies the following important points/guidelines must be followed and checked by everyone:
   a. Family members and health workers should wear gloves and masks and follow hand hygiene.
   b. Relatives may be allowed to see the body of the deceased one time, and the bag should be unzipped by a staff member.
   c. All religious rituals that do not require touching the body can be performed.
   d. The government has not allowed touching, hugging, or bathing the body.
   e. All the family members and the staff at the ground are to perform hand hygiene before and after the cremation or burial.
C. Any misconceptions regarding the handling of COVID-19 dead bodies that people need to know?
   Answer: Most of the public have a common misconception that dead bodies can spread the virus. It is necessary to mention here that once the dead body is packed and sealed in the body bag, there is zero chance of contamination if it is handled properly by the authorities. This is why the medical authorities do not hand over the dead bodies to the relatives. In COVID-19 cases the trained staff of a hospital accompany the dead body up to the cremation ground for proper handling of the dead body so that the people who are not trained in COVID-19 dead body management, like the relatives, do not get infected from the dead body because of mishandling. Even in the case of cremation, there is no risk from ash; therefore it can be collected later. Large gatherings must be avoided, as it is possible that close family members of the deceased may have symptoms of COVID-19 infection or may be shedding the virus.
D. What are the safety measures that everyone must follow while handling and preparing the dead body for burial/cremation?
   Answer: For safety measures, only gloves and masks are not sufficient for handling a COVID patient and one has to use the PPE (personal protective equipment) kit. A PPE kit includes gloves, mask, head-to-foot dress, shoe covers, goggles, and face shield.

Thus from this discussion, it is clear that even in the pandemic caused by COVID-19, a dead person has a right of decent burial/cremation and no one can curtail this right from the COVID-19 dead person as well as from his/her family members for performing the last rituals. The only thing is to strictly follow the guidelines provided by the government, which are for the safety of the family members of the COVID-19 dead person as well as for the society as a whole. A large gathering is not allowed during the funeral at a cremation ground because the family members of the deceased may have symptoms of COVID-19 infection or may be shedding the virus. There is no danger to the life of others if the body is not cremated in an electric crematorium.

According to WHO [18], the dignity, culture, and religious traditions of a person killed by COVID-19 should be taken care of. The families of such people should be respected and they should get protection. It is a tough situation for the administration/government to make a balance the rights of families, the investigation into the cause of death, and the risk of infection relating to COVID-19.

The only thing that is necessary to keep in our mind while dealing with a COVID-19 dead body is that one should strictly abide by the guidelines of the government. This is the only way by which we all will be safe and the dead person will get a decent and peaceful burial/cremation.

5 INCIDENTS OF VIOLATION OF DECENT BURIAL/CREMATION RIGHTS OF A DEAD PERSON DURING COVID-19

Following are some of the incidents that have been reported in various cities in violation of decent burial/cremation rights of a COVID-19 dead person.

1. Recently an incidence was reported from Raipur, the capital city of the state of Chhattisgarh, where cremation of a COVID-19 dead body took place in a local area named Devendra Nagar. As soon as the residents of the said area came to know about it, the residents including women created obstacle to the funeral. Soon after the police intervened and the chaos was settled down after 4 h of conciliation. The public was of the view that cremation of the COVID-19 dead person in the electric chimney will cause health threat in the surrounding localities (Fig. 9.2).

प्रदर्शन : श्मशानघाट में पुलिस ने किया लाठीचार्ज

कोरोना से मृत वृद्धा के अंतिम संस्कार का विरोध, किया हंगामा

हरिभूमि न्यूज ▸▸ रायपुर

कोरोना संक्रमण ने जहां लोगों की दिनचर्या के साथ आदत-व्यवहार को बदल दिया है, वहीं अफवाहों के कारण लोग मानवता का धर्म भी भूल रहे हैं। ऐसा ही एक मामला शुक्रवार को मारवाड़ी श्मशानघाट में देखने को मिला, जहां एक कोरोना संक्रमित बुजुर्ग महिला की लाश को जलाने के विरोध में लोगों का हुजूम उमड़ पड़ा। पांच घंटे की मशक्कत के बाद पुलिस तथा नगर निगम प्रशासन की दखल के बाद देर शाम महिला के अंतिम संस्कार की प्रक्रिया शुरू हो पाई।

नगर निगम के अपर आयुक्त पुलक भट्टाचार्य के मुताबिक 70 वर्षीय महिला की एक निजी अस्पताल में उपचार के दौरान मौत हो गई। परिजन महिला की लाश का अंतिम संस्कार करने जैसे ही मारवाड़ी श्मशानघाट ▸▸ शेष पेज 17 पर

खास बातें

- पुलिस, निगम प्रशासन के दखल के बाद लाश का अंतिम संस्कार हो पाया
- आठ अलग-अलग जगहों पर कोरोना संक्रमित मरीजों की लाश का हुआ अंतिम संस्कार

बल प्रयोग करने के बाद अंतिम संस्कार

जानकारी के मुताबिक कोरोना संक्रमित महिला की लाश को जलाने का विरोध करने तीन सौ से ज्यादा लोगों की भीड़ श्मशानघाट में एकत्रित हो गई थी। भीड़ को देखते हुए तथा समझाने की सारी कोशिश नाकाम होने के बाद मौके पर पहुंचे कोतवाली सीएसपी ने सख्ती बरतते हुए भीड़ को तितर-बितर करने मौके पर उपस्थित पुलिसकर्मियों को बल प्रयोग करने के निर्देश दिए। बताया जा रहा है, मौके पर मौजूद पुलिसकर्मियों ने लाठी चार्ज किया, तब जाकर हंगामा करने वाले यहां से भागने को विवश हुए।

वक्फ बोर्ड ने विरोध नहीं करने पत्र जारी किया

कोरोना संक्रमण से मृत लोगों के अंतिम संस्कार में विरोध के उठते स्वर को देखते हुए मुस्लिम वक्फ बोर्ड ने एक पत्र जारी किया है, जिसमें कोरोना संक्रमण से मृत किसी व्यक्ति की लाश दफनाने को लेकर कोई विरोध नहीं करेगा। विरोध करने पर प्रशासन अपने स्तर पर कार्रवाई कर सकता है।

हल्का बल प्रयोग

मारवाड़ी श्मशानघाट में कोरोना से मृत वृद्ध महिला का अंतिम संस्कार करने के दौरान लोगों द्वारा विरोध किया जा रहा था। इसके चलते लोगों को मौके से हटाने के लिए हल्का बल प्रयोग करना पड़ा।

- देवचरण पटेल, सीएसपी कोतवाली

कार्रवाई करेंगे

पुलिस की मदद से कोरोना संक्रमित मृत महिला के अंतिम संस्कार की प्रक्रिया शुरू हो पाई। विरोध करने वाले लोगों को चिन्हांकित किया गया है। विरोध करने वाले लोगों के खिलाफ थाने में एफआईआर दर्ज कराने को लेकर विचार किया जा रहा है।

- पुलक भट्टाचार्य, अपर आयुक्त

FIG. 9.2 Haribhoomi, Raipur Bhoomi, Saturday, August 15, 2020, 1 p.

2. Two similar incidents were reported from the same state. One involved protest against the burial of a COVID-19 dead body that took place on August 26, 2020, in the Korba district of Chhattisgarh (Fig. 9.3) and another that took place in Raipur on September 7, 2020 (Fig. 9.4).
3. In yet another bizarre and inhuman incident the corpse of a 70-year-old man suspected to have died of COVID-19 was shifted to the burial ground in a JCB vehicle in Andhra Pradesh [19].
4. In Ambala, a village in Haryana [20], villagers protested against the cremation of a patient suspected to have COVID-19 and threw stones at the doctors and police personnel. Police had to open fire in the air to disperse the mob. Only after the crowd dispersed the cremation of the dead body was done.
5. An appalling incident came to light from Bellary [21], Karnataka, wherein a video dated June 30, 2020, went viral where the health workers were seen dumping COVID-19 dead bodies in a large pit. This was clearly in violation of healthcare protocols related to the management of COVID-19 dead bodies. Later the State Health Minister Shri. B. Sriramulu suspended the medical staff while making an appeal to the health department to follow the standard operating procedures (SOP) while conducting funerals for COVID-19 patients.
6. The Nigambodh cremation ground in Delhi [22] refused to perform cremation of a 68-year-old lady who was the second casualty in India because of the novel coronavirus.
7. A shameful and inhuman act took place in Chennai [23] where a reputed neurologist and the chairman of a private hospital, who was suspected to have contracted the infection from a patient he was treating, was denied burial by the residents residing

संक्रमित महिला के अंतिम संस्कार से नाराज ग्रामीणों ने किया विरोध

कोरबा में संक्रमित महिला का अंतिम संस्कार करते। ● नईदुनिया

कोरबा (नईदुनिया प्रतिनिधि)। कोरोना पॉजिटिव दिवगंत महिला का अंतिम संस्कार बाल्को के वार्ड 37 दैहानपारा मोहल्ला और सेक्टर चार के पास खाली पड़ी जमीन पर किया गया। इससे क्षेत्रवासियों में नाराजगी व्याप्त हो गई। वार्ड पार्षदों ने प्रशासन के समक्ष आपत्ति जताते हुए अन्यत्र दाह संस्कार करने की मांग की। बुधवार को बस्तीवासी कलेक्टर को ज्ञापन सौंप कर विरोध जताएंगे।

कोविड-19 के नए नियम के तहत कोरोना संक्रमित मरीज की मौत होने पर उसका अंतिम संस्कार उसी क्षेत्र में किया जाएगा, जिस क्षेत्र का वह निवासी है। अभी तक रायपुर में इलाज के दौरान मौत होने पर वहीं अंतिम संस्कार कर दिया जाता था, पर रायपुर में आपत्ति आने के बाद नए नियम बनाने पड़े। कोरबा पावर हाउस रोड निवासी एक व्यापारी की मां की मौत रविवार की देर रात रायपुर में इलाज के दौरान हो गई थी।

निगम कर्मियों की उपस्थिति में परिजनों ने उक्त स्थल पर अंतिम संस्कार की प्रक्रिया पूरी की। इसकी जानकारी मिलने पर क्षेत्र के लोगों की भीड़ जमा हो गई। बाद में जानकारी मिलने पर नेता प्रतिपक्ष हितानंद अग्रवाल समेत अन्य क्षेत्रीय पार्षद भी स्थल पहुंच गए। निगम अमले से चर्चा के बाद पार्षदों ने वरिष्ठ अधिकारियों से चर्चा कर अंतिम संस्कार अन्यत्र कराए जाने की मांग की। उनका कहना था कि यहां अंतिम संस्कार किए जाने से संक्रमण फैलने का खतरा बढ़ जाएगा, इसलिए बस्ती क्षेत्र से दूर स्थल का चिन्हांकन किया जाना चाहिए। बस्तीवासियों का कहना है कि प्रशासन उनकी मांग पर सकारात्मक पहल नहीं करती है, तो आंदोलन किया जाएगा।

FIG. 9.3 Naidunia, Raipur, Wednesday, August 26, 2020, 7 p.

around the crematorium. When the corporation staff were preparing for the burial of the doctor at TP Chatram burial ground in Kilpauk, the local residents started to protest against the burial of the dead body chanting slogans that the body must not be buried there. After a long discussion between the corporation and the public, when the public did not agree to allow the body to be buried over there, they decided to take the body at Velangadu burial ground. Dr. K. Pradeep Kumar, a colleague and friend of the deceased doctor, said in a statement that 'when they started digging a grave at Velangadu burial ground, a group of people came with sticks and stones to attack them. A JCB machine was used since as per WHO guidelines, COVID-19 patients must be buried at a depth of 12 feet. The ambulance windshield was damaged and they started throwing stones on the body kept in the ambulance. Corporation officials, ambulance drivers and assistants were injured and bled' [23]. Dr. Pradeep, after telling the injured to go back to the hospital, drove the ambulance with broken panes back to the New Hope Hospital of the deceased doctor and roped in two ward boys for help. Later the deceased was buried with strong police protection. Dr. Pradeep, in a heartfelt message, said that 'I dropped sand with my hands hurriedly with tears rolling. This should not happen to anyone in future' [23].

The list does not end here. There are numerous cases where not only the public but also the frontline fighters like doctors have also faced problems for having a decent burial. This incident clearly shows that there is a lack of knowledge regarding the handling of the dead bodies of COVID-19 and about the right of the COVID-19 deceased for a decent burial/cremation. Also, people must understand the actual cause behind the proper handling of the COVID-19 dead bodies. It is not the body that can cause infection to others, but infection can occur only when any blood or body fluids oozing out from the body come into contact with a healthy person. Thus one should avoid coming in direct contact with such fluids. Hence the body is packed in body bags, and unzipping the bag is not allowed, unless to show the face of the deceased to relatives for the last

कोरोना ● होम आइसोलेशन के लिए नए नियम लागू

मोती नगर में संक्रमित के अंतिम संस्कार का विरोध, हंगामा

रायपुर (नईदुनिया प्रतिनिधि)। राजधानी के मोती नगर में रविवार को कोरोना मृतक के अंतिम संस्कार के दौरान लोगों ने हंगामा मचाया। पुलिस और प्रशासन के समझाने के बाद लोग हटे। शहर में आए दिन इस तरह की घटनाएं सामने आ रही हैं, जबकि चिकित्सकों का कहना है कि अंतिम संस्कार पूरी गाइडलाइन से हो रहा है। इसमें खतरा नहीं है। रविवार को शहर पूरी तरह से बंद रहा। दुकानें नहीं खुलीं और बिना मास्क के घर से बाहर निकलने वालों पर प्रशासन ने कार्रवाई भी की।

खबर को विस्तार से पढ़ने के लिए क्यूआर कोड स्कैन करें।

वहीं सोमवार से रायपुर में होम आइसोलेशन के लिए नए नियम लागू होंगे। प्रशासन का कहना है कि कोविड-19 के संक्रमण से पीड़ित बिना लक्षणों और हल्के लक्षणों वाले मरीजों के प्रबंधन के लिए होम आइसोलेशन की अनुमति प्रदान की जाएगी। होम आइसोलेशन के प्रबंधन तथा होम आइसोलेशन के इच्छुक मरीजों की निगरानी एवं समन्वय के लिए सातों दिन, चौबीसों घंटे संचालित होने वाले कॉल सेंटर और कंट्रोल रूम की स्थापना की जा रही है। कोरोना संक्रमित मरीज के रहने के लिए घर में हवादार कमरा और अलग शौचालय होना अनिवार्य है। यदि घर में ऐसी व्यवस्था नहीं है, तो मरीज के इलाज का प्रबंध कोविड केयर सेंटर में किया जाएगा। सामुदायिक शौचालय का उपयोग करने वाले परिवारों को होम आइसोलेशन की अनुमति नहीं दी जाएगी।

मोती नगर में आसपास के लोग विरोध करते हुए। ● नईदुनिया

घर में लगेगा लाल रंग का स्टीकर

होम आइसोलेटेड मरीज के घर के बाहर निर्धारित प्रारूप में लाल रंग का स्टीकर चस्पा किया जाएगा। होम आइसोलेशन की अनुमति प्रदान दिए गए मरीजों से निर्धारित प्रपत्र भरवाया जाएगा। होम आइसोलेशन के लिए उपयुक्त पाए गए मरीजों को उपचार के लिए दवाईयों की एक किट दी जाएगी। होम आइसोलेशन की सम्पूर्ण अवधि के दौरान जिला स्वास्थ्य विभाग द्वारा नियुक्त स्वास्थ्यकर्मी प्रतिदिन चिकित्सकों या उनके अटेंडेंट से फोन के माध्यम से संपर्क में रहेंगे। वे मरीज को सांस लेने में कठिनाई, सीने में लगातार दर्द या दबाव, होंठ या चेहरे के नीले पड़ने, ऑल्टर्ड सेंसोरियम (डिस-ओरिएंटेशन) जैसे गंभीर लक्षण विकसित होने पर इसकी सूचना तत्काल कंट्रोल रूम में दूरभाष के माध्यम से देंगे।

रैपिड एक्शन टीम करेगी काम

विभाग ने ऐसे सभी गंभीर मरीजों के प्रबंधन एवं समन्वय के लिए रैपिड एक्शन टीम के गठन का निर्देश कलेक्टरों को दिया है। गंभीर मरीज की सूचना प्राप्त होने पर मरीज को कोविड केयर सेंटर या विशेषीकृत कोविड अस्पताल आवंटित किया जाएगा। वहां, मरीज अपने अटेंडेंट के साथ स्वयं के वाहन से जा सकेगा। होम आइसोलेशन के निर्देशों का मरीज तथा उनके परिजनों द्वारा अनुपालन सुनिश्चित करने के लिए स्थानीय स्तर पर निगरानी दलों की व्यवस्था सुनिश्चित करने के भी निर्देश दिए गए है। मरीजों द्वारा आइसोलेशन प्रोटोकॉल के किसी भी निर्देश की अवहेलना करने पर उन्हे तत्काल कोविड केयर सेंटर शिफ्ट कर दिया जाएगा। साथ ही अंडरटेकिंग की अवहेलना करने एवं महामारी अधिनियम के उल्लंघन के लिए उन पर कार्यवाही करने कहा गया है।

FIG. 9.4 Naidunia, Raipur, Monday, September 7, 2020, 2 p.

time. Till now, there is no scientific report to prove that burial/cremation can spread the infection to another healthy person.

6 SUGGESTIONS

1. Whatever is written in the guidelines for the management of the COVID-19 dead bodies must be strictly followed.
2. Media should take initiative to spread awareness to the public regarding their rights and duties toward the burial/cremation of COVID-19 dead bodies.
3. Proper training must be given to the frontline workers such as doctors, nurses, other medical staffs, staffs of burial/cremation ground, police personnel, and municipal staff regarding the COVID-19 dead body management.
4. One has to understand the actual reason behind the restrictions imposed on the gatherings during paying funeral rites and on performing religious ceremonies like bathing of dead bodies. The restrictions are actually for the prevention of direct contact of any fluid oozing out from the body due to post mortem/infection.

5. General awareness relating to the misconception of burial/cremation of COVID-19 dead bodies is necessary among the public in order to protect the right of a deceased person so that the deceased person can have a decent burial/cremation. Only through this awareness can the general public understand that infection can spread only when the body is mishandled by the cremators.
6. Transparent body bags can be used to pack COVID-19 dead bodies for preventing them from mixing up with others. Also then there will be no need to unzip the bag for showing the face to the relatives for the last time.
7. Everyone must follow the precautionary measures for the prevention of COVID-19 as per the WHO and Indian government guidelines.

7 CONCLUSION

'Everyone who has died from the corona virus is a person who had a life, family, and friends. The victims of this crisis deserve dignity in their death' [24].

FRANCES EVE, DEPUTY DIRECTOR OF RESEARCH AT CHINESE HUMAN RIGHTS DEFENDERS.

The decent burial/cremation right is not just only a right for the dead person but it is the last rites performed by the loved ones for the dead person and failing to perform this right can actually cause long-term psychologic trauma to the family of the dead person. Actually by denying one the right to decent burial, we are hurting the sentiments of the loved ones left behind by the dead person. Itis a well-established law all over the world for granting a decent burial/cremation right to dead persons in accordance to their religious beliefs in any circumstance. Burial/cremation of a dead person is an emotional time for the family members; even the government understands this psychologic aspect and that is why even in this pandemic situation the government has allowed performing last rituals. The only exception is that one should not perform any rituals where touching the dead body is required. Without touching the dead body, one can perform the last rituals of the funeral.

There is a knowledge gap among the public regarding (1) the COVID-19 dead body management and (2) why it is important to follow those guidelines. Making guidelines for funeral rights and applying restrictions on hand-over of COVID-19 dead body to the relatives is done as a preventive measure from spreading the virus to a healthy person. One must have proper knowledge about the guidelines for the safety of the public at large. The proverb *A Little Knowledge is a Dangerous Thing* actually fits in the current situation where people do not know how to deal with the pandemic situation. They just follow the rumors without applying their mind.

Prevention is Better Than Cure

ACKNOWLEDGMENT

The author is thankful to Dr. S. Khan, M.D. (Medicine), S K Hospital, Gangapur, Dist- Sawai Madhopur, Rajasthan, for giving his valuable time and for sharing his esteemed knowledge and insight on this important subject.

REFERENCES

[1] Management of dead bodies in disaster situations. Pan American Health Organization; 2004 [Internet]. [cited 2020 Sep 9] Available from: https://www.who.int/hac/techguidance/management_of_dead_bodies.pdf.
[2] Universal Declaration of Human Rights. United Nations [Internet]. [cited 2020 Sep 10] Available from: https://www.un.org/en/universal-declaration-human-rights/.
[3] S.Sethu Raja vs The Chief Secretary on 28 August, 2007. Indian Kanoon [Internet]. [cited 2020 Sep 10] Available from: https://indiankanoon.org/doc/273832/.
[4] [Internet] Madras HC: Suo Moto cognizance of opposition by mob to the burial of body of doctor, died due to Covid-19 - issues notice, observes Art. 21 includes right to decent burial. Advocate Retainer; April 20, 2020 [cited 2020 Sep 11] Available from: https://www.advocateretainer.com/blogs/news/ty.
[5] Common cause (A Regd. Society) vs Union of India on 9 March, 2018. Indian Kanoon [Internet]. [cited 2020 Sep 11] Available from: https://indiankanoon.org/doc/184449972/.
[6] Mahajan VD. Jurisprudence and legal theory. 5th ed. New Delhi: Eastern Book Company; 2005. p. 380−1.
[7] Gaur KD. Textbook on Indian penal Code. 6th ed. Haryana: LexisNexis; 2016. p. 535p.
[8] Unclaimed dead bodies be given decent disposal: Supreme Court [Internet] The Economic Times November 25, 2013 [cited 2020 Sep 18] Available from: https://economictimes.indiatimes.com/news/politics-and-nation/unclaimed-dead-bodies-be-given-decent-disposal-supreme-court/articleshow/26370200.cms.
[9] [Internet] Fundamental right of decent burial, cremation can't be denied: SSC. Daily Excelsior; July 16, 2020 [cited 2020 Sep 18] Available from: https://www.dailyexcelsior.com/fundamental-right-of-decent-burial-cremation-cant-be-denied-ssc/.
[10] Bindra J. Letter to CJI: take suo motu cognizance of manhandling of covid-19 dead bodies [Internet] Mint June 20, 2020 [cited 2020 Sep 18] Available from: https://www.livemint.com/news/india/letter-to-cji-take-suo-motu-cognizance-of-manhandling-of-covid-19-dead-bodies-11591790376216.html.

[11] Thakur A. "Covid-19 patients are treated worse than animals", says Supreme Court [Internet] HW June 12, 2020 [cited 2020 Sep 13] Available from: https://hwnews.in/news/national-news/covid-19-patients-are-treated-worse-than-animals-says-supreme-court/139093.

[12] Singh B. Meghalaya HC orders action against those hindering burial, cremation [Internet] The Economic Times April 18, 2020 [cited 2020 Sep 18] Available from: https://economictimes.indiatimes.com/news/politics-and-nation/meghalaya-hc-orders-action-against-those-hindering-burial-cremation/articleshow/75209882.cms?from=mdr.

[13] Kadam S. Do the dead have rights in India? [Internet] Sabrang April 21, 2020 [cited 2020 Sep 18] Available from: https://sabrangindia.in/article/do-dead-have-rights-india.

[14] COVID-19: guidelines on dead body management. Government of India Ministry of Health & Family Welfare Directorate General of Health Services (EMR Division) [Internet]. [cited 2020 Sep 19] Available from: https://www.mohfw.gov.in/pdf/1584423700568_COVID19GuidelinesonDeadbodymanagement.pdf.

[15] Loiwal M. West Bengal: son forced to pay Rs 51,000 to see body of father who died from coronavirus [Internet] India Today August 10, 2020 [cited 2020 Sep 18] Available from: https://www.indiatoday.in/india/story/west-bengal-son-forced-pay-rs-51000-see-father-s-body-died-coronavirus-1709674-2020-08-10.

[16] Thaver M. Thane hospital hands over wrong body of Covid patient, family conducts last rites twice in a week [Internet] The Indian Express August 10, 2020 [cited 2020 Sep 18] Available from: https://indianexpress.com/article/cities/mumbai/thane-hospital-hands-over-wrong-body-of-COVID-patient-family-conducts-last-rites-twice-in-a-week-6496925/.

[17] Infection Prevention and Control for the safe management of a dead body in the context of COVID-19 [Internet] World Health Organization March 20, 2020 [cited 2020 Sep 18] Available from: https://apps.who.int/iris/bitstream/handle/10665/331538/WHO-COVID-19-lPC_DBMgmt-2020.1-eng.pdf?sequence=1&isAllowed=y.

[18] Chopra V. Can coronavirus spread by dead bodies of people that have died due to coronavirus? [Internet] Inventiva May 04, 2020 [cited 2020 Sep 18] Available from: https://www.inventiva.co.in/trends/vanshika/can-corona-virus-spread-by-dead-bodies-of-people-that-have-died-due-to-coronavirus/.

[19] Body of 70-year-old shifted in JCB on suspicion of him being COVID-19 victim in Andhra [Internet] The Indian Express June 27, 2020 [cited 2020 Sep 20] Available from: https://www.newindianexpress.com/states/andhra-pradesh/2020/jun/27/body-of-70-year-old-shifted-in-jcb-on-suspicion-of-him-being-covid-19-victim-in-andhra-2162033.html.

[20] Dey S. Coronavirus: residents protest cremation of COVID-19 suspect, throw stones at Haryana cops, doctors [Internet] NDTV Coronavirus April 28, 2020 [cited 2020 Sep 20] Available from: https://www.ndtv.com/india-news/coronavirus-ambala-residents-protest-cremation-of-covid-19-suspect-throw-stones-at-haryana-cops-doctors-2219350.

[21] Agarwal P. No dignity in death: dead bodies of COVID-19 patients thrown into large pits in Karnataka, staff suspended [Internet] The Logical Indian July 01, 2020 [cited 2020 Sep 20] Available from: https://thelogicalindian.com/news/video-on-dumping-of-covid19-victims-bodies-in-karnataka-22016.

[22] Gothi C. Delhi crematorium refused to perform last rites of woman who died of coronavirus, says family [Internet] India Today March 14, 2020 [cited 2020 Sep 20] Available from: https://www.indiatoday.in/india/story/delhi-crematorium-refuses-to-cremate-woman-who-died-of-coronavirus-1655434-2020-03-14.

[23] COVID-19: Chennai neurologist denied dignified burial as mob vandalises ambulance, injures staff [Internet] The Indian Express April 20, 2020 [cited 2020 Sep 20] Available from: https://www.newindianexpress.com/cities/chennai/2020/apr/20/covid-19-chennai-neurologist-denied-dignified-burial-as-mob-vandalises-ambulance-injures-staff-2132770.html.

[24] Standaert M. China virus funeral order fuels upset as death toll rises [Internet] Aljazeera February 09, 2020 [cited 2020 Sep 22] Available from: https://www.aljazeera.com/news/2020/02/china-virus-funeral-order-fuels-upset-death-toll-exceeds-sars-200209010139644.html.

CHAPTER 10

Effects of Information Communication Technology on Da'wah Activities Amidst COVID-19 Pandemic

FOLAMI AHMADU BOLANLE • BALOGUN MUHSIN ADEKUNLE • MUSOLIHU MAJEED OLAYORI

1 INTRODUCTION

The recent use of Information Communication Technology (ICT) in da'wah activities took a new dimension during Coronavirus disease 2019 (COVID-19) and it has been generating a lot of debate. Most of our du'at have been versatile with the new trend and it has helped them so much in conveying and impacting knowledge to their listeners on different internet handles, while the traditional scholars who are non-ICT compliance are finding it difficult to advance their da'wah activities. To this end, and since the global outburst of the COVID-19 pandemic on December 2019, ICT has continued to be the focus of most ICT compliance du'at in propagating da'wah activities. This is due to the role it could play in the areas of Facebook, E-mails, LMS, Zoom, Google Meet, etc., which serve as an avenue for listeners to log via mentioned ICT handles in the comfort of their homes to listen and watch religious lectures and teachings live.

The introduction of this new avenue in propagating da'wah has always been a welcome development but not without some hindrances and implications. Da'wah as an effective tool for teaching and conveying Islamic beliefs and practices since ancient days has survived the storm of extinction up till date; many intellectuals in the world have come to be conversant with its impact and regarded it as one of the internationally recognized means employed by Islamic scholars in propagating Islam due to its viability as an authentic avenue for teaching people the basic tenets of Islam and its richness in quality, which made Muslims and non-Muslims to acquaint themselves with the study of Islamic practices and culture.

Da'wah activities in Nigeria have undergone many developments in recent times, not minding the misrepresentation of its course being orchestrated by the westerners who believed it is an avenue to make the world and its inhabitants practicing one religion (Islam). Many du'at (da'wah personnel) and experts in this field have been improving in the manner they invite people to Islam with substantial methods and suggestions in salvaging the phenomenal from being downtrodden. This new development has made da'wah activities to trend worldwide especially during COVID-19 Pandemic via information communication technology. The method adopted by traditional du'at in Nigeria to exhort human (listeners') minds into good act with the intention of their accepting the truth and abiding it by their actions as the best means of propagating Islamic religious practices, but most of them are yet to reckon to the new method due to the lack of ICT knowledge [2].

However, with the introduction of ICT, da'wah activities have been well structured in terms of delivery either on the internet, with or without the du'at, thereby gradually exposing the listeners to learning Islamic beliefs and practices via this ICT handles. Through this means, the listeners will have the opportunity of interacting and asking the scholars questions on the religion, and this will enable them to have intrinsic knowledge about Islam and gather more wealth of experience from other listeners worldwide. Internet then becomes a good teacher which provides opportunities for Arabic teachers and learners than we had before and serves as a good companion for modern users of Arabic [6].

Our flair for this study was borne out of significant decrease of da'wah activities due to the presence of COVID-19 Pandemic in different parts of Nigeria. This unwarranted disease has denied most du'at who are not ICT compliance the engagement of da'wah activities

COVID-19: Tackling Global Pandemics through Scientific and Social Tools. https://doi.org/10.1016/B978-0-323-85844-1.00008-8

owing to the guiding rules stipulated by the NCDC which warns against gathering more than 20 people, with this restriction, most non-ICT compliance du'at could not communicate to their listeners face to face in point of the fact that most mud'u comes from near and far to come and listen to their message. Specifically, this study is needed to acquaint du'at to the new trend in da'wah activities in the country and how to change the traditional narrative at any point in time when such disease come into limelight. The study will also abreast them to the fact that da'wah activities can be largely disseminated to the world within a short space of time through the help of ICT.

In this regard, the researcher intends to find out both negative and positive impacts of ICT on da'wah activities during COVID-19 pandemic. The study will investigate on the impact of ICT on da'wah activities during COVID-19 pandemic. It is specifically aimed at finding the gap of "Traditional Du'at" who are not acquainted with the use of ICT, which denied them the access of conveying religious teachings to the populace during COVID-19 pandemic and to also intimate those who are ICT compliance on how to use them during this period.

2 RESEARCH OBJECTIVES

2.1 To find out the extent of the usage of ICT in da'wah activities during COVID-19 pandemic

2.2 To study the impacts of ICT on da'wah activities during COVID-19 pandemic.

2.3 To be able to engage all stakeholders on the usage of ICT for da'wah activities during COVID-19 pandemic.

3 RESEARCH QUESTIONS

3.1 What is the extent of the usage of ICT for da'wah activities during COVID-19 pandemic?

3.2 What are the impacts of ICT on da'wah activities during COVID-19 pandemic?

3.3 How can we engage all stakeholders on the usage of ICT for da'wah activities during COVID-19 Pandemic?

4 CONCEPTUAL ANALYSIS

4.1 Da'wah

The term *Da'wah* is derived from the verb *Da'aa* (to call/ to invite), *yad'u* in the past continuous tense and the root is *Du'a* or *Da'wah* [8]. It is an attempt or endeavor in the form of speech or activity for inclining or bending people to believe and have faith in a certain idea, thought, or opinion which one is calling them to (Poston 5). *Da'wah* technically is motivating and directing people to performance of good deeds in the right path [3]. According to Ammarah, it requires one to inform another way of God and, accordingly, *Da'wah* comprises teaching and educating and all forms of communication as well [4]. Al-Ilori defined it as drawing people's attention and rationality toward an ideology or belief which is of benefit and advantage to them. It is also rescuing people from going astray or from tribulations which beset them (Al-Ilori n.d). Adetona described *Da'wah* through different terminologies, and these include the following:

Al—wa'z means the exhortation of human minds into good act with the intention of accepting the truth and abiding by it in their actions.

Al-Irshaad—The guidance of men toward the right path and sticking onto goodness.

At-Tadhkir—The making known to men Allah's favors and warning them against worshipping other but Allah.

Al-Bashaarah—The giving of glad tidings about a blissful entrance [1].

All of the above definitions simply stage *da'wah* as inviting, calling, informing, admonishing, educating as well as rescuing people (Adetona).

4.2 Information Communication Technology

Information and Communication Technology, as a disseminating source, has been defined by many scholars in different connotations. According to Wikipedia, the term Information and Communication Technology covers all advanced technologies in manipulating and communicating information. Olagunju opined that electronic systems can be used for broadcasting telecommunication and all forms of computer-based communication. Information and Communication Technology—centered education, therefore, covers the use of computers, online self-learning packages, Satellites, Radio, Optical fiber technology, i.e., Tele presence system, and all types of information technology, hardware, and software [5]. The concept of information and communication technology also covers the application of modern technologies which allow various forms of information to be processed, transmitted, manipulated, and retrieved with speed and accuracy. These include Data Communication on Digital, Telecommunication by using wired or wireless telephone, and Computer hardware and Software which allow computer to function according to predetermined instructions.

4.3 COVID-19 Pandemic

COVID-19 is defined as an illness caused by a novel coronavirus now called severe acute respiratory syndrome coronavirus 2 (formerly called 2019-nCoV), which was first identified amid an outbreak of respiratory illness cases in Wuhan City, Hubei Province, China [1]. It was initially reported to the World Health Organization (WHO) on December 31, 2019. On January 30, 2020, the WHO declared the COVID-19 outbreak a global health emergency [2,3]. On March 11, 2020, the WHO declared COVID-19 a global pandemic, its first such designation since declaring H1N1 influenza a pandemic in 2009 [4].

5 RESEARCH DESIGN

The researcher proposes to use the descriptive survey design to gather data relating to a study on impacts of ICT on da'wah activities during COVID-19 Pandemic. Rice (2001) says that any researcher who adopts the descriptive research design attempts to produce data that are holistic, contextual, descriptive in depth, and rich in detail. Descriptive survey designs are used in preliminary and exploratory studies to allow researchers to gather information, summarize, present, and interpret it for the purpose of clarification [7]. The descriptive survey research design was suited to the study because it allowed the researcher to study the state of the usage of ICT by religious scholars to navigate da'wah activities during COVID-19 pandemic in Nigeria.

5.1 Study Population

The study was to target du'at (religious scholars) in Lagos State out of other states in Nigeria, because most of them are Lagos based. Below is the tabulated population of religious affiliations selected for the study. The researcher will draw study sample from this pool of religious affiliations available which is estimated at 330 population.

5.1.1 Religious affiliations of participants

Religious Affiliations Selected	Frequency	Percentage (%)
NASFAT	100	26.6
LEKKI MUSLIM UMMAH	70	12.4
ANSARUDEEN	100	24.7
MICA	40	22.7
TMC	20	13.6
Total	330	100

5.2 Data Analysis and Presentation

The results of the study were presented and discussed in relation to the research questions outlined for this study above. This study aimed at investigating the impacts of Information Communication Technology on da'wah activities during COVID-19 pandemic in Lagos State. Data were analyzed using triangulate method of enquiry. Quantitative data were analyzed by frequency tables and percentages. Qualitative data were analyzed qualitatively using content analysis based on analysis of meanings and implications emanating from respondent information and comparing responses to documented data on the impacts of Information Communication Technology on da'wah activities during COVID-19 pandemic in Lagos State and presented thematically in line with the objectives of the study (Table 10.1).

TABLE 10.1
Feedback on ICT Involvement in Da'wah Activities.

S/N	Questions	Yes	No	Total
	Have you ever read religious lessons via ICT during COVID-19 pandemic?	283 46.7	57 10.3	**330** **100%**
	Have you had addiction or tending to reading religious lectures via ICT during COVID-19 pandemic?	197 56.4	123 14.2	**330** **100%**
	Have you ever noticed if ICT influenced your religious learning during COVID-19 pandemic?	217 5	113 7.7	**330** **100%**
	Did you post religious questions to scholars via ICT after reading their teachings during COVID-19 pandemic?	168 54.5	162 23.3	**330** **100%**
	Was it interesting using ICT to read Islamic teachings during COVID-19 pandemic?	209 16.8	121 42.8	**330** **100%**

Source: Field Work, 2020.

5.3 Response Rate

It is calculated by the number of questionnaires collected or the number of people with whom interviews are completed divided by the number of the entire sample. In this study, data were collected from five sampled religious organizations. A total of 50 questionnaires were distributed and 45 were returned. This posits that a response rate of above 50% is adequate for analysis, and, therefore, 91.25% response rate was considered as being very good for analysis (Table 10.2).

5.3.1 The extent of the usage of ICT for Da'wah activities during COVID-19 pandemic

Here, analysis from the questionnaire is tabulated below on participants' involvement, age, and level of education in categories and finally on the way they used ICT in getting religious messages from religious scholars to see the magnitude and reality of how ICT has been helpful to them in advancing their da'wah activities during COVID-19 pandemic and how it is has improved their religious lives positively (Table 10.3).

5.3.1.1 Summary of the study. Study objectives were formulated and research questions were derived from the objectives. The instruments used for the study were questionnaires which were sent to respondents via Google link form, and afterward, the responses were used in the collation of data. Data collected were coded and inputted using statistical (SPSS). The results gotten from the responses were carefully analyzed and presented to showcase the actual situation.

5.3.1.2 Findings of study. It was gathered from the study most da'wah activities were done through ICT during COVID-19 pandemic whereby % said that they communicated most of their lectures to the listeners,

TABLE 10.2
Negative and Positive Impacts of ICT on Da'wah Activities During COVID-19 Pandemic.

S/N	Questions	Yes	No	Total
	I was able to advance my da'wah activities via ICT during COVID-19 pandemic.	233 46.7	97 10.3	**330** **100%**
	Da'wah activities via ICT during COVID-19 pandemic reduce stress of movement and cost of spending.	274	56 14.2	**330** **100%**
	Noncompliance of ICT reduces my chances of da'wah activities during COVID-19 pandemic.	182 52.5	148 7.7	**330** **100%**
	Delivering religious lectures via ICT during COVID-19 pandemic improves my knowledge of communication technology.	229 54.5	101 23.3	**330** **100%**
	Da'wah activities via ICT during COVID-19 were the best.	264 16.8	66 42.8	**330** **100%**

Source: Field work, 2020.

TABLE 10.3
Engagement of all Stakeholders on the Usage of ICT for Da'wah Activities During COVID-19 Pandemic.

S/N	Questions	Yes	No	Total
	Most du'at need to be trained on how to use ICT for da'wah activities during COVID-19 and beyond.	283 46.7	57 10.3	**330** **100%**
	Du'at stakeholders' forum should be created to do checks and balances on principles to be used in da'wah activities through the ICT.	275 56.4	55 14.2	**330** **100%**
	Du'at need to be trained on better usage of ICT to advance da'wah activities during COVID-19 pandemic.	303 52.5	29 7.7	**330** **100%**
	Establishment of unity forum among du'at in order to desist from using the ICT negatively.	313 54.5	17 23.3	**330** **100%**
	Da'wah activities require acquisition of profound knowledge of religion, da'wah methodology, and ethics of ICT.	291 16.8	39 42.8	**330** **100%**

Source: Field Work 2020.

% said they have been unable to reach out to their listeners during COVID-19 pandemic owing to the fact that they are non-ICT compliance.

Fig. 5.3.1.2 shows that % of respondents believed that making use of the ICT to advance da'wah activities has improved their competencies in the knowledge of internet and it has helped them a lot in impacting most of their da'wah officers. It was also gathered from the study that % of the respondents agreed that using ICT for da'wah activities during COVID-19 pandemic has reduces for them the stress of moving from one lecture venue to another which sometimes cost them a lot in terms monetary and health wise.

Fig. 5.3.13 shows that % of the respondents believed that the possible way of engaging stakeholders to train du'at on how to use the ICT to advance da'wah activities. % suggests that du'at should also be enlightened on the methodology of da'wah and positive usage of the ICT.

5.4 Recommendations from the Study

1. The du'at need to be trained on better usage of the ICT for da'wah activities during and after COVID-19 pandemic so as to minimize negative attitudes displayed by them via these handles and reduce online risks associated to communication technology.
2. Creation of du'at stakeholders' monitoring forum discussion on how to checkmate da'wah activities via ICT.
3. Development of ICT programs for positive use to advance da'wah activities during COVID-19 pandemic and in the future.
4. Establishment of unity among du'at in order to curb them from introducing provocation into their da'wah activities.
5. Du'at must also be enlightened on how to exercise high level of moral standard in carrying out da'wah activities via ICT during COVID-19 pandemic and in the future.

5.5 Conclusion

In view of the above summary, it was evident that ICT played a pivotal role in da'wah activities of Islamic scholars during COVID-19 pandemic. The du'at mostly used ICT for propagating the religion and communicating religious messages to their followers in order for them to learn more about the religion they practiced and to deviate them from being idle during the pandemic.

It was gathered from respondents that most of these activities occurred during the Ramadan period whereby various religious activities are being exercised by all and sundry, but for the fact that this deadly disease surfaced before this holy period and within which denied du'at from visiting religious centers due to the guidelines given by the NCDC on social distancing and minimum number of people to converge at a venue for any function.

The fact that ICT is part of avenues where information can be passed across to the populace, most du'at felt it is also the fastest means to throw shade at one another forgetting the power of ICT and how information can be spread within a short of second.

REFERENCES

[1] Adetona LM. The impact of modern technology on da'wah in the third world nations. J. Arabic Rel. Stud. 1997;16(3): 13–5. University of Ilorin.

[2] Folami A. Dawah activities among notable scholars in Ojo. Int J 2019.

[3] Mahfuz A. Hidayah al-Murshidna ila tariqal-wa'az wal Khitab. Al-Qahirah Dar al-Islam. 1979.

[4] Muhammad R. Lilam Ar-Rabawiy wa duarihi fi Khidmati al-Dawatil Islamiyyah Mutaaghayiratil Asr. Ghaza: Kulliyati Usul al-Din Islamiyyah. 2005.

[5] Olagunju AM. An investigation into teachers' awareness and essence of utilization of information and communication technologies for effective science education. In: Olagunju AM, editor. Journal of computer literacy, vol. 5. Ibadan: Franco Printers; 2003.

[6] Oloyede IO. Islamic studies in the world of information and communication technology: opportunities and challenges. In: Imam YO, editor. Public administration and and information and communication technology (ICT) through the lenses of Arabic islamic studies. Ijebu Ode, Nigeria. Sebootimo Pub; 2009.

[7] Orodho JA. Techniques of writing research proposals and reports. Nairobi: Masolo Printers; 2002.

[8] Poston L. Islamic da'wah in the west: muslim missionary activity and the dynamics of islam. New York: Oxford University Press; 1992.

FURTHER READING

[1] Al-Ilory, A. Tarikh al-Da'wati ila Allah bayn al-Ams wa al-Yawm. Maktabah Wahibah n.d.

[2] CDC. How corona virus spreads. 2019. https://www.cdc.gov/coronavirus/2019-ncov/preventgettingsick/howcovidspreads.html?CDC_AA_refVal=https%3A%2F%2Fwww.cdc.gov%2Fcoronavirus%2F2019-ncov%2Fabout%2Findex.html.

[3] Lawal IA. Using ICT for teaching Arabic as a foreign language in Nigeria: issues and challenges. 2017. The IAFOR international conference on language learning, official conference proceedings, Dubai.

[4] Ramzy A, McNell DG. W.H.O declares global emergency for novel coronavirus. 2019. https://www.medscape.com/viewarticle/924596.

[5] Rice A. Students push their facebook use further into course work. The Chronicle of Higher Education; 2011. Retrieved from, http://chronicle.com/blogs/wiredcampus/students-push-theirfacebook-use-further-into-academics/33947.

CHAPTER 11

The Impact of Compensation Practices on Employees' Engagement and Motivation in Times of COVID-19

HADEYA JEHA • MOHAMAD KNIO • GEORGES BELLOS

1 INTRODUCTION

A reward system is considered one of the crucial issues in managing organizations. It is a tool to achieve organizations' objectives. We really believe in the reward system, incentives, and compensation when they are managed fairly and equitably. This is why strategic human resources management (HRM) practices appear in the literature and are very important for employee empowerment and motivation [1]. Human resources (HR) departments are organizations' basic resources to achieve their failure or success. Compensation packages in all organizations are made up of salaries, wages, bonuses, and commissions. Employers always take into considerations the role of rewards and their benefits because their effect will play a fundamental role in sweetening the contract between both parties with the priorities that most employees need [1].

HR departments in all organizations try their best to put themselves in a good financial position because they believe that they deserve a worthy salary; consequently they always look forward for a position that pays according to their efforts. Right compensation for the right employees shows that employees are really appreciated as an integral part of the company and as human beings. When employees feel they are appreciated, they feel happy coming to the job, the morale of the job in role will increase, and they will be engaged and motivated in their work to give the best results [1].

Employees' engagement and motivation toward the job they are performing are being affected by the compensation package they receive; rewards, bonuses, and commissions are all the focal point of a company's success. Compensation managers in all organizations must take into consideration a couple of factors to reward their employees in order to ensure that all things are on the right track. It has been argued that compensation packages, in addition to the reward systems in all organizations, reflect the degree of an employee's engagement, motivation, job satisfaction, and willingness to stay in the organization [1].

All organizations seek to achieve their goals by creating happy and satisfied personnel among their workforce. So for every single organization to achieve its strategic goals, it will highly depend on its capacity to grab, keep, and maintain competent pleased staff into its employment [2].

1.1 Research Question

RQ1: To what extent compensation (extrinsic and intrinsic) is related to employee engagement and motivation? This question addresses the correlation between an organization's reward system and an employee's motivation and engagement. This question was previously answered and it was noticed that in times of pandemics, the intrinsic (i.e., noncash) rewards were preferred, so this issue needs to be investigated further here as to whether or not it is the case now, with coronavirus disease 2019 (COVID-19) or not.

As such, and to properly address and answer this question, the research will focus on both extrinsic and intrinsic motivational factors such as salaries, wages, recognitions, job satisfaction, and appreciation. Then the researchers will dig deeper to interpret the relationship between compensation and employee engagement. For this purpose, data will be collected from three nongovernmental organizations (NGOs) through the selection of a sound research methodology that will structure a good research design to guide in collecting and interpreting the data.

COVID-19: Tackling Global Pandemics through Scientific and Social Tools. https://doi.org/10.1016/B978-0-323-85844-1.00004-0

1.2 Research Aim and Objectives

The research objectives (ROs) are as follows:

RO1: To determine the degree at which packaged compensation affects employees' engagement and motivation.

RO2: To evaluate the relationship between the reward system and employee engagement.

RO3: To explore the relationship between compensation and high performance.

RO4: To identify the compensation packages available at NGOs.

RO5: To identify the ways to increase employee engagement and motivation.

RO6: To identify the strategies used by NGOs to increase the engagement and motivation of their employees.

1.3 Research Hypotheses

As we present our research methodology, we will refer to monoqualitative research methodologies and will not formulate any hypotheses. But we will discuss the findings based on the research questions and answer them with "yes" or "no".

2 LITERATURE REVIEW

Researches on the effect of compatible compensation packages on an employee's motivation and engagement attracted academics over the past three decades. Compensation packages represent one of the most complex issues that face HR managers when it comes to motivating and engaging employees. Many studies demonstrated the effects of reward system on the motivation of employees because the level of motivation is linked to the outcomes. Hence several compensation plans and strategies were introduced over the years to make employees more productive. In this literature review the researchers extracted data on all the dimensions of motivation and what causes motivation and engagement at workplace, in addition to the theories attached to the assigned topic.

2.1 Compensation

Compensation is a word that represents the overall exchange between the organization and its HR—the former party gives something to the latter in return for something else. A couple of years ago, reward systems were confidential. However, in today's competitive world, compensation has become more transparent and employees have started to take their rights depending on a compensation package system. Thus creating this system becomes a critical issue for the organizations [1].

Compensation packages need compensation management because it helps the organizations to better achieve their strategic goals by taking into consideration the effective implementation of a reward system in order to keep all employees engaged beyond expectation. This issue is important for both parties, that is, the employee and the employer [1]. Yet the implementation of compensation packages will not only decrease employee turnover rates but also stabilize or even decrease them.

However, compensation can be translated as a form of financial and tangible benefits that staff can receive as a part of employment relationship [1,2]. Various cases on several sectors, including NGOs, were explored. NGOs comprise groups that are independent of the local government and have a cooperative or a humanitarian goal and provide their employees with compensations [1,3]. Compensation packages tend to create an atmosphere of job satisfaction for the employees; the package includes daily allowance, bonuses, salaries, and incentives, as well as recognitions and promotions. This package plays an important role in an employee's performance, and consequently, it is mandatory for the organization to clarify its compensation system based on job measurements to avoid a misunderstanding [1].

An employee's engagement and motivation as independent variables can be directly affected by other factors such as daily allowance, salary, bonuses, and other benefits. This research will be considered as a test to measure the role of compensation packages on an employee's engagement and motivation in NGOs in Lebanon [1].

Compensation is defined as a reward that is given to employees for a service they did [4]. It is linked to the direct and indirect rewards. HR professionals define compensation as salaries, wages, commissions, bonuses, insurance, and other types of nonfinancial rewards [5]. Employees sink deeper when they define compensation as the salary they receive for the service or work they have done at their workplace.

Moreover, compensation is defined as the "overall reward package whether it is tangible or intangible, monetary and non-monetary, physical and psychological" that the organization gives to its employees for a work or service they performed [6]. There is a fundamental goal behind formalizing a reward system in every single organization and usually these purposes can be listed as (1) to grab professional and qualified employees in order to fit some positions, (2) to stabilize them so the turnover and the quit rates will be under control, and (3) to motivate the employees so that they will be fully engaged in their jobs and give the maximum of their knowledge, skills, and abilities.

2.2 The Importance of Engagement and Motivation

Can you imagine the staff being not motivated toward their jobs? Definitely they will start to waste their time by surfing the web for personal issues or they will be job hunters at another competitive organization, consequently they are wasting both the time and the resources of the organization.

Motivation at work creates the sense of commitment, so all employees will do their best in order to produce the best outcome [7]. It is also a step to increase an employee's job satisfaction, which will reflect positively on the growth of the organization and improves an employee's development so that he/she will easily reach the goals. Once employees notice that they are achieving their goals through the growth of the organization, automatically they will realize the connection between their efforts and the results, which will help them to stay in progress. This increases the efficiency of the employees. An employee's efficiency does not necessarily depend on his/her knowledge, skills and abilities (KSAs), so for the organization to achieve its goals, the staff must have a good balance between their abilities and their willingness to perform the assigned tasks; therefore this kind of balance will reflect positively on the productivity and efficiency [8].

2.3 Effective Compensation Packages

When formalizing an effective compensation package, organizations take into consideration three major ingredients, namely, psychologic satisfaction, monetary rewards, and nonmonetary rewards, as there are various types of compensation packages that exist and experts differentiate effective compensations from ineffective ones [1,3,9,10].

2.4 Motivation and Employee Engagement

Motivation can be defined as the enthusiasm to perform a certain action, and it is a result of intrinsic and extrinsic factors [11]. In his theory of motivation, Frederick Irving Herzberg stated that responsibility and opportunity for an employee to grow in a job increase motivation; therefore motivation can be defined as the self-awareness that drives an employee to act in a way that helps him/her achieve specific goals [12]. Motivated employees are required in all organizations because they help in achieving targets.

Actually there are five theories that talk about motivation and they will be explained later in the chapter. In this section the "McGregor theory" (or the "Theory X and Theory Y") will be discussed, as it can be applied to any organization [13]. When applying Douglas McGregor's theory (or the Theory X, Theory Y, and Theory Z), organizations are encouraging the employees to express their self-motivation [14]. In Theory Y, there is a hierarchy of chain in which the top management must encourage the subordinates through punishment, as a form of reward. This also helps them to participate in different ways, which creates responsibilities [13,14,40]. Motivation factors under the abovementioned theories provide a chance for the employees to increase their self-esteem; they look for chances that make them more responsible in their jobs and achieve higher levels of job satisfaction, which facilitate the way in front of the organization to achieve its goals [13,14].

The more psychologically empowered employees are, the more engaged they will be. Motivation and engagement are two criteria that every employee must enjoy to contribute to the success of an organization [1,3]. In general, engagement is broad and is important for the overall success of the organization, whereas motivation is narrow and is necessary for accomplishing challenging tasks in the job. Rewards may play an important role to attract and motivate employees in all departments [1,3]. But a major issue in employee motivation is massive employee turnovers, and the more the employees refaced with high turnovers, the less they tend to be satisfied, engaged, or motivated at work, as the data in Table 11.1 shows [13].

TABLE 11.1
Employee Numbers Immediately Before and After the Massive Turnover at the American University of Berut (AUB) and the American University of Berut Medical Center (AUBMC) Took Place (see Chapter 16).

Professorial staff	1200
Nonacademic staff	4340
Laid-off employees (June 2020)	1500
Laid-off employees from AUBMC (July 2020)	850

2.5 Understanding Motivation Through the "Needs Theories"

Many studies have been conducted to discover facts related to employee motivation. Here we provide a brief explanation of compensation from different perspectives because it can act as a stimulator for building a positive, competitive, and cooperative environment at workplace [1]. These are briefly discussed in the following in terms of rewards, incentives, and compensation. Some of these theories were developed over the year by many researchers; Chapter 2 discusses these theories in detail.

2.5.1 The 1943 Maslow's hierarchy of needs

In 1943, A. Maslow, proposed the Maslow's hierarchy of needs and operated it in the fields of business and HRM and education (Fig. 11.1). It states that people are motivated by five categories of needs that are organized hierarchically such that once any of the lower needs is met, the others are considered [15–17]. Recently, the need of safety has increased when employees fear for their well-being amid COVID-19, which includes over and above their personal safety and their job security, once experts noted how governments reacted to the pandemic [18].

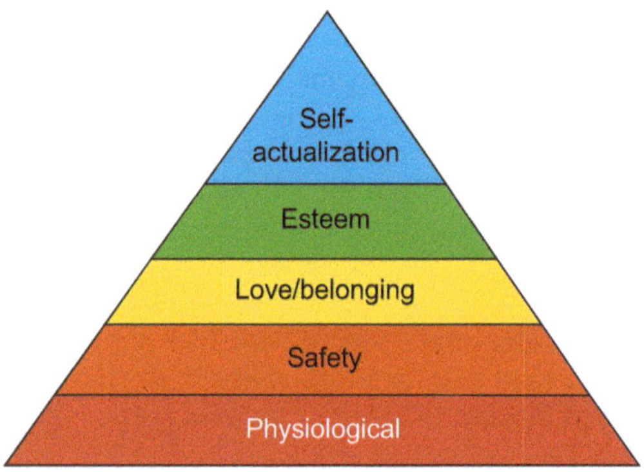

FIG. 11.1 Pyramid of needs. (Adapted from Maslow's hierarchy of needs Jeha H. The impact of compensation practices on employees' engagement and motivation. MBA Thesis submitted to the School of Business, at the Lebanese International University; 2020. Summer 2020.)

2.5.2 The 1959 Herzberg dual factor theory

The Herzberg dual factor theory or hygiene theory was developed by an American psychologist, F. I. Herzberg (Fig. 11.2). According to this theory the motivator and the satisfier are the two factors that create any employee's satisfaction at work [12,17,19,20]. It is either enhanced or weakened when economic crises, COVID-19, and other external factors affect employees' job security; studies were done to measure these effects collectively in Lebanon and globally and how governments responded to the pandemic [18].

2.5.3 The 1969 Clayton Alderfer theory

The Clayton Alderfer theory, or the ERG theory, is an extension of the Maslow's hierarchy of needs theory (Fig. 11.3). It suggests that there are three basic needs that motivate human beings [15,17,21,22]. Therefore this theory could also be used to explain how these needs can affect employees' motivation at work when the world faces a global pandemic [18].

2.5.4 The 1985 self-determination theory

The self-determination theory proposes that people always look toward growth and that they are motivated when they feel that their actions will affect the outcomes (Fig. 11.4). This theory can be used to measure or observe how COVID-19 can influence this [17,18,20,23].

2.5.5 The 2002 four drive theory

The four drive theory, as its name clarifies, states that there are four main driving agents that motivate

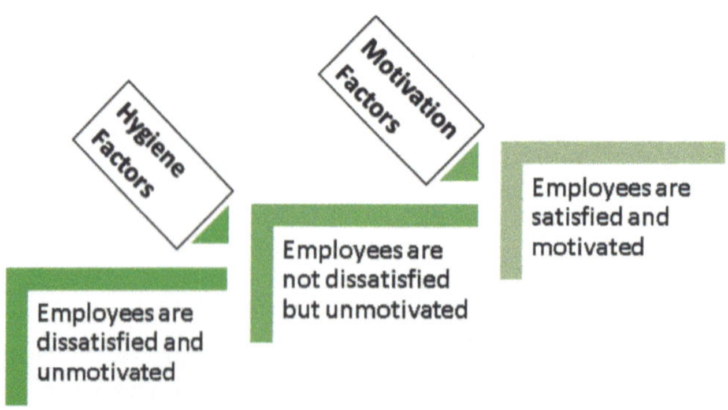

FIG. 11.2 The two factor theory of motivation. (Adapted from Herzberg's theory Jeha H. The impact of compensation practices on employees' engagement and motivation. MBA Thesis submitted to the School of Business, at the Lebanese International University; 2020. Summer 2020.)

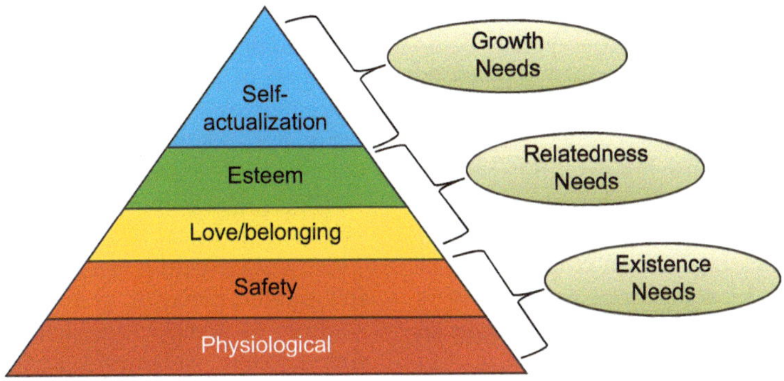

FIG. 11.3 The motivation pyramid. (Adapted from Alderfer's and Maslow's theories Jeha H. The impact of compensation practices on employees' engagement and motivation. MBA Thesis submitted to the School of Business, at the Lebanese International University; 2020. Summer 2020.)

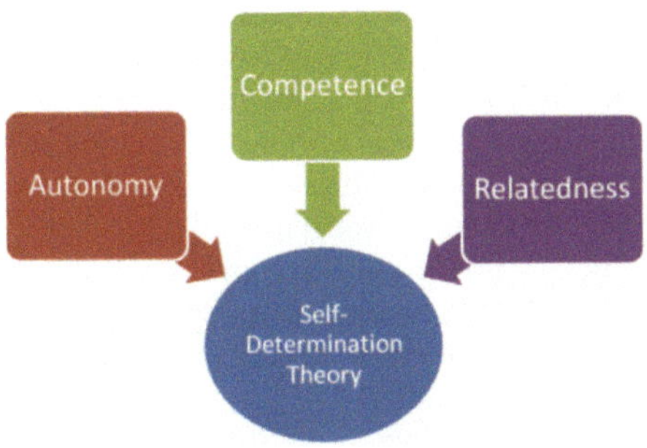

FIG. 11.4 The self-determination theory (figure traced according to [13]).

employees, even in times of pandemics (Fig. 11.5). This theory, which encompasses the drive to acquire, the drive to bond, the drive to learn, and the drive to defend, is interesting to apply to the fields of HRM and education and could be useful to investigate in the current case of coronavirus pandemic [11,18,24,25].

2.6 Employee Engagement Definition

Nowadays there is no single definition for employee engagement. HR managers define it as a state of being, and authors spend a lot of time trying to create a universal definition for this concept [5]. Yet "William Kahn", founder of the "personal engagement and disengagement" concept, wrote in his article that "the psychological factors that stands behind engagement and disengagement within the job they are performing" [13,26,53]. These factors lead researchers to answer the question "what is employee engagement?" and several sources were used to answer this question [1,3,9,10]. These cited sources define engagement as "a workplace approaching resulting in the right condition for all members to give their best every day, be committed to their organizational goals, and value, be motivated to contribute to the organizational success and with an enhanced sense of their own well-being" [1,3,9,10,13,27].

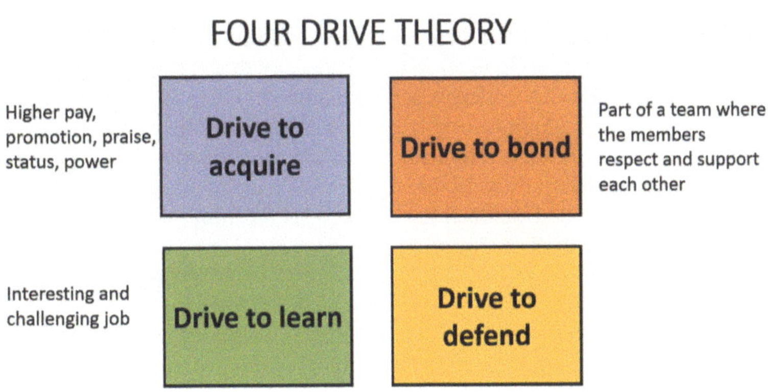

FIG. 11.5 The four drive theory (figure traced according to [13]).

While the Chartered Institute of Personnel and Development (CIPD) defines this concept as "being positively present during the performance of work by willingly contributing intellectual efforts, experiencing positive emotions and meaningful connections to the others" [28]. Sinking deeper into this definition clarifies that intellectual thinking means thinking deeply about the job and finding some ways to perform it in a good way. Thus effective engagement is the feeling of satisfaction about a job, while social engagement is finding a room to negotiate with other colleagues to discuss work improvements [13].

In addition to the previous definitions, the Institute for Employment Studies (IES) has set a comprehensive definition of employee engagement that it is the positive attitude an employee has toward the job, where an employee cooperates with his/her colleagues so that he/she can achieve better outcomes [28]. Consequently, it is the employer's duty to enhance this relation and help in creating an engaged employee [13,28].

Some researchers went beyond this to analyze whether an employee's engagement is an attitude, a behavior, or an outcome to the situation [1,3,9,10].

There are several arguments in which some of the researchers see it as a positive attitude toward the job, the organization, and the colleagues, and some employees feel supersatisfied because they are working at a good reputable organization, with a professional manager and some senior colleagues. Other researchers described engagement as a behavior, like when an employee recommends his/her organization to others. Lastly, another group of researchers link engagement to the outcomes [1,3,9,10].

Based on all these definitions, engagement can be defined as the degree of interest in work. It affects both the organization and the employee, and this is why compensation managers must appreciate the HR as a valuable asset in the organization because they are the only ones who create the strategy to make the organization a profitable one.

When employees feel engaged, they try to express themselves "physically, cognitively, and emotionally during role performance" [29]. According to Kahn's report, engaged employees must be physically available [13,26]. So based on the abovementioned definitions, engagement can be summarized as the level of commitment and participation for organizational value [26].

2.7 Employee Engagement Types

In 2004, Mike Johnson wrote a book entitled *The New Rules of Engagement* in which he stated that the ability to engage employees in work will be one of the hottest topics in the upcoming years. Nowadays it is becoming the greatest organizational battles and still grabbing the attention of all HR professionals; this can show how this emerging topic makes HR professionals to be proactive and seek to enhance the engagement level in order to achieve business goals [30].

Disengaged employees are those who are not satisfied by their work, or with the organization they work for; disengagement is always being discussed as a negative attitude, behavior, or outcome to the company. According to Kahn, "personal disengagement is the simultaneous withdrawal and defense of a person preferred self in behavior, physical, cognitive and emotional absence and passive, incomplete role performance" [26].

Disengaged employees are not motivated to their job, and they are not satisfied with their work and with their position. As a result, they feel stressful all the time, which affects the organization; always tend to be inefficient and disloyal to their organization; are always ready to go where the grass is greener; and see no room for career advancement [1,3,9,10,13]. They are not poorly preforming employees, but they just do the necessary steps to perform their job, arrive to and leave work on time, and do not think of tomorrow's tasks [31].

They resist all things related to change, as they are task-oriented and not goal-oriented, they do not have work relationships with the organization, and they do not care about the mission and the vision of the organization they are working in [32]. Researchers clarify the cause behind being disengaged as an employee's loss of trust in his/her supervisor, employer, or head of the department, which leads to lack of communication with all of them. This phenomenon can be summarized as lack of passion toward work and overload of work responsibilities [1,3,9,10,13].

To face all the challenges due to this phenomenon the organization, the employer should try hard to find the causes that stand behind their active disengagement to prevent things from being worse. This phenomenon has a huge negative impact on the reputation of the overall organization. When examining its source, it will be easy to protect the employees and try to find ways to set things back on track [1,3,9,10,13,31,32].

2.8 Performance and Engagement Strategies

There is no doubt that the engagement of employees is linked to their performance; those employees who are hired in the right position with a very good job description are likely to do their best to perform the tasks and to stay as long as they can in the organization. Performance influences an employee's attitude; therefore HR managers should perform a survey to evaluate the performance of employees and should do a performance appraisal as a tool to measure their performance at the end of the year [1,3,9,10,13,18].

Employee engagement should be the easiest and the best way for taking corrective actions, learning, and improvement. Employee engagement will create better results if the management measured it regularly [1,3,9,10].

2.9 Summary of the Literature Review

This literature review answers the question what is compensation and the importance of compensation on employee engagement as well as its types. The researchers evaluated the theories that are linked to motivation as an important part of the research. This critical literature review clarifies the importance of compatible packaged compensation on employees' engagement and motivation amid times of COVID-19 [13,18].

So the next section will pick up from the definition of compensation, provide its historical background of extrinsic and intrinsic factors, and clarify the strategies that can be used as a tool to increase performance engagement, by electing the correct research methods to use [33]. Therefore one needs to think deeply to interpret the link between rewards and compensation with employee engagement and work motivation.

3 METHODOLOGY

3.1 Selecting the Research Design

Fig. 11.6 presents the research methodology process was undergone to select the best research design for conducting this research. The research approaches typically used to best fit the selected research design are also explained and shown in Figs. 11.7 and 11.8. The data collection method suggested a monoqualitative

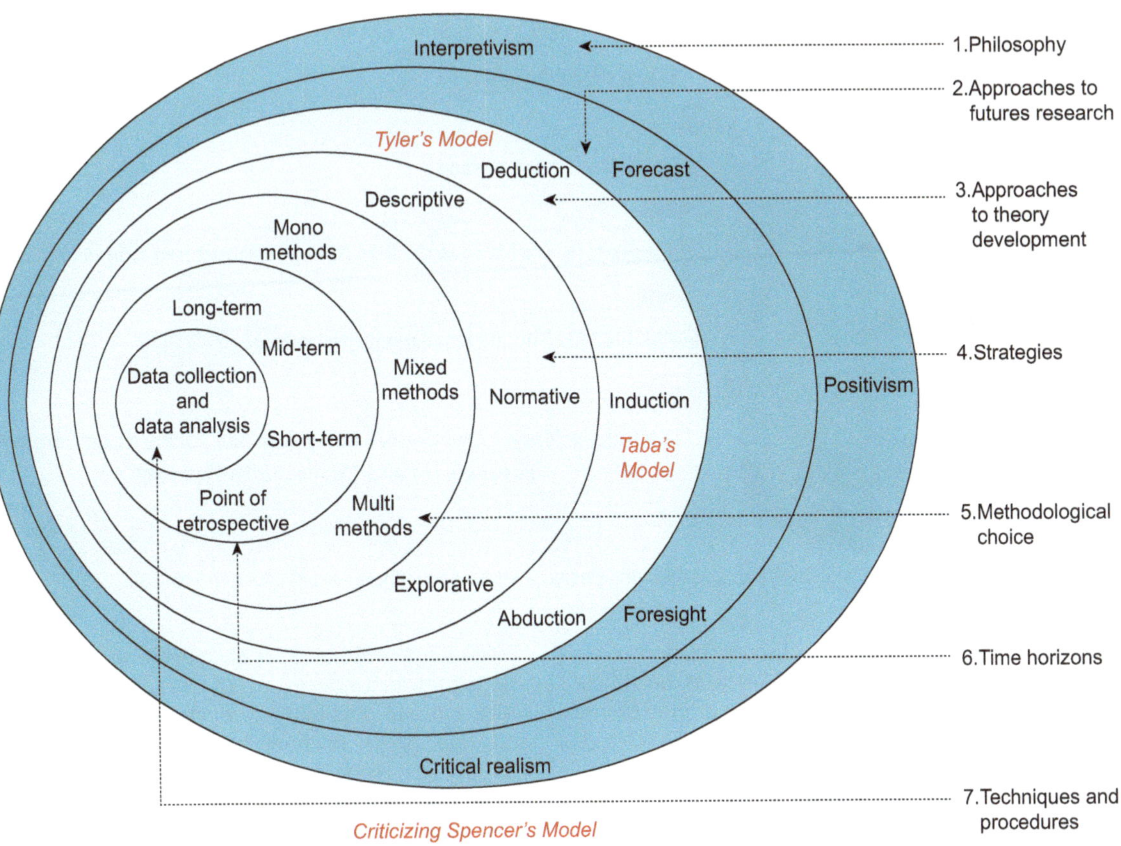

FIG. 11.6 Research process. (Adapted from Saunders' Research Onion Jeha H. The impact of compensation practices on employees' engagement and motivation. MBA Thesis submitted to the School of Business, at the Lebanese International University; 2020. Summer 2020.)

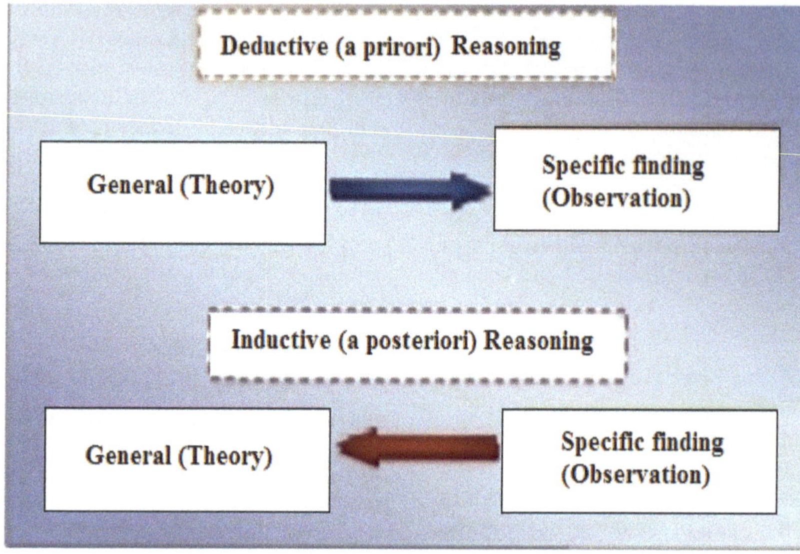

FIG. 11.7 The deductive and inductive approaches explained (figure traced according to [13]).

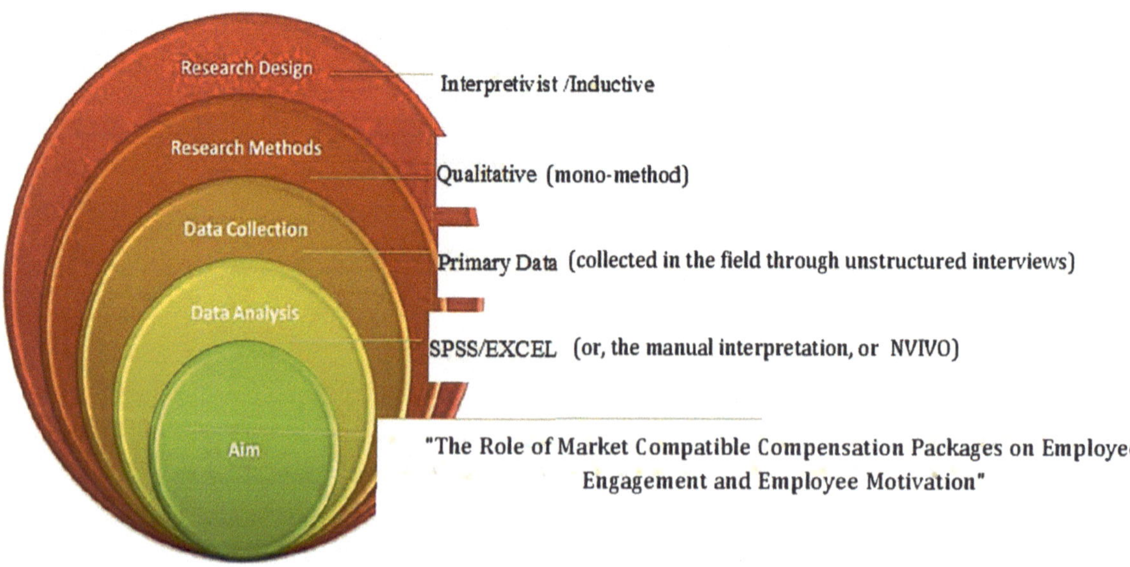

FIG. 11.8 Data collection and research design methodology (figure traced according to [13]).

strategy, where the gathering and interpretation of information was adapted accordingly, as both the philosophy adopted and the approaches used allowed the researchers to reach their conclusions [33]. The research approaches (shown in Fig. 11.7) will be explained further, as they help in electing the research strategy and methodology needed to develop the final research design (Figs. 11.8–11.10). Fig. 11.6 shows the research process as utilized by social scientists [54].

Sampling techniques, as such, were adapted to the selected research strategy, which was surveying through using a mono-qualitative research design (Fig. 11.8), and one needs to understand the whole process [33,34]. This research proposes to use sampling techniques that are adopted for qualitative methodologies, which urged scholars to select the convenient technique over the snowball, or judgmental ones if surveying was to be considered [35]. Techniques and procedures for

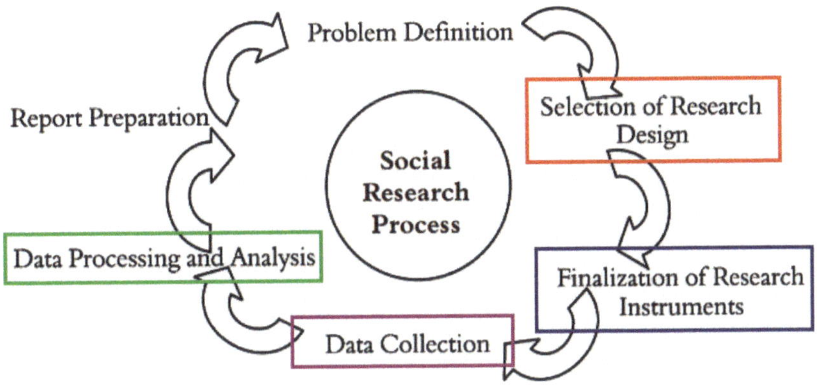

FIG. 11.9 Research process as per exploratory methods (figure traced according to [13]).

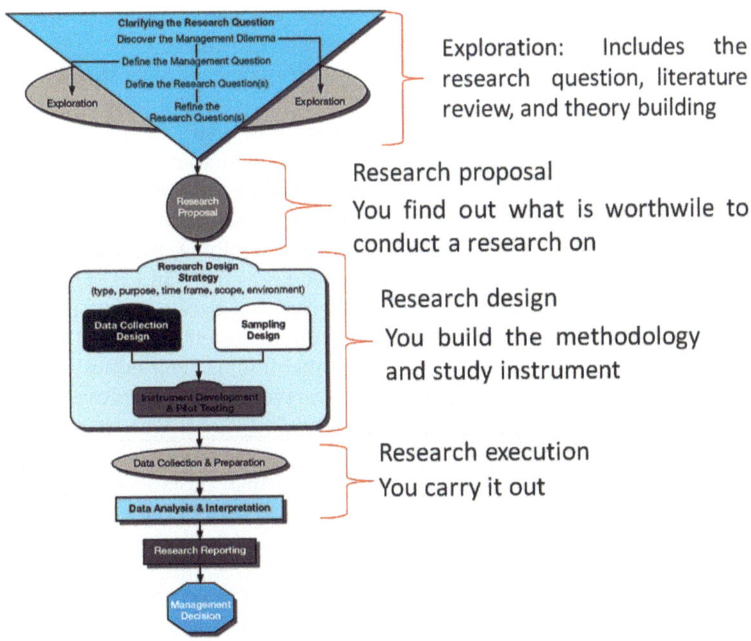

FIG. 11.10 The research design process. (Compiled from various sources Jeha H. The impact of compensation practices on employees' engagement and motivation. MBA Thesis submitted to the School of Business, at the Lebanese International University; 2020. Summer 2020.)

data analysis are based on the types of data collected; for instance, for quantitative data one could use Excel or SPSS (Statistical Package for the Social Sciences), while for qualitative data one can do interpretation by hand, use statistical tools (such as Excel or SPSS), or rely on NVIVO, among other techniques [34,35].

Fig. 11.8 shows the outcomes of the selected research design based on the selection of the research philosophy, approach, strategy, methodology choice, and time horizon. This is why the following research design that encompasses Interpretivist philosophies and inductive approaches was adopted [34,35]. The selected research methods were monoqualitative research methods. Data collection methods include collecting primary data through unstructured surveys. The data analysis stage involves manual interpretation of interview data, but we will consider digital methods, if time permits. Here, and based in Fig. 11.8, these methods were selected based on the data by several experts in research methodology development [34–36].

Fig. 11.8 shows the list of the abovementioned research design to fulfill the study aim in the best way possible using U. Sekaran's and R. Bougie's arguments on quantitative methodologies [34]. We can consider triangulating the data with the literature review and pilot studies; however, as this is a method that requires mixed research methods (combining qualitative and quantitative methods), it does not suit the scope of this research, but could be considered in a future study [37]. Finally, we show the major difference between confirmatory (quantitative) and exploratory (qualitative) studies in Fig. 11.9, while in Fig. 11.10 the full research design from start to finish is presented to illustrate the point further.

3.2 Outcomes

After finishing with the selected mode of data collection (or sampling, if applicable), through the proper selection of the research philosophy, research approaches, research strategy, and methodological choice in details, one needs to refer to Fig. 11.7 to better understand the research approaches that exist in research methodology. Following the best research methodology to explore the relationship between employees' motivation and compensation practices, quantitative or confirmatory methods cannot be relied on (Fig. 11.9), as the study is merely descriptive or exploratory [33].

Therefore, and for the above-stated reasons, the Interpretivist philosophy was used, as the study relied on unstructured interviews as a mode of data collection in order to get the sought answers. As a result, and aligned with this study's viewpoints, an inductive research approach was followed, combined with the monoqualitative methodological choice (Figs. 11.9 and 11.10). All necessary data for accomplishing this research were collected by conducting interviews with three managers from three different NGOs. Finally, the research design, which is shown in Fig. 11.9, follows the research methodology choice that was proposed before [33,36,38].

4 FINDINGS AND ANALYSIS

This section presents the results and the findings of this research based on the research question and other interview questions. It also presents an overview of the interview results, together with some interpretations following recent trends in researches attempting to analyze the effects of COVID-19 on rewards and compensation schemes, through a descriptive analysis on the role compensations play on employees' engagement and motivation in three NGOs, while attempting to interpret the responses for the current situation at hand [13,18].

4.1 Findings

The five managers who were interviewed represented the three NGOs that were approached for the study.

4.1.1 Answers of the first three managers

In their institution, responding to **question 1**, the managers said that they mostly have volunteers and that they recently got accredited, explaining why they relied on volunteers. As such, intrinsic motivation works best at this association, as they are not on the payroll. The managers, in their answer to **question 2**, reiterated what they mentioned in question 1. Yet they added that although they mostly rely on intrinsically motivated volunteers, they do provide rewards for the most dedicated ones.

The managers, raise an issue in **question 3**, which is donor fatigue. This concept is essential because they want to diversify funding sources, to avoid just that. Once they do, they want to propose more serious packages; as for now, the association only provides food and transportation allowances, based on project funds.

For **question 4**, the managers agree that they should get to know the wants and needs of their employees in the long run because now, the association's volunteers are the management's main concern. The managers go on mentioning their ethical duties toward all subordinates (volunteers, employees, and so on) following Mazjoub (2018). Part of their duties is to empower the task force and provide it with job security, in order to boost their morale.

For **question 5** the managers state that at their association, they don't distinguish with age but with performance and reward accordingly, and on a sidenote they told us that "it is common practice, in times of crises to rely on performance appraisals to fire all the bad employees first." The managers told while responding to **question 6** that this is a typical case for internal (i.e., intrinsic) drives. Yet, of course, financial compensations will not be refused by anyone, but, by nature, at NGOs like theirs, cash bonuses are rarely, if never, disseminated. Also, in Lebanon, the labor law and ethical considerations apply [47,48].

For **question 7**, the NGO managers told us that their dedicated volunteers will want to continue with them, so this alone is an encouragement. While, for **question 8**, the managers told us that they did not reach this stage yet, as they barely passed the first 6 months of operations. For **question 9** the managers indicated they did not reach this stage yet because they are only working with volunteers now. But when they start hiring they will ensure that they feel wanted, welcome, and empowered. So they will definitely work on enhancing a family environment. For **question 10**, the managers concluded that as they work only with volunteers now, they can only empower them now to feel they belong to their NGO.

4.1.2 Answers of the fourth manager

This manager represents the second NGO we approached. He, through answering **question 1**, told us that at his association, motivation is on multiple perspectives, as it is based on the institution's values in its operations in the humanitarian sector. He also adds that teamwork is essential as well as rewards through achievements are considered. For **question 2**, the manager told us that there are both internal and external influences. In his words, he defined that they keep their employees to be "highly engaged through monetary permits such as salary raises, bonuses, rewards ... symbolic benefits such as badges, and so on, as the leader exerts 'reward' power."

The manager also details another form of appraisal, which I am particularly fond of myself, "reward by punishment". Although he noted that his institution's leaders rarely exert to their "coercive" power, it is important to mention it here. Note that the managers refer to "reward by punishment" from well-referenced and quite well-documented sources, and many HR specialists disagree because, for instance, some may never consider punishment as a reward, like many communist leaders considered in the days of the Ex-Soviet Union [1,32,41].

The manager went on explaining that intrinsic factors may be harder to observe and trigger and include autonomy, which is greatly appreciated and encouraged by the top management. Also, belonging is very important for them, while they take many initiatives such as celebrating personal events, organizing gatherings, and so on, as they work in the humanitarian field, where "finding meaning" is often a great motivator in itself.

The manager pursues the discussion while responding to **question 3**. He told us that at his institution, different compensation packages are available depending on organizations, departments, and positions. He goes on saying that there are benefits such as vacation days, sick days, personal days, free medical care given by the company's own medical center, discounted dentist care, and bonuses [1].

Addressing the issue, while answering **question 4**, the manager told us that they use two types of motivators: a standardized promotional system that is based on common factors (i.e., years of work, key performance indicators (KPIs)...) and a more personalized or profile-based approach (as mentioned previously). He thinks that both types are important and have different impacts on employees, where one needs a degree of standardization when it comes to motivation, and this statement agrees with organizational behavior (OB), Leadership, among other HRM specialists [1,3,10,31].

As such, the manager addressed the following points in **question 5**. As stating his points, he told us that in his association, age is always a debatable issue when it comes to motivational systems. Here, we quote him directly, "Of course age affects how you approach employees. To motivate employees, you can either approach them individually, in teams, or comprehensively. Also you can motivate them either pre or post task fulfillment".

The responses for **question 6** seem to be in line with many cases seen in the current situation in Lebanon (and the world) affected by the COVID-19 pandemic-induced recession [18]. So the association's manager said that "unlike what may be expected in an economically challenging environment and work conditions, the most appreciated and effective motivational systems are non-monetary". This agrees with many ways that HRM specialists consider what firms can do to keep good employees [9].

In citing the manager's statements for **question 7**, we notice an almost full agreement with HRM specialists [1,3,9,10]. He responded by saying that "planning for a motivation campaign or system in the company, surely you would like to sense whether its fruitful or not." This means that measuring motivation depends on the aim behind these benefits and if they aim to boost employee morale or increase productivity.

The manager goes on saying that they always track the effects of the association's promotional and motivational systems and recognize the proportional relation between them. Finally, he concludes his answer by saying that some ways (i.e., monetary vs. abstract) to motivate staff have greater effects over others.

For **question 8**, here is the manager's answer—he fully agrees in the fact that older staff show greater loyalty, different expectations, and different interaction with motivation. Newer employees, in general, want more materialistic benefits, whereas older employees tend to seek appreciation and recognition.

However, the direct effect on newer employees is more obvious and instantaneous. From a managerial perspective, older employees are fixed and essential for the company and thus they may receive more promotions, while newer employees are yet to be proven essential and worth investing in.

The **penultimate question** was answered, as the manager omitted the **last question**, with regret, as he told us that it did not really apply to his association. So we end this section with the manager's response for **question 9**. Although the manager "personally thinks they should experiment the effect of giving employees the chance of choosing or modulating their reward packages", he says that there is no such practice yet. But, in either case, the upper management should be very attentive to their staff's wants and needs, irrespective of whether they were empowered to participate in decision-making or not.

Finally, according to him, the most impactful way is when the management solely decides but deeply studies and personalizes the system to meet employees' desires. This does not fully agree to or other HRM or OB specialists who do encourage participatory decision-making as a form of motivation [1,10,31]. Last but not least, the manager concluded the interview by thanking us and by saying that the last point, addressed in **question 10**, does not apply to his association (i.e., there are currently no strategies).

4.1.3 Answers of the last manager

These responses are from the last institution's manager we approached. He told us, while responding to **question 1**, that the following incentives are proposed: insurance, extra transportation, incentive loans, trainings, appreciation, employee of the year award, and summer and Ramadan activities. He stresses that all salaries are paid in fresh dollars for the relief unit, and that the National Social Security Fund (NSSF) is provided, as well as marriage leaves and flexible work time.

For **question 2**, the manager said that his foundation provides capacity building training, dollar discounts at the health department, vocational training programs, and flexible microloans. For **question 3**, he told us that his institution does not have any compensation packages available.

In his response to **question 4**, the manager details that they use performance evaluation (template), NSSF, promotion, and salary scales, among other incentive profiles [1]. For **question 5**, his response is clear, as they do not usually do that. Yet, while addressing **question 6**, the manager reminds us that they rely on the reward systems, as per employee preference, linked to performance appraisal.

For **question 7**, the manager reminded us that the incentive systems did instigate more staff engagement, motivation, working extra hours, and more productivity. Yet, by tackling **question 8**, the manager told us that there are extra benefits for our senior-level staff, such as insurance and priority in promotions, as they show more dedication than the junior-level staff.

The manager confirmed to us that, while answering **question 9**, the staff are part of the evaluation process (linked to employee empowerment). The manager wrapped up, while answering **question 10**, by saying that further employee incentives can be seen through options for volunteering, Ramadan village, crisis responses, flexibility, general meetings, and awareness sessions.

4.2 Analysis

Due to the primary scope of the research was to conduct unstructured interviews with the five managers we approached in three different NGOs, using monoqualitative research methods, we will no longer use triangulation, but just mention that it is used as a way to validate and to ensure the data we collected had we used quantitative or mixed methods could be correlatable, internally reliable, or both [13,33–38].

Here we will analyze the responses of all five managers by comparing and contrasting them. We will also attempt to tie their responses to existing literature in order to see if they are coherent.

4.2.1 Question 1: How do you motivate your employees? How do you perceive them to be motivated?

The first three managers stated that they only have volunteers at their NGOs. As such, only intrinsic motivation for encouraging field operations is possible, or considered, as there are no employees under a payroll, so monetary rewards are not applicable. Such findings agree with other researches stating that employees prefer noncash incentives over financial ones in times of pandemics [13,18].

As the first NGO is recently accredited, it explains why there are no paid staff yet. The last two managers, at their respective institutions, point out that they use both cash and noncash rewards, having different types of work available (e.g., internships, voluntary jobs, and various contractual jobs). So, basically, the last two NGO managers said there are various incentives or rewards that attract new employees and retain old ones, irrespective of whether they are new hires or not because of the current times [13,18,39].

4.2.2 Question 2: How are they motivated? Are they motivated by internal or external factors?

The first three managers, like with question 1, said they only work with volunteers for now, so they focus on intrinsic motivators. But the last two managers admit that they, in their respective institutions, provide both cash and noncash rewards [13].

Strikingly interesting is that the fourth manager refers to punishing as a way of reward, which many HRM experts disagree with. But if we take the classroom management perspective of reinforcement as stipulated by educators, we realize that rewards (behavior reinforcement) can be positive or negative, just as punishment [1,40].

Punishment whether it is seen as a reward or a reprimand ultimately results in behavior termination, or modification; experts agree that punishment can be used as a reward, which is strange according to some other experts [1,40]. This is the trade-off of leaders using "reward" versus "coercive" power, where in both cases the leaders exert "legitimate" power [1,32,41].

The last two managers agree with regard to rewards. But the last manager seems to not discuss any use of coercive power [1,32,41]. They agree that leaders who use reward power in a way favor employee motivation, irrespective of whether it is through material or nonmaterial incentives, and this is an essential function of reward management that may help in times of crises, such as the COVID-19 pandemic [13,18].

4.2.3 Question 3: What are the compensation packages available at your organization?

The first three and the last manager may agree on the nature of compensations present, or absent, in their institutions. At their NGOs, the first three managers were highly concerned about donor fatigue, if they keep relying on the same sources for funding, so the only compensation scheme currently operational is food and transportation [13].

This means that they cannot sustain a steady compensation scheme, as they ultimately want job safety and stability for their staff, which will not be currently feasible. For different reasons the manager of the last NGO also says that there are no compensation packages available. However, according to the manager of the second NGO, at his association, there are various compensation packages available, unlike the last NGO, which has no such plans. The first four managers, however, could agree on the basic needs as food and transport, which to some extent may be covered in their respective institutions [13].

4.2.4 Question 4: Do you use an incentive profile (security, recognition, rewards, responsibility, and so on) in order to know what can motivate the employees?

The first three managers highly consider empowering all people working for them and by doing the right thing as leaders, in terms of ethical practices and especially in times of major recessions, pandemics, and economic meltdowns, once mass turnovers are not considered [1,13,42].

The fourth manager, similar to the first three managers, wanted to motivate and empower employees through rewards and other incentives, as he believes that it is best to motivate employees by following a standardized way. The last manager, in agreement with the fourth manager, says that they use performance assessment templates to evaluate employee performances (Fig. 11.11). Aside from the incentives, they

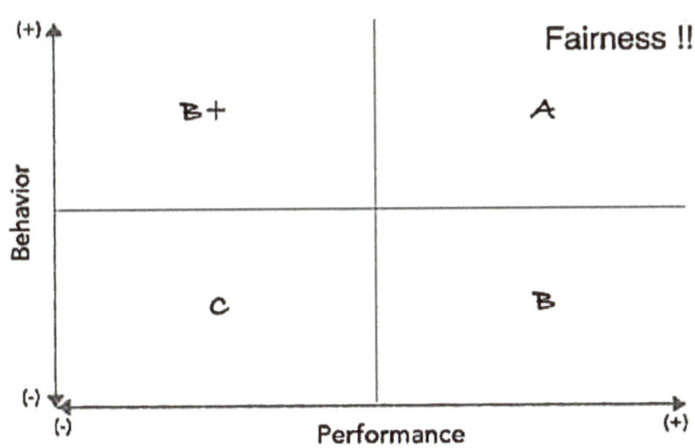

FIG. 11.11 Standard performance appraisal graph used for keeping good staff and firing the bad ones (see: [13]).

also provide securities such as the NSSF, promotions, and salary scale incentives, irrespective of the COVID-19 pandemic [1,3,9,10,13].

4.2.5 Question 5: How do you approach employees concerning motivation, do you approach them differently because of age?

The first three managers said that they do not distinguish by age but they "distinguish by performance, and reward accordingly". But for the fourth manager, age is always a debatable issue when it comes to motivational systems, and it may or may not be related to pandemics [13,18].

For them, age affects how they approach their staff, and they are approached individually, in teams, or comprehensively in terms of rewards and incentives. However, the last manager, agreeing with the first three, stated that they do not usually distinguish employees by age, but may agree [1,10].

4.2.6 Question 6: Which reward system do you believe the employees prefer? Why?

The first three managers stated that in their case, they definitely consider intrinsic rewards, as NGOs do not focus on material rewards, although monetary compensations can be given, but it is not common. The fourth manager seems to agree as well [13].

In their association, the fourth manager seems to consider that according to the staff the most appreciated and effective motivational systems are nonmonetary. As such, the association's staff tend to focus on noncash rewards, in spite of the hard times faced by Lebanon and the world [18]. Finally, the last manager believes that in his institution, employees prefer the rewards linked to performance appraisals (Fig. 11.11).

4.2.7 Question 7: Do you believe the reward system at your organization has increased employees' motivation and performance? Why or why not?

The first three managers stated that most, if not all, volunteers who once worked with them will always want to be working with them. This also encourages program continuity, by which staff will be internally motivated to perform well as long as the projects are ongoing, irrespective of issues such as the COVID-19 pandemic, the successive governments dealing with the various economic drawdowns, and the recent Beirut Harbor blast [13,18,43].

The fourth manager stated that while planning for a motivation campaign or system in his association, the leader needs to ensure that the benefits proposed actually align with the institutional goals. Examples like increasing staff loyalty, enhancing staff productivity, and encouraging staff initiatives can be useful to construct a way to measure how effective these benefits are in measuring motivation. He also adds that employee motivation increases productivity and that they always track the effects of their promotional and motivational systems as they recognize the proportional relation between them. Finally, the fourth manager concludes that some ways to motivate employees (i.e., monetary vs. abstract) have greater effects than others when it comes to improving organizational performance [6,13,27,31].

The last manager, in his institution, agrees that this increase has resulted in more engagement, motivation, working for longer shifts, and increased employee productivity. Here we can say that all three managers somehow agree that increased motivation enhances productivity, irrespective of whether the rewards are cash or noncash, but may need to be well designed [13].

4.2.8 Question 8: Is there any difference between the motivation of employees who have been working for your organization for a long period and those who have just started?

The first three managers told us that they did not reach that stage yet, as they started being fully operational about 6 months ago. But the last two managers do show a difference in seniority or tenure in their respective institutions.

So according to the fourth manager, of course, a great difference is noticed. For him, "older employees have greater loyalty, different expectations, and different interaction with motivation". He adds that "older employees are fixed and essential for the company and thus they may receive more promotions". This point is corroborated by the last manager who stated that at his institution, extra benefits are available for their senior employees, such as insurance and priority in promotions. Tenure and seniority are discussed in various aspects by several experts in the field [1,10,13,31,44].

He says, "newer employees, in general, want more materialistic benefits while older employees tend to seek appreciation and recognition." Yet, he adds that "the direct effect on newer employees is more obvious and instantaneous" and that they "are yet to be proven essential and worth investing in" [13]. The last manager seems to agree as well and probably so do the HRM experts that new hires need to still prove themselves once they start a new job [1,3,9,10,31]. On a sidenote, the same point is stated by the first three managers while working in the humanitarian field, as the entire association needs to prove itself able to all stakeholders.

4.2.9 Question 9: Are your employees included in the decision-making process of the reward system?

The first three managers do not have that, as there are only volunteers now. But they will highly consider their subordinates as being part of a family. Moreover, they will be feeling wanted, welcome, and empowered. So definitely, the association's management, when they start hiring staff, will empower them. The last manager agrees and says that in his institution, the subordinates are included in the evaluation process, as it is linked to employee empowerment [13].

However, the fourth manager stipulates that at his institution, currently the employees are not involved in the decision-making per se. But he thinks that they should experiment the effect of giving their staffs the chance of choosing or modulating their reward packages. In both cases, the management should be very attentive about their staff's wants and needs, irrespective of who decides on these incentives. Finally, he adds that at his institution, "the most impactful way is when the management solely decide but deeply study and personalize the system to meet employees' desires" [13].

4.2.10 Question 10: What strategies do you currently use to increase employees' engagement?

The first three managers stated that with their volunteers now, they can only empower them to feel they belong to their institution. But the fourth manager stated that at his association, there is no such strategy. Finally, the last manager at his organization stated that options to show employee involvement, such as volunteering, crisis management, work flexibility, general meetings, and awareness sessions, among other incentives, are available for employees to sign up for [13].

4.3 Summary of the Findings and Analysis

The data was collected from three NGOs (whose names are not disclosed for confidentiality issues) in the form of interviews. Interview questions were designed based on the literature review. Therefore this research confirmed that there is a strong positive correlation between the compensation practices and employees' motivation.

The findings simply stated that are various ways to motivate employees. Both known ways are intrinsic (internal) and extrinsic (external) means. In terms of motivation as such, we refer to both, but in terms of incentives and rewards management, we refer to key sources to discuss this issue, pertaining to how organizations deal with COVID-19 [1,18,31,32,43]. All HRM, Leadership, and OB specialists argue that intrinsic (noncash) and extrinsic (cash or status) benefits are used interchangeably in the workplace, and employees sometimes prefer noncash over cash rewards, especially in the humanitarian sector. In some cases, this facilitates employees' feelings of job safety in times of major pandemics [1,9,10,13,18,31,32,41].

The first three managers have a plan to fully empower employees and provide them with stable and secure jobs once the funding is made sustainable. The strategy is planned in such a way to rely as much as they can on the first phase of operations [13]. Interchangeably the last two managers, agreeing with the first three in terms of intrinsic or basic compensations, provide various cash and noncash benefits for their employees. Finally, and in the light of the above, we noticed that most managers noticed that employees (those performing well at least) prefer stable securities and noncash incentives over cash bonuses, especially in these hard times, which help them stay motivated on the job [13,18].

The reverse case of employee motivation, perhaps if punishments are involved, or simply high turnovers are conducted, was observed in reviewing the literature if one took firms that exerted massive sackings [45,46]. Such management teams will be sure to have demotivated employees at hand, will have to deal with very high feeling of job dissatisfaction, will have to witness quite unsafe job conditions, and will have a harder time to demotivate their staff in order to make them perform better or to increase productivity if the company itself behaves unethically or illegally [13,18,42,45–47]. This is estimated especially in terms of major pandemics, when turnovers increased in these firms, irrespective of whether they helped businesses increase productivity, or not, based on how turnovers are defined [1,3,9,10,13,18,42,45–48].

5 CONCLUSION, LIMITATIONS, AND RECOMMENDATIONS

5.1 Conclusion

In this section, the researcher will discuss the findings of the previous chapter and the conclusion derived from this study, in addition to some proposed recommendations deemed fit on the effect of compensation practices on employees' engagement and motivation. Finally, the answer to our research question, *"To what extent compensation (extrinsic and intrinsic) is related to employee engagement and motivation?"* is *"to a great extent,"* as there is a link.

In response to the crisis Lebanon is facing, HR managers are concerned about a very critical issue that is keeping employees engaged and motivated, after they have suffered from hyperinflation, downsizing, losing bonuses, and account freezing. Motivating the employees during these crises while recognizing that once they have the opportunity senior talents may leave Lebanon to a place where grass is greener has created a new battle for HR managers.

The results have answered the research question, which were initially discussed, and all objectives were met. The goal behind this research study is to detect the relation between compensation practices (extrinsic and intrinsic) and employees' engagement and motivation. For starters, we see that most NGOs believe that the intrinsic and noncash benefits (or rewards) are what that drive their employees to increase productivity, through internal motivation. This is especially the case with interns and volunteers, provided they are not aiming to get a job once the term ends.

Yet, ultimately all workers in the humanitarian sector have also external motives. Yes, as some of the institutions we sampled said that some staff, indirectly at least, go for the high salaries, good packages, and incentives, which neither the public sector nor the private sector have. This is especially true in larger international NGOs such as the UN (which the UNICEF, UNESCO, UNRWA, UNDP, and so on are part of), Mercy Corps, and the ICRC.

As we previously mentioned, compensation practices play an important role in all organizations, and this is why HR managers adopt different strategies that help them to increase the motivation, in addition to the level of engagement, at work. Both extrinsic and intrinsic factors are the fundamentals used by the organization for motivation in order to achieve better outcomes.

This study shows that reward system managers are the motor of increasing employee engagement and motivation at workplace. Engaged employees need continuous care and support from their managers and a safety environment that helps them to be innovative. It is very important for the managers at all levels to use several strategies that help them to increase employee engagement, and not in those that hurt it [13,18,43,45,46].

Findings show that monetary compensations are not the primary and the only motivator for all employees, especially at NGOs. However, managers play an important role in employees' engagement in the workplace, and the interviewees of this study indicated that they are a vital component in increasing the level of motivation. Engaged workers need continuous support from their managers and supervisors and a safe atmosphere in the workplace.

Motivation for HR managers is an easy concept to understand but one of the difficult issues to implement. Engaged employees help the organization to achieve its mission, implement its strategy, and achieve better outcomes. This research has shed the light on several HR practices, such as training and reward systems, that help the organizations to enhance employees' engagement. It also shows that engagement and motivation at workplace are complex than they appear on the surface. The decision to increase the level of engagement depends on an organization's policy and the structure of its workforce.

For these reasons, it is important for the managers to create an engagement culture, to discuss the value of engagement through organizational mission statement, to follow up with all the employees at different departments to emphasize that they are implementing the engagement action plans, and to adjust their plans, policies, and strategies as needed.

5.2 Limitations

Limitations are "factors known at the inception of the study that may affect the outcome of the study" [49]. Doing any research project or thesis dissertation during these tough times is bittersweet, and the authors will explain why below, following Dr. K. Rogoff's and Dr. F. Gerges' arguments [13,18,43]. The main limitation of this research was a major time constraint, and because of the COVID-19 lockdown, government management of the pandemic, and the Beirut Harbor blast, the data that was collected was gathered just before the event on August 4, 2020, that transformed Beirut for good [13,18,43,50]. As a result, this research was carried out it in a period when nobody was in a mindset to consider answering survey or interview questions. These are hard times and the available data we collected had to fulfill our primary constraint, that is, time, as we had at most 6 weeks to complete the entire research, excluding the harbor blast, which made interpreting the results a harder task.

It is "bitter" due to the great displeasure for what is happening in Lebanon and for the confinement that prevents us from conducting face-to-face interviews, as an online option was mandatory, after getting the managers' approval, which was not easy [13]. As the August 4, 2020, Beirut Harbor massive explosion destroyed some of the main offices of the initial NGOs that were targeted, a contingency plan had to be adopted to target other NGOs, which made the task more daunting [13,18,43,50]. In addition, it was impossible to collect data through semistructured questionnaires with their respective employees, owing to the confidentiality and secrecy policies some of those NGOs had. Moreover, some of the NGOs that were initially approached refused to take part in the research because of their internal policies.

Even though the research had come up with preliminary answers, while addressing the three NGOs that the study was able to reach out to, none of the responses disclosed any confidential information about the detailed compensation practices in detail. Yet, the interviewees did their best to answer all the questions in the best of their abilities [13].

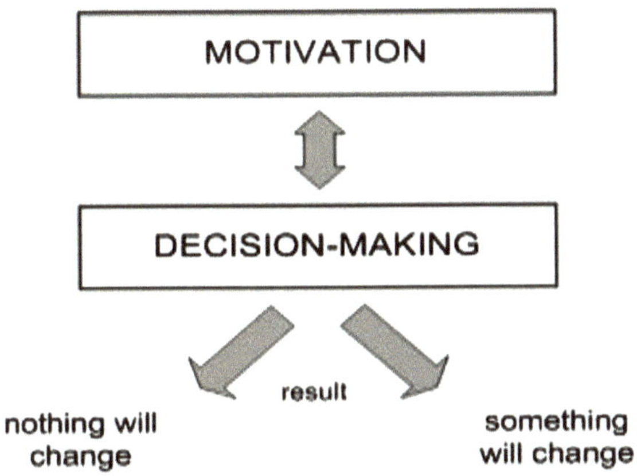

FIG. 11.12 Results of decision-making, linked to employee benefits and compensation, showing a change [13].

But it is "sweet" because I will be an HRM graduate soon. Consequently the journey of thousands of miles will start soon. I will definitely work on these incentives in the place I will be responsible for, that is, the HR department, which is why the lessons learnt and takeaways from this project will be applied to my next job as an HR manager [51]. Hence it is not fully disturbing. In the last section, I lay down my suggestions, as well as a few recommendations, which will be shared in the next section. The research proposes also to define a simple way to design a straightforward relationship with administrative decisions and employee behavior changes, which is shown in Figs. 11.12 and 11.13 and explained in the next section in detail [1,3,9,10]. Fig. 11.12 briefly lists the outcome of the literature review through the decision-making scheme.

5.3 Recommendations

Before ending this project, I took the liberty to trace a path for future HR managers to consider. As mentioned earlier, Fig. 11.13 details an important issue that HR managers working with coworkers in the benefit and compensation units need to be aware of, that is, which decisions positively affect employee motivation and which will hamper their motivation.

This point is not directly related to the current study, but lays the ground for argument. It could be the grounds for establishing a grounded theory for a future HRM research on decision-making in times of crises and

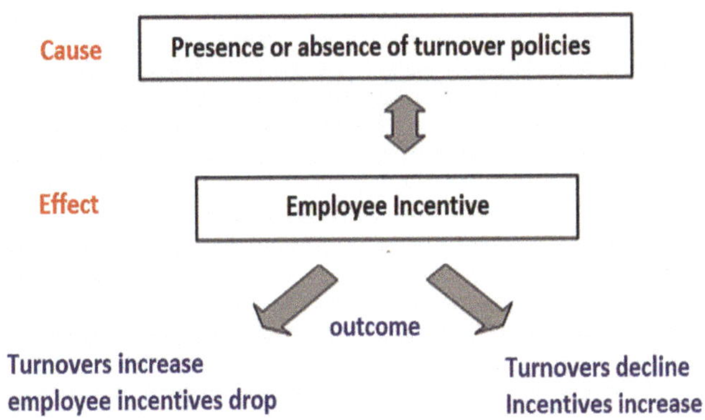

FIG. 11.13 Established relationship between employee turnover and employee incentives, with mapped outcomes. (Adapted from Jeha H. The impact of compensation practices on employees' engagement and motivation. MBA Thesis submitted to the School of Business, at the Lebanese International University; 2020. Summer 2020.)

seeing which decisions are best to boost employee morale and which are terrible, especially when rewards and incentives are involved; with the current pandemic crisis, this can be in fact challenging.

Many staff prefer either noncash rewards or strictly to benefit from incentives tied to performance appraisals (Fig. 11.12). How is this important is what other firms can benefit from the takeaways and lessons learnt [51]. Well, rewards boost employee morale during crises, such as major economic recessions, financial collapses, and pandemics; with massive employee sackings in companies that face economic drawdowns, they need to boost employee morale with noncash benefits, fire all the bad employees first, and work on strategies to fairly and equally reward the good ones [1,3,9,10,13,18].

The results of this study confirms that HR managers need to create engagement strategies that help them to meet the desires and needs of the employees at their organizations in order to implement the best and effective engagement and motivation strategies that fit with the organization's mission and vision.

The recommendations for future studies require addressing some of the limitations of this thesis, especially the limited number of interviewees and the political and health issues that our country faced while the thesis was in progress. Future theses should study the impact of compensation practices on employees' engagement and motivation in a broader sample, with more HR managers and program managers, in a larger geographic region with international NGOs because they might have different strategies that help them to achieve engagement and motivation among their employees.

This qualitative study is somehow limited because it was for a couple of NGOs, so for future studies, I recommend exploration into a variety of case studies on engagement and motivation. Suggestions could be to establish grounded theories for contributing positively to the literature.

Future studies must include quantitative and mixed research approaches when trying to investigate engagement and motivation, which could ensure a heightened level of information. Definitely, those studies will need critical realist view points and abductive approaches and will simultaneously use deductive and inductive approaches to build theories and test them such that they validate their data and ensure they are internally correlated and reliable once the study needs to be repeated again. Figs. 11.12 and 11.13 illustrate a potential cause-and-effect relationship with decision making and employee motivation (expressed as a change in employee behavior). Recall that we discussed that rewards and incentives are methods of benefits and compensation that managers can use to praise good performing employees. This is referred to as leaders exerting their "reward" power, as opposed to the leaders relying on their "coercive" power to punish employees for bad behaviors [31,41].

Fig. 11.12 links the summary the literature review with the summary findings, in a way to review the research purpose, in terms of COVID-19 [18,52]. This relationship was also noticed throughout this research, and Fig. 11.13 shows a hint of what decisions managements take to control the rewards and benefits in a way that benefits and compensations could prevent or accentuate turnovers [1,3,9,10,13,18,52].

REFERENCES

[1] Martoccio J-J. Strategic compensation: a human resource management approach. 7th ed. Prentice Hall; 2010.

[2] Lawler EE. Rewarding excellence. Wiley; 1999.

[3] Mondy R-W. Human resource management. 12rh ed. Pearson; 2012.

[4] Wertz K, Bryant J-J. managing workers' compensation: a guide to injury reduction and effective claim management. 1st ed. 2000 [Boca Raton].

[5] Gomez-Mejia L, et al. managing human resources. 8th ed. Pearson; 2016.

[6] Sirkin MS. Executive compensation. 11th ed. 2006 [New York].

[7] Ganta VC. Motivation in the workplace to improve the employee performance. Visakhapantnam: Andhra University Department of Commerce and Management Studies; 2014.

[8] Skirpak SJ. Motivating employees. In: Fundamentals of business; 2016 [Virginia].

[9] Gomez-Mejia L, et al. Managing human resources. 7th ed. Pearson; 2011.

[10] Mondy R-W, Martoccio J-J. Human resource management. 14th ed. Pearson; 2016.

[11] Kahn W, Yawar I. An investigation of the relationship between work motivation (intrinsic & extrinsic) and employee engagement: a study on allied bank of Pakistan. Sweden: Master's Thesis Submitted to the Umeå School of Business, Economics and Statistics; 2013.

[12] Riley S. Herzberg's two- factor theory of motivation applied to the motivational techniques within financial institutions. Eastern Michigan University; 2005.

[13] Jeha H. The impact of compensation practices on employees' engagement and motivation. MBA Thesis submitted to the School of Business, at the Lebanese International University; 2020. Summer 2020.

[14] Hattangadi V. Theory X & theory Y. Int J Rec Res 2015;2:20–1.

[15] Maslow A. Motivation and personality. 1st ed. Harper & Brothers; 1954.

[16] Maslow A. Motivation and personality. 5th ed. Harper & Brothers; 1987.
[17] Bloisi W, Cook C-W, Hunsaker P-L. Management and organisational behaviour. 2nd ed. Maidenhead: McGraw-Hill Education; 2007.
[18] Rogoff K. Mapping the COVID-19 recession. 2020 [Online] Available at: https://www.project-syndicate.org/commentary/mapping-covid19-global-recession-worst-in-150-years-by-kenneth-rogoff-2020-04. [Accessed 3 March 2021].
[19] Hertzberg F, et al. The motivation to work. John Willey & Sons; 1959.
[20] Deci EL, Ryan RM. Intrinsic motivation and self-determination in human behavior. New York: Plenum Press; 1985.
[21] Alderfer C-P. Existence, relatedness, and growth; human needs in organizational settings. New York: Free Press; 1972.
[22] Salisu J-B. "The impact of compensation on public sector construction workers in Jigawa state of Nigeria". A thesis submitted in partial fulfilment of the requirement of the University of Wolverhampton for the degree of doctor of philosophy (PhD). Nigeria: University of Wolverhampton; 2016.
[23] Ryan RM, Deci EL. Self-determination theory and the facilitation of intrinsic motivation, social development, and well-being. Am Psychol 2000;55:68–78.
[24] Nohria N, Lawrence P-R. "Driven: how human nature shapes our choices"?. 1st ed. Wiley; 2002.
[25] Malik K. Management: an integrated approach. 1st ed. 2007 [Manchester].
[26] Kahn W. Psychological conditions of engagement and disengagement at work. Acad Manage J 1990;33: 692–724.
[27] MacLeod D, Clarke N. Engaging for success: enhancing performance through employee engagement. 2009 [London].
[28] Chartered Institute of Personnel and Development. Developing managers to manage sustainable employee engagement, health and well-being. 2017. Research report phase 2. London.
[29] Lai E-R. Motivation a literature review. New York: Pearson; 2011.
[30] Johnson M-O. The new Rules of engagement: life-work balance and employee commitment. The Chartered Institute of Personnel and Development; 2004.
[31] Robbins S-P, Judge T-A. Organizational behavior. 17th ed. Penguin; 2012.
[32] Northouse P-G. Leadership: theory and practice. 6th ed. Thousand Oaks: SAGE; 2013.
[33] Saunders M, Lewis P, Thornhill A. Research methods for business students. 5th ed. Pearson; 2009.
[34] Sekaran U, Bougie R. Research methods for business: a skill-building approach. 7th ed. NY: John Wiley & Sons; 2016.
[35] Saunders M, Lewis P, Thornhill A. Research methods for business students. 7th ed. Pearson; 2015.
[36] Cooper D-R, Schindler P-S. Business research methods. 12th ed. McGraw-Hill; 2014.
[37] Creswell J-W. Educational research: "planning, conducting, and evaluating quantitative and qualitative research". 4th ed. Pearson; 2012.
[38] Edson MC, Henning PB, Sankaran S. A guide to system research. 2nd ed. USA: Springer; 2017.
[39] Melkonian R. Measuring the consequence of administrative decisions making on employee motivation. Unpublished MBA Thesis presented to the School of Business at the Lebanese International University; 2020. June, 2020.
[40] Cangelosi J-S. Classroom management strategies: gaining and maintaining students' cooperation. 7th ed. Wiley; 2014.
[41] Hughes RL, et al. Leadership: enhancing the lessons of experience. 8th ed. International Edition: McGraw Hill; 2012.
[42] Mazjoub Z. Business ethics essentials. 2018 [Beirut].
[43] Gerges F-A. "Will Beirut's blast be a catalyst for change"?. 2020 [Online] Available at: https://www.devdiscourse.com/article/politics/1159968-will-beiruts-blast-be-a-catalyst-for-change. [Accessed 12 October 2020].
[44] Martoccio J-J. Employee benefits: a primer for human resource professionals. 5th ed. McGraw Hill; 2014.
[45] Azzi B. AUBMC lays off hundreds of staff members. Beirut Today; 2020. July 18th, 2020.
[46] Zaazaa B. Hundreds of Lebanon's prestigious AUB medical centre staff vow to fight back after mass layoffs. 2020 [Online] Available at: https://english.alaraby.co.uk/english/indepth/2020/7/30/lebanon-medical-staff-fight-back-after-shameful-mass-layoffs. [Accessed 14 August 2020].
[47] Al-Khatib K, Abdallah F. Introduction to business law. 2017 [Beirut].
[48] Yassine D, et al. Principles of accounting. 2nd ed., vol. I; 2015 [Beirut].
[49] Zikmund WG, Babin BJ, Car JC, Griffin M. Business research methods. 9th ed. USA: Erin Joyner; 2016.
[50] El-Bacha F. Coronavirus: L'erreur libanaise. 2020 [Online] Available at: https://libnanews.com/coronavirus-lerreur-libanaise/. [Accessed 14 August 2020].
[51] Youness H. Trends of business. 1st ed. 2017 [Beirut].
[52] Pohanková A. Motivation and decision-making process in managing change within the organization. Hum Res Manage Ergon 2010;4:1–9.
[53] Purcell J. Disengaging from engagement. Hum Resour Manage 2014a;24:241–54.
[54] Oliva PF. Developing the curriculum. 7th ed. New York: Pearson; 2012.

FURTHER READING

[1] Bryman A. Social research methods. 4th ed. Oxford University Press; 2012.
[2] Rebore R-W. The essentials of human resource management". Pearson; 2012.
[3] Everard K-B, Geoffrey M, Wilson I. Effective school management. 4th ed. SAGE; 2004.

CHAPTER 12

Examining the Impact of COVID-19 on Today's Businesses

MOHAMAD KNIO • ALI-ERIN BALIKEL • GEORGES BELLOS

1 INTRODUCTION

Following the recent WHO research, coronavirus disease 2019 (COVID-19) is considered as a world pandemic. This triggered widespread and major employee sackings worldwide, as well as major, never seen before economic depressions forcing many firms globally to downsize over and above the sanitary crises [1]. According to Dr. Kenneth Rogoff, this major recession has had catastrophic implications for the world of business. Similar arguments on the trends of business from Dr. Hasan Youness go hand in hand with strategic planning, leadership, and international business, a few years before this major pandemic started, if one choses to describe or link this period both in Lebanon and worldwide [1,17].

1.1 Research Question

This study ought to respond to this question: *"Does the COVID-19 pandemic harm organizational performance"*? This research question, as it was expressed from the literature, needs to be answered by the end of the study. It will explore the correlation between COVID-19 as a worldwide pandemic and world economies, employment, and ultimately business performances. If the abovementioned relationship is treated in terms of variables, one can assume that COVID-19 is an independent variable, while the business performances can be treated as dependent variables. Potential mediating variables, which will be tested on the degree to which COVID-19 affects company performances, are dire economic situations, employee turnover, and employee motivation. All of which could be related to each other as well [2].

1.2 Aims and Objectives

This study intends to investigate how COVID-19 currently influences business productivity. Internationally, all countries faced problems due to the COVID-19 pandemic [1]. Firms have globally been affected by its impact through massive sackings and major economic drawdowns. As such, a large number of firms were compelled to shut down. Aside from a pervasive sanitary problem, COVID-19 directly caused high rates of unemployment, economic recessions, and even loss of staff work incentive, as well as the lack of mental health awareness [1,3,4].

This study will be done by examining the global impact of COVID-19 on international business performance in order to describe the influence of the coronavirus epidemic on the administrative decision-making procedure. This study links economics, business management, and human resources to explain the economic changes, causing the drop in business performances that ultimately lead to employee demotivation, through terrible economic conditions [1,5].

For Lebanon, both political corruption and the COVID-19 pandemic led to the dramatic drop in the local currency, but the economic indicators showing a defaulting system were observed decades before [1,3,6,7]. As the economic indicators, which were formerly discussed in the McKinsey 2018 report [6,7], were listed, many experts have set off the alarm [1,5–7].

The above respectively discuss major factor that have caused high causes of stress in world populations, triggered by these economic indicators and the coronavirus on business performances [1,4,8].

A study linking dummy variables to political events in light of examining how these phenomena affect business practices could be considered for a future study [9]. For now, these events, similar to those considered in that study, could be studied further in an attempt to link them to COVID-19, as they deserve serious attention. Hence, this study intended to investigate the effect of the link between political events and COVID-19 on business performances [9]. Following James S. Cangelosi's arguments on stressors, worldwide stress levels have

COVID-19: Tackling Global Pandemics through Scientific and Social Tools. https://doi.org/10.1016/B978-0-323-85844-1.00001-5

increased because of the COVID-19 led-economic depressions [1,3,4,8]. This may have caused increased levels of depressions and anxieties. The outcomes of the COVID-19 pandemic led to economic recesses that were much worse than the US 1929 Great Depression and the 2008–10 world recession [1].

1.3 Overview

Unprecedented economic drawdowns have occurred globally since December 2019. In many parts of the world, there were recessions that occurred before the COVID-19 epidemic was declared world pandemic by the WHO. Also, in many countries, as a result of the existing recessions and those accentuated by the coronavirus pandemic, major sackings and unprecedented recessions occurred [1,3,5,10,11].

COVID-19 has forced many major companies to downsize, to lay off staff, or to shut down operations permanently as a result of the COVID-19-imposed fiscal drawdowns, triggered by such unprecedented economic downturns. Dr. Kenneth Rogoff, a Harvard University professor, quoted that these economic drawdowns were never seen before [1]. Although never seen before, the economic depression triggered by the epidemic was so widespread that its impact on global trade may have caused irreversible effects [1,3,4,8].

Practices such as employee turnovers and massive sackings, as well as cutting down on task forces or on allocated wages, regardless of whether they occur in Lebanon or in the rest of the world, are both illegal and unethical as per Lebanese and International Labor Laws [12–17,71,73]. Both Dr. Hasan Youness and Dr. Zeina Majzoub have respectively added arguments pertaining more to the trends of business and to business ethics practices, in Lebanon, and have illustrated various cases as well [17,18].

Because of the global coronavirus pandemic, terrible epidemic international unemployment rates dramatically increased and presented figures as high as 50% unemployment rates in nations worldwide [1,10,11]. Recent figures that appeared on CNN declared that about 35,000,000 COVID-19 cases were registered globally, with a death toll of over 7,000,000 people. In the same broadcast, it was mentioned that both President Trump and the First Lady were tested positive to the coronavirus infection. The BBC, on a daily basis, updates the latest COVID-19 recommendations, like storing shoes at the entrance of the house.

Famous news broadcasts reported (either on BBC or on CNN) that COVID-19 will have a vaccine by mid-2021 and that the world economy will be fully recovered only by 2025. If one stays pragmatic once a global pandemic killing several tens of thousands of people and severely influencing most sectors pervasively, especially tourism and hospitality, then there should be dim hope for an immediate economic boom [1,5]. Moreover, once the tourism sector got affected a worldwide deficit of $1 trillion was reported [1,19,42]. So if one reverts back to the tourism potential of any coastal site (such as that of the Levantine coast) that attracted tourists for site seeing, or for luxury holidays, it will definitely be severely affected by the COVID-19-induced recession, as travels to destinations were halted [19].

2 LITERATURE REVIEW

Today, some of the economic issues are rooted in banking issues seen since 2011 in the Arab world, explaining some economic hardships faced by Lebanon [20]. These arguments need to be considered to see how far back we can trace the defaulting mechanisms. In Lebanon, the poor management of COVID-19 by the authorities is quite evident [21]. So Lebanon faces several economic problems and the banks exert capital control. This does not help now, with the coronavirus pandemic [21].

Recent research on the Coronavirus points out that COVID-19 began in 2019 in Wuhan, according to some sources [22]. Yet there are some controversies about this disease that still have puzzled experts. Issues pertaining to COVID-19-induced economic recessions and unemployment have never been seen before [1]. Very pertinent arguments and indicators are shown in the Middle East, Europe, and the United States [1,2,9].

On one hand, job satisfaction is a measure of the workplace satisfaction of staff [23]. At the same time, it is imperative for firms to understand the importance of human capital, as their survival and their professionalism are contingent on staff, which need to be motivated and satisfied at the workplace [24]. On the other hand, work motivation is a measure of the interaction of internal and external forces that trigger work-related behavior and regulate its shape, course, concentration, and extent [25].

The phenomenon of job mobility in turnover studies was studied before [2,26]. The link between job mobility and job turnover intentions was analyzed [26]. One of the ways to cater for employee empowerment (or motivation) is reward management [2]. However, in general, rewards enhance staff motivation (extrinsic) [27]. Work motivation is to have well-motivated employees and having managers with the possibility to motivate others. This can help in cases of pandemics like now with COVID-19 [28]. Low job satisfaction will increase voluntary job terminations, absenteeism, unpunctuality, and even weak mental health. These were all caused by the COVID-19 pandemic globally [29].

These two theories can be argued independently from each other, as it is questionable that the extent to which an employee is satisfied at work is dictated by the presence of factors and circumstances that motivate an employee [30]. Yet, given today's financial situation, motivated employees characterize both a competitive advantage and a major strategic asset in any work environment [31].

The studied impact of strategic human resource management (HRM) on organizational performance can prove useful in these times [32]. However, decisions are important in measuring motivation. There is a direct link between decision-making and productivity (or job performance) [33].

Contradictory arguments, stressing on motivation and job satisfaction follow each other and respond to various firms and job-related variables, could be raised [34]. Decision-making affects employee motivation, which is useful to investigate now [35]. There are alternatives to attrition as specialists in HRM stipulate. Their arguments stress that there are alternatives to attrition, as for them, mass layoffs are bad for employee motivation and company productivity [13].

Performance appraisals are key in decision-making. It is believed that it is not normal to massively fire staffs just to cut costs. This is counterproductive [73], as issues pertaining to major recessions or depressions were also raised and were pertinent to COVID-19 cases [1,3].

Companies need to consider motivating employees to enhance productivity by various means through a confirmatory study assessing the role of employee motivation. A study currently wishes to see how these results can be applicable to the current COVID-19 situation [36]. Employee motivation through different phenomena was also observed in a study analyzing this behavior in the public sector. But COVID-19 knows no distinction between sectors [1,37].

A study was conducted on low-skilled workers' intrinsic motivation. It measured the link among intrinsic job quality, intrinsic motivation, skill use, low-skilled work, participative decision-making, and independence [38]. The following research, investigating employee incentives, job satisfaction, competitive advantage, human capital, and strategic HRM, provided information that is not always possible to explore in times of crises, if the results were to be applied in the COVID-19 period [1,39].

Empowering women and treating women equally in the workplace to enhance motivation is the current trend in education and in HRM practices; however, it was proven that women are better leaders than men. Furthermore, female employees should be empowered for enhancing their motivation in the workplace, under any situation [40,52].

Employee participation in decision-making can also be considered as a way of employee empowerment [41]. Employees can get empowered once they participate in companies' decision-making processes. Empowerment enhances employee motivation (intrinsic), as also mentioned by other specialists, even in the COVID-19 situation when employees choose intrinsic rewards over cash rewards, as evidenced in the humanitarian sector [2,27,42].

The difference between constructional and expansionary policies is important to be mentioned, as these details help in the studies of how respectively world states and central banks ought to get us out of recession and to go back to equilibrium conditions [43]. The link between motivation and organizational performance is important because as firms keep employees motivated, their performance will increase irrespective of the macroeconomic factors [43,44].

The relationship between both variables which are related to each other, was also measured. This may be useful in these tough times, i.e., with COVID-19 [1,45]. The relationship between employee motivation and work performance, which are related to each other, was also measured. This link demonstrated that as motivation increases, so does performance [1,46].

Employee motivation and satisfaction stipulated that these studies are linked to studying the dynamics of motivation in general, as they could be useful for studying the case during COVID-19 [1,47]. A research on the link between strategic HRM and company productivity has shown important results for application during the COVID-19 pandemic lockdown period [1,48]. Studying the strategic approach to HRM during times of crises is important to manage in times of crises or now with COVID-19 [1,49].

The decision-making map helps the management to take the right decision. The detailed motivational model can be replicated for the current research [50]. Decisions affecting employee motivation and other variables, related to them, could be influenced by the COVID-19 pandemic [1,51].

The following study also believes that motivation can be enhanced through empowerment, as it enhances employee motivation [52]. Employee motivation was studied in an exploratory analysis of the part that motivation plays in the workplace. The study indicates that motivation is a very essential part of HRM practices [53]. Motivation is affected by performance appraisals, depending on the type of motivation and if performance evaluations are used to fire people or not [54].

Yet, companies need to consider motivating employees to enhance productivity through various means. This is related to studying motivation by proposing circular economic systems, unlike the Lebanese rentier economic systems [55].

The impact of employees' motivation on business efficiency was also studied in relation to company productivity. This research could link organizational effectiveness with the current times under COVID-19 [1,56].

Company decision-making is linked to employee motivation and enhances empowerment [57]. This, on the other hand, discusses issues that can undermine employee empowerment. As unethical procedures occurred in Lebanon during COVID-19 and since the October 17, 2019, anticorruption revolution, employee job safety and satisfaction have dropped a great deal [1,17].

Dr. Zeina Majzoub's arguments are pertinent, and Chapter 8 in her book is fully dedicated to employment ethics [17]. Dr. Louis Hobeika presents discussions about countries having high-interest rates and believes that it means the countries have weak currencies, as an elevated interest rate can be linked to a bribe for citizens to opt for the national currency [7]. His arguments are pertinent and link to the contractionary monetary policies, a fact that rentier (and not cyclical) economies rely on, which are deemed as unethical practices [7,17,55].

Dr. Jad Chaaban's article presents important arguments that align with the arguments of Dr. Louis Hobeika, another economics expert who also agrees that high interest rates lead to weaker economies [6,7]. This is especially true when saying that haircut and capital control are illegal and unethical practices [6,7,16,17]. This is iterated by who advises that legal issues need to be considered. The capital control and haircut mechanisms should be further reviewed in this case and are measures of heavy capital defaults [16,17,43].

Career advancement is a way to measure performance or productivity in the workplace. This is a form of motivation, or a measure of job safety/satisfaction, as this study shows [57,86]. Career fulfillment and work motivation are also linked. The more an employee is motivated in the workplace, the better his/her job fulfilment will be. This relationship links career fulfillment, self-esteem, and work motivation together [58]. Employee motivation can be measured by how discourteous behaviors can affect job satisfaction. The study analyzing this link checks how much workplace incivility affects motivation [59].

2.1 COVID-19-Induced Recessions and Layoffs

Major fiscal drawdowns were faced both in Lebanon and worldwide anywhere from Summer 2019 to January 2020 [1,10,11]. COVID-19 enhanced the economic drawdowns, by causing major firms to totally shut down and major employee sackings, among other crises that were observed across various studies from multiple industries [1]. Yet, if one goes back stating the past three decades, one can understand by looking "for the optimal blend in turnover" what may happen once one observed what went on in both the private and the public sectors in such recession situations [3,5,26,60−65].

What happens once a company undergoes a mass employee turnover has been well studied in previous research [1−3,5,10,11,63]. So, as studied before, the loss of key staff and the resulting drawdown in productivity are major effects of such turnovers [2]. Similarly, the fiscal problems that arose in the past are also a major trigger for massive turnovers [63].

As this research will take a few examples from various countries to demonstrate how the coronavirus pandemic affected business performances, examples from Lebanon, France, and the United States will be compared on how, across continents, institutions faced the economic hardships as affected by COVID-19 [1,10,11,20]. So some of the cases that this research will examine will help us better understand the situation. According to Dr. Kenneth Rogoff, and others, several privately and publicly owned institutions, which faced several hardships that were triggered by the coronavirus epidemic, had to take very hard managerial decisions, such as massive sackings [1,5,10,11].

Although the medical sector suffered from extremely high turnovers, the employees are still needed as frontline workers, if one sees what happened in Lebanon [1,10,11]. Worldwide, hospital employees, street sweepers, and garbage collectors faced extremely harsh working conditions that have caused increased burnouts, leading to increased job terminations in these industries, [1,5,10,11].

Way before the coronavirus epidemic, the existing socioeconomic conditions were already very harsh [1,5,10−12,26,64,66−71]. Galletta et al. [70] Many additional external threats, also triggering massive employee sackings in some sectors, which could even include loss of staff, or key knowledge because of death, among other causes, may have influenced the administrative decisions [5,10,11].

Several cases involving universities, banks, and several institutions in the hospitality sector, which all replied differently to the economic hardships triggered by the coronavirus epidemic as related to managerial decisions, were investigated [1,5]. In most cases, the sector that was the most affected by COVID-19 was the hospitality and the tourism sector, with losses of up to $1 trillion recorded during the COVID-19 period [1,19,42].

Other research discussing similar cases related to the unethical mass layoffs that occurred during the pandemic were studied as well [1,5,17]. It is to no surprise that major economic drawdowns forcing firms to downsize, fully shutdown, or cut down on labor costs will actually make them decide to mass-fire employees in such cases, irrespective of whether it will negatively harm employee motivation or organizational performance [1,5,10,11]. Both economic drawdowns and worldwide pandemics lead to substantial employee dismissals [1,2].

Finally, through a previous study showing the relationship between taking the right decisions and enhancing employee empowerment and motivation was investigated. It provides the same message of the "optimal turnover blend" that was referred to [63,72]. The takeaways one can get after conducting this research is that it is not the best option to categorically lay off workers especially in times of crises, if the firms considering mass firings as a way to survive amid the COVID-19 times [1,2,5,10,11].

A good way to fire workers is through performance appraisals and by firing only the poorly performing ones [13,73]. This would be in lieu of the pervasive staff attritions as was the case with the Al Habtoor Hotel and American University of Beirut Medical Center (AUBMC) or any other place that had to fully shutdown or downsize its operations due to the COVID-19 pandemic [3,5,10,11].

2.2 Types of Employee Layoffs

Employee layoffs are categorized as either voluntary or involuntary, based on whether the employee choses to leave the job or is forced to do so, respectively [71]. According to previous studies on employee layoffs, the types of employee turnovers that have occurred during the COVID-19 pandemic are involuntary, and basically are permanent attritions, as firms massively firing staffs will not renew their task force. Yet, this occurs when companies oblige a member of the workforce to depart, and these types of turnovers usually occur due to poor performance, hiring errors, team incompatibility, and conflict in the workplace, but are not necessarily caused by major pandemics [13,73].

In this research, the concept of employee turnover, as induced by the COVID-19 pandemic, will be referred to here as the nonrenewable turnover. Turnovers are defined in terms of sales, accounting, or even staff and involve the degree of staff replacements within a period [13,14,73]. In general, staff layoffs occur once members of the workforce depart from their jobs and get replaced by others [73]. But employee attritions involve the nonrenewal of departing staffs, where alternatives to mass layoffs exist for decreasing labor costs [13].

Within the scope of this study, during major pandemics, the staff firings may be pervasive and nonrenewable, and this is what the research wishes to study. This can be expressed by the mass firings in the case of AUBMC, in the early Summer 2020, as it nearly laid off 50% of all of its employees [10,11]. Such massive attritions have also begun during the Lebanese economic crisis, which increased employee dissatisfactions and job unsafety [10,11]. These were caused by improper administrative decisions seen since October 2019 in Lebanon. As such, the study needs to analyze the effect this crisis has on business performance and to see if it plays a role in employee motivation and how it affects staff turnovers [2].

Companies that must operate in periods of crises need to keep their employees motivated [13,73] One of the ways they can do so is by taking the right management decisions and that these work best to keep their workforce satisfied and motivated, irrespective of whether the world is currently facing an unprecedented pandemic or not [5]. Therefore it takes the right management decisions, in spite of the situation an institution faces, to keep or to lose employees and to measure its effects on organizational performance [5].

2.3 So Which Decisions can be Considered Good in Such Terrible Cases?

Irrespective of the situation, employers and company decision makers must empower their employees to keep them motivated, especially in times of pandemics [1,5]. This is essential for a company's success in such times as the decisions considered need to be catering more on its employees' needs than on their discipline and guarantee them job safety [5]. Finally, yet importantly, employers ought to provide benefits and rewards to good employees, to keep them satisfied on the job, and ought to fire the bad ones, as layoffs need to be solely dependent on performance [13].

With the COVID-19 crisis, massive firings cannot be omitted, but a company's top managers can consider smarter ways to cut down on losses and still keep the productivity going [10,11]. In spite of the tough times that are witnessed globally, staff firings may be needed [5]. Yet the employees who must be fired ought to be laid off based on performance and not solely on cost-cutting schemes. This is mainly because companies that consider downsizing need to perform optimally. If not, this could cause employee demotivation [73]. This is why, before that, performance appraisals must be carried out [13].

3 RESEARCH METHODOLOGY

Several important research methodologists clearly define business research techniques and how one can use monoqualitative, monoquantitative, or mixed research methods [74–78]. Fig. 12.1 shows the steps to be taken to select a business analyst philosophy, approach, or method choice. When breaking down the process into layers, the selected structure becomes subjective and descriptive after descriptive thought and induction mechanisms [71].

Some organizations use descriptive research to explore areas that have not previously been investigated using standard methods, mostly working with mixed research methodologies [77,78]. Statistical (or numerical) studies usually involve descriptive research using causal relationships or experiments, which will be explored through the surveys we will be running [76]. Fig. 12.1 shows the summary discussion of the selection of the research methodology that was used to design our research model.

As the initial methodology stipulates descriptive analyses of secondary data, neither statistical software, for interpreting quantitative data (SPSS or Excel), nor NVIVO, for interpreting interview questions, will be used, as they apply for primary data [78]. Although we will be running a survey (quantitative) as well as collecting secondary data (qualitative), we will still be mostly relying on qualitative methodologies, with a numerical component [77].

As we did a descriptive study, sampling was not used because we did neither an interview nor a questionnaire to disseminate [78]. Yet, as a possible outcome of our research, our conclusions and recommendations need to be representative of the data we had collected from secondary sources and online, so we are considering a pilot study of 10 questions.

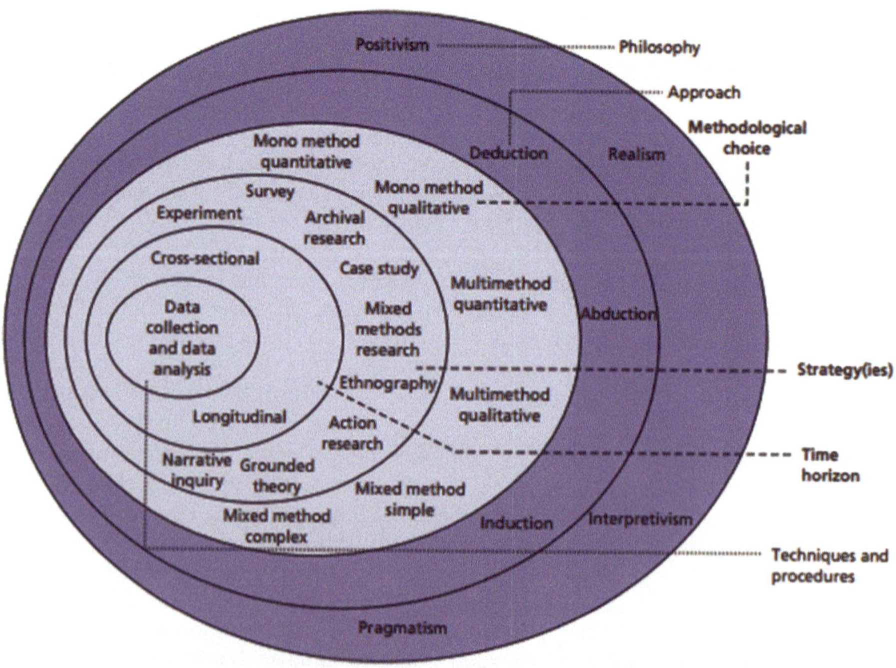

FIG. 12.1 Research onion. (Saunders M, et al. Research methods for business students. 7th ed. Pearson; 2015.)

The various data, which were collected from various sources from the literature review, provided us some insights including HR-specific findings regarding HR best practices in crises [13]. Fig. 12.1 shows our research design that will be recapped at the end of this section. The research design basically relies on monoqualitative research methods, but with a small 10-question survey to test the results, which is used as a pilot study for such purposes [76,78].

So this section reviews in brief the methodology that will be implemented for collecting and analyzing data by using monoqualitative research methodologies using the following model. This proposed model, shown in Fig. 12.1, provides an interpretive relationship of the factors affecting organizational performance, which was treated through various means. Employee layoffs (treated as either attritions or turnovers, based on if the workforce gets renewed or not) with respect to the observed economic indicators that eventually led to the 2019 economic crisis were measured [5]. Yet we will not examine any of those and just interpret these parameters to see how they are influenced based on the relationship between COVID-19 and the managerial decisions.

As stated earlier, we chose to bring about the findings from previous research, which investigated similar phenomena, to properly understand how exactly the COVID-19 pandemic-induced managerial decision-making affects employee motivation, which shows the same relationships [1, 5]. Our motive is to describe and to understand the relationship between the potential identified interdependent variables (the COVID-19 pandemic) and (managerial decision-making) and dependent variables (organizational performance) that could be measured quantitatively in the future, if we study the case of AUBMC in more detail [1,5,10,11].

Our research design favored descriptive analysis and narrative inquiry (interpretivism), literature review through data analysis (positivism), and an approach used to criticize what we studied (critical postmodernism). All these views lead to the qualitative research, as is shown in Fig. 12.2 [79,80].

4 FINDINGS

For the purpose of this research, we divided the chapter into three parts. First, and following a comprehensive review of the literature, where the selected variables we needed to investigate were clearly defined, we found some cases that need further investigation [1,10,78]. Then several secondary data were collected and interpreted following recent trends in HRM practices following these cases [2,71]. Finally, through the selection of the descriptive methodology (after selecting the narrative inquiry and case study as a research strategy, as we are going for a multimethod qualitative study), we decided to describe and interpret the patterns we saw from the literature.

As we had to rely on qualitative research deigns, we found out that there was direct link between the COVID-19 pandemic and managerial decision-making [79]. The decision-making process in turn seemed to affect employee motivation, job satisfaction/safety, turnover, productivity, and organizational performance [81]. Employees and managers alike lose job satisfaction and motivation once their jobs are no longer secure, and in these times, just as what happened to AUBMC, or at other institutions in Lebanon, across sectors [10,11].

4.1 Cases

This research selected as a main case study the situation of mass turnovers that occurred at an institution in the educational sector in Lebanon, which describes the mass sackings at the AUBMC [10,11]. Although in Europe, the United Kingdom, and the United States,

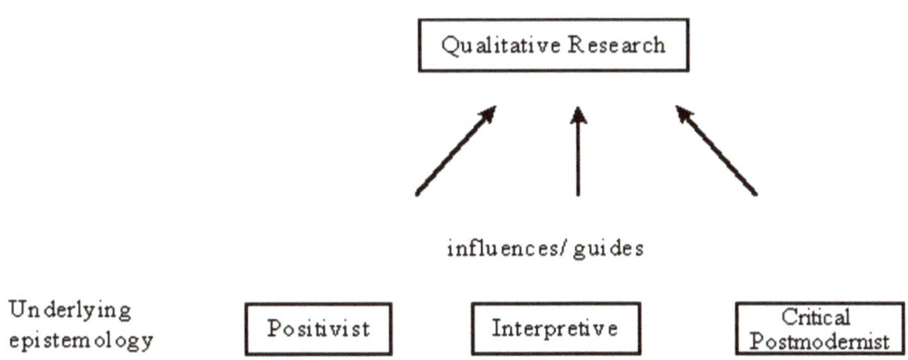

FIG. 12.2 All views contributing to qualitative research designs [78].

unlike the Lebanon government bailout plans helped firms cut down on losses, in Lebanon, the successive crises, government mismanagements, and corruption gave no room for subsidies to assist firms facing losses in times of COVID-19.

4.1.1 TheAmerican University of Beirut Medical Center mass firings

This case illustrates a company in Lebanon from the educational sector that massively fired its staff in Summer 2020 due to COVID-19 and the dire world economic cases [1,10,11]. Here we are presenting the finding of the descriptive statistics from this case. We are just showing the chart representation of all the cases we chose to discuss, through narrative inquiry (and case studies) [10,11]. For this purpose, some statistics were collected from the official American University of Beirut (AUB) website and from two articles on the case of the firings (Table 12.1). These illustrated what happened during Summer 2020, discussing the plan of the AUB administration to conduct a massive attrition on its staffs in order to cut off on labor costs [10,11]. Table 12.1 shows this rapid statistic about employees.

A good theoretic base to consider while looking at the AUBMC case is like a PhD thesis looking at similar variables for the selection of the variables [2]. This was used in order to shed light on some of the parameters we will be observing ourselves, namely, through the sound methodology, which were used to collect data and interpret them, in terms of their relationships to COVID-19, and which parameter influences the other [1,2].

Even though, in a previous study we checked during our literature review, the educational sector was not hard hit as the hospital industry had been, this shows different evidence.[1] Our findings stipulate similar results to the previous study as it concluded that there is an unquestionable doubt that a link exists between the decisions that encourage employee turnovers will harm staff motivation and after such a case of mass firings, the remaining employees will lose job satisfaction, no longer feel secure, and lose all/most workplace motivation [81].

This mass sacking of the AUB and the AUBMC employees totaling to 2350 employees in mid-2020 was a case that has been well documented when the institution took the decision to deal with massive layoffs [10,11].

TABLE 12.1
Staff Numbers Before and After the Massive Sackings at AUB and AUBMC Took Place.

Academic staff	1200
Administrative staff	4340
Fired staff (June 2020)	1500
Fired staff at AUBMC (July 2020)	850

AUB, American University of Beirut; *AUBMC*, American University of Beirut Medical Center.

4.1.2 COVID-19 cases in Europe, the United States, or other countries affecting company performance

As a massive turnover case was discussed in a developing country (Lebanon), the AUBMC was selected to present this situation [10,11]. So an attempt was made to see if the recent mass sackings in the AUBMC in Lebanon can also be observed in a developed nation. As such, after collecting interview results from managers as part of the interpretation of secondary data collection, the researchers chose to sample nations by interviewing managers and we selected two cases from France [1,5]. With the Summer 2020 that started, many beaches, hotels, and restaurants had to struggle with the pandemic to stay in business [1].

Remember that the Lebanese government proposed no subsidies for industries to restructure, as opposed to the French government or other nations that provided government aids to help businesses rebuild. For example, France received part of the €600 billion aid for relaunching its economic activites.[2] The same applied to the United Kingdom, the United States, and the rest of Europe.

4.2 Interview Results

Here we are affixing the interview themes table (see Table 12.2) collected from a previous study that was used for building the arguments that were interpreted following the methodology carried out in a research done in Ireland [40]. The study, using interview themes coding, was relied on to interpret our own data a second time, as they had to be linked to COVID-19 [1,40]. So a simpler form for the coding was used, based on the research we investigated on the effects of the harsh economic crises on

[1]The AUBMC case has been documented in both the July 20, 2020, and July 24, 2020, l'Orient le Jour issues. We chose to mention it here because these issues have been widely documented as we were collecting our own evidence.

[2]In the early phases of the lockdown, the European bailout money was €600 billion. By June 2020, the amount was raised to €750 billion, €250 billion in cash and €500 billion in grants

TABLE 12.2
Preliminary Themes Table Collected From One of the Previous Studies We Consulted for the Research (see [40] for the Methodology Used).

Questions	THEMES		
1	No panic	Slight/some panic	Full panic
2	No loss		
3	Productivity unaffected		
4	Trainings online	Postponed	Not applicable
5	Management helping		Unethical
6	Goodwill		No goodwill
7	Management helping		Management not concerned
8	Safety procedure with fumigation		Safety procedure without fumigation
9	Helping and HR present	NO HR team/ethical	NO HR/unethical
10	Postponed/canceled	New hires	No new recruitment

HR, human resources.

management decisions, employee motivation, and ultimately on employee turnover [1,2,40].

These results constituted the initial study that was studied using the interpretation of secondary data. These data were then tested on a random population through a pilot study, the results of which will be discussed in the next section.

The data related to COVID-19 were collected based on the arguments of the research we analyzed that provided the secondary data we analyzed for our study [1,10,11,21,22]. Below are the question themes addressed, and Table 12.2 presents the general idea/themes, which were also sampled in a pilot study survey, conducted during the research, and where the answers of the first 10 respondents were considered for the pilot study.

Question 1: Addressed if there was panic in the institution or not.

Question 2: Addressed how institutions deal with loss of staff or knowledge.

Question 3: Addressed if there are losses do they affect productivity at the organization.

Question 4: Addressed if the institution considered canceling trainings.

Question 5: Addressed if the organization prevented turnover and if management helped.

Question 6: Addressed if the institution helped its staff cope with depression and anxiety.

Question 7: Addressed if the organization had high job insecurity, if management acted.

Question 8: Addressed if the institution had measures against the spread of COVID-19.

Question 9: Addressed if the organization had any difficulty if the management handled it.

Question 10: Addressed if the institution canceled any new recruitments during lockdown.

4.3 Pilot Study Results

A 10-question pilot study survey was conducted on a random population sample composed of mid-aged professionals. Question 1, "*there was panic in the institution or not,*" yielded 60% of the respondents to be neutral in regard to the statement, whereas 30% agreed and 10% disagreed. Note that the interview themes results are shown in Table 12.2, as similar results were displayed.

For question 2, "*The institutions has dealt with loss of staff or knowledge,*" 70% of the respondents agreed and 30% disagreed. Note that the interview themes results are shown in Table 12.2. A similar finding was recorded for question 3 "*If there are/were losses, they affected productivity at the organization*". The only difference being that 20% disagreed and that 10% were neutral. Note that the interview themes results are shown in Table 12.2. This shows a link between questions 2 and 3, showing that loss of staff do affect company productivity, and the results differ based on sampled sectors, as employees saw the issues differently than the managers.

Question 4, "*The institution considered canceling trainings,*" showed 20% of the respondents to agree, 70% to disagree, and only 10% to be neutral. Note that the interview themes results are shown in Table 12.2, as similar results were shown.

Question 5, "*The organization prevented turnover, and the management helped to cope with it,*" showed 70% of the respondents to agree, 20% to disagree, and 10% to be neutral. Note that the interview themes results are shown in Table 12.2. Results differ based on sampled sectors, as employees saw the issues differently than the managers.

Question 6, *"The institution helped its staff cope with depression and anxiety,"* showed 40% of the respondents to agree and 60% to disagree. Note that the interview themes results are shown in Table 12.2. Results differ across sampled sectors. Results differ based on sampled sectors, as employees saw the issues differently than the managers.

Question 7, *"The organization had high job insecurity, and the management helped to cope with it,"* showed 80% of the respondents to agree and 20% to disagree. Note that the interview themes results are shown in Table 12.2, and similar results were shown.

Question 8, *"The institution has/had measures against the spread of COVID-19,"* showed 56% of the respondents to agree, 33% to be neutral, and 11% to disagree. Note that the interview themes results are shown in Table 12.2. Results differ based on sampled sectors, s employees saw the issues differently than the managers.

Question 9, *"The organization has/had any difficulty if the management handled it,"* showed 70% of the respondents to agree, 10% to be neutral, and 20% to disagree. Note that the interview themes results are shown in Table 12.2, and similar results were shown.

Question 10, *"The institution canceled any new recruitments during lockdown,"* showed 50% of the respondents to agree, 30% to be neutral, and 20% to disagree. Note that the interview themes results are shown in Table 12.2, as similar results were shown.

4.4 Analysis

Before we start to analyze the collected data from the literature in Lebanon (and France or the rest of the world), panic cases were seen worldwide about employees fearing for their jobs [49]. Societies today highly depend on good HRM practices, especially in terms of crises and panic situations, as they are primordial to properly deal and manage crisis situations, like the COVID-19 pandemic [1].

So proper HRM practices greatly impact the workforce. In other words, the decisions company managers take will greatly affect its workforce and consequently will affect employee morale and job satisfaction [49,82]. Accordingly, this research highlights why we chose to focus on how a company management effectively dealt with crisis management, now very much on the table with world economic depressions triggered by the COVID-19 pandemic and related mass sackings, if there are any [3,11,49,83].

In other words, from the outlook of this research, even though the likely occurrence of future crises may exist, we must prevent them at all costs (at least those we can stop) or cope with them in the best ways possible to minimize the negative effects on the task forces if the top management fails to identify the issues at hand and do nothing to deal with them [1,11,49].

Moreover, this study has collected data by analyzing the cases of two companies in Lebanon and two in France how they dealt with the case at hand [1,11]. The various ways that institutions at company (or government) levels use HRM policies are setup in ways to truly cope with these hard cases, through comparing classical (Lebanon) and modern (the developed world) ways to handle the effects of crises on any company's labor force [1,11,49].

Last but not least, within the last year, several specialists, including Dr. Ken Rogoff, among the experts interviews we collected as secondary data, in the research we based our interpretations from, refer to the hard conditions businesses faced due to COVID-19 in their answers. They present the economic cases tied to corruption-induced financial crisis in Lebanon[3] and to the major COVID-19 pandemic.

Yet several articles greatly criticize how the successive Lebanese governments since 1990 managed the resources in Lebanon, which include the latest Beirut Harbor blast that led to a major catastrophe, on August 4, 2020, at 6:00 p.m [20]. We refer back to the 2018 McKenzie Report and the data available from the www.worldbank.org official website for details on the economic indicators that led the way to the current economic case. We decided to mention them as part of the problem that led to the COVID-19 recession in Lebanon, and worldwide, by looking at several variables [1,83].

We highly encourage our readers to read more on those issues, as they are beyond the scope of our research, but they MUST be raised for any interested reader to dig deeper to the possible root cause of the misery Lebanon has been in since 1990. Recent estimates stipulated our debt to GDP ratio to be around 454% by the end of 2020, yet this value is questionable.

It is suggested to fire poorly performing staff before the good ones, de facto, as we must consider sacking based on performance only, especially in times of crises, to fire the bad and to keep the good [54,71,84]. Thus an existing performance appraisal scheme is adopted by each HR department similar to the one in Fig. 12.3

[3]The articles referred to are those that criticized the absence of government reforms or actions that all governments failed to take in order to avoid all the crises.

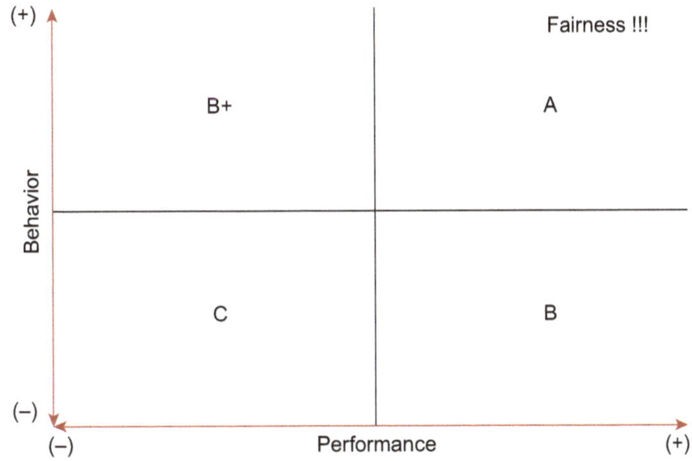

FIG. 12.3 An alternative to massive sackings can be utilized in times of crises (see [5]). The best performing staff are graded as "A" and the worst as "C" if we exclude the origin (i.e., the point of junction of the abscissa and ordinate), companies based on performance evaluations for the bad employees, displaying terrible performance and behavior before firing the good ones. All experts agree that the best performing ones must be fairly treated and kept at all costs especially in times of crises (cf. [71]).

[13,73]. This can also be recommended for future HR practices in case firms are not relying on performance evaluations as a means to lay off poorly performing employees.

5 CONCLUDING REMARKS

5.1 Conclusion

There were several factors that affected industrial performance, including the Lebanese October 2019 anticorruption revolution and the COVID-19 pandemic [1,2,5,22]. Even the educational sector faced massive sackings, when the AUB and its medical center laid off nearly half of its workforce [10,11].

The effects of the financial crisis (induced by the COVID-19 pandemic) forced institutions like AUBMC to massively fire half of their workforce by taking a decision that is highly negative for the remaining employees, as their job safety, satisfaction, and motivation will wane as a result [1,2,5,10,11,83]. These are capital to comprehend staff demotivation, and the resulting voluntary staff departure, which will happen henceforth [10,11].

In the long run, one means to get somewhat good incomes to develop entrepreneurial skills and to establish a business that can flourish in times of crises. In the literature, some of the respondents in the surveys agreed that entrepreneurship can help secure incomes as well as working job and income security [5].

All targeted stakeholders have been identified objectively. They have been approached due to the outcomes of both the 2019 Lebanese October 17th Revolution and current COVID-19 pandemic lockdown that have raised awareness of increased cases of depression, anxiety, lack of mental health awareness, and other stressors that led some employees to terminate their jobs, or worse in certain cases [5].

Answering the research question(s), *"Does the COVID-19 pandemic harm organizational performance?"* was the purpose of the study. After we studied real cases in Lebanon, like with the AUBMC, where massive sackings occurred, we believed that yes, the coronavirus pandemic did cause this institution (and many others too) to perform less optimally [1,10,11].

As the AUB and its medical center has to lay off nearly half of its staff, the institution had to downsize, de facto. Although the AUB faced a problem forcing it to downsize, many others had to shut down operations totally, as other studies have investigated.

As such, the research objectives denote a link between the COVID-19 pandemic and organizational performance both in Lebanon and in France. The case in France, however, is different, unlike Lebanon, in both its response to the coronavirus outbreak and coping with it.

Unlike the Lebanese government, the French government subsidized the industries and all major sectors not to go bankrupt. This is why there were not as much institutions in France that were compelled to shut down.

As a result, our conclusions concur with our literature review findings. The study's results are clear and *'add to the understanding of the issue portrayed in the literature review'*. Although there are no major new discoveries, this study presents a new outlook on the impacts of the coronavirus epidemic on business performance, both worldwide and in Lebanon.

Future research implications are important. We need to focus on the lessons learnt and in the takeaways from both Lebanon and the rest of the world and clearly show that we need to increase employee motivation, empower them, keep their job environments safe in order for them to be satisfied to work, and for them to perform better, as was demonstrated in past research [1,2].

Once you do that, your employees will want to work for you even in times of major pandemics, and this is how major companies were able to keep their task force unlike others who had to massively sack their workforce, in order to downsize. Mind you, this is irrespective of pandemics or economic crises [1].

As a result, the decision-making process was in many cases a response to the crisis at hand, without considering the damaging effect it had on employee motivation. Yet some mangers made it a point to work on employee motivation to keep them satisfied such that productivity does not drop, whether at a government level like in France or like several companies in Lebanon that refused to lay off a single worker, even though the COVID-19 economic crisis makes life difficult [1].

5.2 Recommendations

The first point to consider is limitations. The initial study took five weeks to complete, so it is of utmost importance to consider a future study that has a longer time span, which can generate a good dataset to be validated and reliable. So a larger sample is suggested for the quantitative data collection, relying on the current pilot study results [78].

As we are entitled to criticize the entity in charge of solving the COVID-19 economic crisis case (or not), we provide a word of appreciation for France, and its President, Emmanuel Macron, as it is attempting to help its people through these tough times. As such, we suggest the best course of action based on the findings is to provide the Lebanese Government with a useful guide that helps it to finally bail out its industries from bankruptcy, just as France and other countries in the world did.

The terrible economic situation in Lebanon and the COVID-19 pandemic have caused massive employee turnovers across multiple sectors, both in Lebanon and worldwide. Therefore we recommend that firms exposed to such harsh strains on their operations to reconsider mass sackings of workers, unlike what the AUBMC did [1,5,10,11].

As massive employee attritions harm, instead of aid, organizations regardless of the harsh financial situation, we need to promote various modes of easing to alleviate the strains or by amortizing the costs. We need to treat the best behaving and preforming employees fairly and attend to keep them if you do not want to lose them, as it is highly unfair to depart from such outstanding members of your workforce [2,17].

Per se, it is better to use any performance analysis chart in order to select which employees to fire first and which ones to fire last (see Fig. 12.3). This makes sense, as the situation does not permit firms that have problems in operations to stay in business without shutting down and to retain all their workforce, regardless the situation.

Therefore in such situations, they can keep operating, shut down unproductive branches, reallocate the current staff from the branches that were closing to the active ones, and thus keep their task force (after laying off the bad employees).

Definitely, entrepreneurship is an option to consider, especially in cases like these, where employee motivation is at stake [5]. The company may want to propose special classes for its best performing staff to get the chance to become entrepreneurs and may be to help them getting promotions or by increasing incentives and bonuses.

As indicated, mass firings (at institutions such as AUBMC) harmed staff job satisfaction a great deal. Consequently, unsatisfied workers will lose motivation and perhaps add to more turnover when they willingly terminate their jobs. This will truly hamper a firm's productivity, revenue, and organization performance. Hence, we recommend any firm's management to focus on investing in its workforce. It should also keep the employees motivated, such that industrial output can improve, especially in times of crises.

As some people believed that massive turnover did not hurt the economy, affect employee motivation, and lead to poorer performance, we had a hard time to understand why they act distracted. We urge company owners and senior managers to train these people in the harmful effects that a massive turnover has on sacking company task forces and the negative effects it has on the existing employees who fight to keep their jobs [85].

This simply reiterates that indeed massive layoffs harm employee motivation, affect profits, and ultimately impinge on company productivity, especially in times of major pandemics. So it is suggested to seek for alternatives for mass sackings if the main strategy is to increase organizational performance, as with the performance appraisal graph shown in Fig. 12.3 [1,13,73,86].

REFERENCES

[1] Rogoff K. Mapping the COVID-19 recession. 2020 [Online] Available at: https://www.project-syndicate.org/commentary/mapping-covid19-global-recession-worst-in-150-years-by-kenneth-rogoff-2020-04. [Accessed 13 August 2020].

[2] Brooks A-M. It's all About the motivation: factors that influence employee motivation in organizations. PhD thesis. University of Tennessee; 2007.

[3] Stefaniak J, Baaki J, Blake AM. An examination of the decision-making process used by organizational leaders during the great recession. Perform Improv Q 2012;24: 81–102.

[4] Feldman. Development across the life span. 2nd ed. Pearson; 2016.

[5] Beam J. What is employee turnover?. 2014 [Online] Available at: http://www.wisegeek.org/what-is-employee-turnover.htm. [Accessed 14 August 2020].

[6] Chaaban J. Lebanon needs a budget for its people, not is ruling class. Beirut Today 2019 [Online] Available at: https://beirut-today.com/2019/04/23/lebanon-budget-people/ [Accessed August 2020].

[7] Hobeika L. High-interest rates mean a weaker currency. 2019 [Online] Available at: https://en.annahar.com/article/1039542-highinterest-rates-mean-a-weaker-currency. [Accessed August 2020].

[8] Cangelosi J-S. Classroom management strategies: gaining and maintaining students' cooperation. 7th ed. Wiley; 2014.

[9] Zach T. Political events and the stock market: evidence from Israel. Int J Bus 2003;8(3):243–66.

[10] Azzi B. AUBMC lays off hundreds of staff members. Beirut Today 2020. July 18, 2020.

[11] Zaazaa B. Hundreds of Lebanon's prestigious AUB medical centre staff vow to fight back after mass layoffs [Online] Available at: https://english.alaraby.co.uk/english/indepth/2020/7/30/lebanon-medical-staff-fight-back-after-shameful-mass-layoffs. [Accessed 14 August 2020].

[12] Kim S. Factors affecting state government information technology employee turnover intentions. Am Rev Publ Adm 2005;35:137–56.

[13] Gomez-Mejia L, et al. Managing human resources. 7th ed. Prentice Hall; 2011.

[14] Yassine D, et al. Principles of accounting I. 2nd ed. 2015 [Beirut].

[15] Martocchio JJ. Strategic compensation: a human resource approach. 9th ed. Pearson; 2016.

[16] Al-Khatib K, Abdallah F. Introduction to business law. 1st ed. 2017 [Beirut].

[17] Majzoub Z. Business ethics essentials. 1st ed. 2018 [Beirut].

[18] Youness H. Trends of business. 1st ed. 2017 [Beirut].

[19] Corluka G, et al. Classification of tourist season in costal tourism. UTMS J Econ 2016;7:71–83.

[20] Gerges F-A. "Will Beirut's blast be a catalyst for change"?. 2020 [Online] Available at: https://www.devdiscourse.com/article/politics/1159968-will-beiruts-blast-be-a-catalyst-for-change. [Accessed 14 August 2020].

[21] El-Bacha F. Coronavirus: L'erreur libanaise. 2020 [Online] Available at: https://libnanews.com/coronavirus-lerreur-libanaise/. [Accessed 14 August 2020].

[22] Huang C, et al. Clinical features of patients infected with 2019 novel coronavirus in Wuhan, China. Lancet 2020; 395.

[23] Spector P-E. Job Satisfaction: application, assessment, cases and consequences. Thousand Oaks, CA: Sage Publications; 1997.

[24] Finck G, Timmers J, Mennes M. Satisfaction vs. motivation. In: Across the board, vol. 25; 1998. p. 55–6.

[25] Pinder C-C. Work motivation in organizational behaviour. Upper Saddle River, NJ: Prenctice Hall; 1998.

[26] Sousa-Poza A, Henneberger F. Analyzing job mobility with job turnover intentions: an international comparative study. J Econ Issues 2004;38.

[27] Zubair A, et al. Employee's participation in decision making and manager's encouragement of creativity: the mediating role of climate for creativity and change. J Serv Sci Manage 2015;8:306–21.

[28] Edhund M, Nilsson H. Employee motivation in medium-sized manufacturing enterprises: to case studies from northern Sweden. Master's Thesis in Marketing presented at the Lulea University of Technology; 2007.

[29] Hellriegel D, Slocum J-W. Fundamentals of organizational behavior. Mason: South-Western/Cengage Learning; 2007.

[30] Furham A, Eracleous A, Chamarro-Premuzic T. Personality, motivation and job satisfaction: hertzbrg meets the Big Five. J Manage Psychol 2009;24:765–79.

[31] Temblay M-A, Blanchard C-M, Taylor S, Peltier L-G. Work extrinsic and intrinsic motivational scale: its value for organizational psychology research. Can J Behav Res 2009;4:213–26.

[32] Nemli-Çalişkan E. The impact of strategic human resource management on organizational performance. J Naval Sci Eng 2010;6(2):100–16.

[33] Pohanková A. Motivation and decision-making process in managing change within the organization. Hum Res Manage Ergon 2010;4:1–9.

[34] Ayub N, Rafif S. The relationship between work motivation and job satisfaction. Pak Bus Rev 2011;13:332–47.

[35] Blašková M, Blaško R. Decision taking in motivating employee. Hum Res Manage Ergon 2011;5:1–14.

[36] Flynn R-D. An exploratory analysis of the role of motivation within the workplace. A dissertation submitted in partial fulfilment for a MA in Human Resource Management. National College of Ireland; 2013. September, 2013.

[37] Paillé P. Do coworkers make the service customer? A field study in the public sector. Rev Public Person Adm 2013;33:28–57.
[38] Janssen A. Intrinsic motivation in the context of low-skilled work: the influence of intrinsic job quality and demands-abilities fit. Masters thesis in Human Resource Studies presented to the University of Tilburg; 2014.
[39] Khosorowshahi S, Nejad I-A. The relationship between the employees' motivation, job satisfaction and financial development of parsian hotels chain in Tehran. Sweden: Masters Thesis presented to the Luleå University of Technology; 2014.
[40] Nava M. Exploratory study of leadership styles and gender: an analysis of female HR managers styles over their career trajectory in Ireland. Master of Arts in Human Resources Management. Submitted to the National College of Ireland School of Business; 2014 (2014).
[41] Irawanto D-W. Employee participation in decision-making: evidence from a state-owned enterprise in Indonesia. Management 2015;20:159–72.
[42] Chaterjee S. COVID-19: tackling global pandemic through scientific and social tools. Elsevier; 2020.
[43] Farnham P. Economics for managers. Global Edition. Pearson; 2015.
[44] Jalloh AAZ, Jalloh A-U. The effects of motivation on employee performance: a strategic human resource management approach. Int J Manage Sci Bus Res 2016;5(12).
[45] Kuranchie-Mensah E-B, Amponsah-Tawiah K. Employee motivation and work performance: a comparative study of mining companies in Ghana. J Ind Eng Manage 2016;9:255–309.
[46] Lee MT, Raschke R-L. Understanding employee motivation and organizational performance: arguments for a set-theoretic approach. J Innov Knowl 2016;1:162–9.
[47] Maksic F. Employee motivation and satisfaction: the case of Clinica Centre University of Sarajevo. Master's Thesis presented to the University of Ljubljana, Faculty of Economics; 2016.
[48] Peregrino de Brito R, Barbosa de Oliveira L. The relationship between human resource management and organizational performance. Braz Bus Rev 2016;13:90–110.
[49] Vardarlıer P. Strategic approach to human resources management during crisis. Procedia Soc Behav Sci 2016;235:463–72.
[50] Yasodara Jayarathna SMD, Weerakkody WAS. Impact of decision making, reward management on job performance: mediation of job satisfaction: a case of a private banks in Sri Lanka. Eur J Bus Manag 2016;8.
[51] Ceschi A, et al. Decision-making processes in the workplace: how exhaustion, lack of resources and job demands impair them and affect performance. Front Psychol 2017;8.
[52] Devi S. Women empowerment: role and importance. Int J Adv Edu Res 2017;2:232–3.
[53] Hadziahmetovic N, Sait Dinc M, Jahic K. A research review of motivation enhancing of human resource practices. J East Eur Centr Asian Res 2017;4.
[54] Idowu A-O. Effectiveness of performance appraisal system and its effect on employee motivation. Nile J Bus Econ 2017;5:15–39.
[55] Kontakos P. A circular economy roadmap for Cyprus by 2020. Thessaloniki: Proceedings of the 8th International Conference on International Business; 2017. May 2017.
[56] Nguyen M-L. The impact of employees' motivation on organizational effectiveness. Finland: Masters thesis presented to the Vaasan Ammattikorkeakoulu University of Applied Sciences; 2017.
[57] Blašková M, et al. Responsible decision making for sustainable motivation. Sustainability 2018;10:1–23.
[58] Ayankola I-A, Popoola S-O. Effects of work motivation and self-esteem on career fulfillment among health information management personnel in federal teaching hospitals in southern Nigeria. J Hum Resour Manag 2020;8:85–95.
[59] Jamal R. The effects of workplace incivility on job satisfaction: mediating role of organizational citizenship behavior, intrinsic and extrinsic motivation, emotional exhaustion. Int J Hum Resour Stud 2020;10.
[60] Baldwin JN. Public versus private employees: debunking stereotypes. Rev Public Person Adm 1991;11:1–27.
[61] Babin B, Boles J. The effect of perceived co-worker involvement and supervisor support on service provider role stress, performance, and job satisfaction. Journ. Retail.; 1996. p. 72.
[62] Adkins CL, Werbel JD, Farh J-L. A field study of job insecurity during a financial crisis. Group Organ Manage 2001;26:463–83.
[63] Siebert WS, Zubanov L. Searching for the optimal blend of turnover. Acad Manage J 2009;52.
[64] Battistelli A, Portoghese I, Galletta M, Pohl S. Beyond the tradition: test of an integrative conceptual model on nurse turnover. Int Nurs Rev 2013;60.
[65] Wynen J, Op de Beeck S. The impact of the financial and economic crisis on turnover intention in the U.S. Federal government. Publ Person Manag 2014;43.
[66] Becker B. Hospital unionism and employment stability. Ind Relat 1978;17:96–101.
[67] Dalton DR, Todor WE. Turnover turned over: an expanded and positive perspective. Acad Manage Rev 1979;4:225–35.
[68] Brannon D, Zinn JS, Mor V, Davis J. An exploration of job, organizational, and environmental factors associated with high and low nursing assistant turnover. Gerontologist 2002;42:159–68.
[69] Buelens M, Van den Broeck H. An analysis of differences in work motivation between public and private sector organizations. Publ Adm Rev 2007;67.
[70] Galletta M, Portoghese I, Battistelli A. Intrinsic motivation, job autonomy and turnover intention in the Italian healthcare: the mediating role of affective commitment. J Manage Res 2011;3:1–19.
[71] Gomez-Mejia L, et al. Managing human resources. 8th ed. Prentice Hall; 2016.

[72] Ismail A, Sieng LLC, Abdullah MMB, Francis SK. Linking supervisory role in training program to motivation. Intang Cap 2010;6:15–32.
[73] Mondy R-W. Human resource management. 12th ed. Pearson; 2012.
[74] Sekaran U, Bougie R. Research methods for business: a skill-building approach. 7th ed. NY: John Wiley & Sons; 2016.
[75] Cooper D-R, Schindler P-S. Business research methods. 12th ed. McGraw-Hill; 2014.
[76] Greener S. Business research methods. London: Ventus Publishing; 2008.
[77] Creswell J-W. Educational research: "planning, conducting, and evaluating quantitative and qualitative research. 4th ed. Pearson; 2012.
[78] Saunders M, et al. Research methods for business students. 7th ed. Pearson; 2015.
[79] Hossain D-M. Qualitative research process. Postmod Openings 2011;2:143–56.
[80] Scotland J. Exploring the philosophical underpinnings of research: relating ontology and epistemology to the methodology and methods of the scientific, interpretive, and critical research paradigms. Engl Lang Teach 2012; 5:9–16.
[81] Carsten J-M, Spector P-E. Unemployment, job satisfaction and employee turnover: a meta-analytic test of the muchinsky model. J Appl Psych 1987;72:374–81.
[82] Lambert E, Hogan N. The importance of job satisfaction and organizational commitment in shaping turnover intent: a test of a causal model. Crim Justice Rev 2009;34.
[83] Mobley W-H. Intermediate linkages in the relationship between job satisfaction and employee turnover. J Appl Psychol 1977;62:237–40.
[84] Van Kleef GA, Homan AC, Beersma B, Van Knippenberg DV, Van Knippenberg BV, Damen F. Searing sentiment or cold calculation? The effects of leader emotional displays on team performance aligned on follower epistemic motivation. Acad Manage J 2009;52.
[85] Meier KJ, Hicklin A. Employee turnover and organizational performance: testing a hypothesis from classical public administration. J Public Adm Res Theor 2008;18: 573–90.
[86] Perumal G, et al. Human resources management practices and affective commitment among financial industry employees in Malaysia. Int J Eng Adv Technol 2019;8.

CHAPTER 13

How COVID-19 Will Change the Future of Tourism Industry

VIANA HASSAN • MOHAMAD KNIO • GEORGES BELLOS

In the past decades, the world has experienced significant epidemics such as AIDS, H1N1 (hemagglutinin type 1 and neuraminidase type 1), and New Delhi metallo-β-lactamase 1 (NDM1). Traveling had a role in contributing to the transmission of those diseases by affecting public health negatively, but those diseases never had implications on the global economy and on the tourism sector as the coronavirus (COVID-19) pandemic did. Due to the epidemic, a deficit of US$ 1 trillion in export revenues from tourism was reported, which helps better understand how COVID-19 will change the tourism sector.

The research is based on different source materials (reports, blogs, articles, papers, and statistics from various sources such as the United Nations World Tourism Organization [UNWTO], the World Travel & Tourism Council [WTTC], and so on), besides using the qualitative method through two interviews and 100 questionnaires targeting leaders and employees, respectively, in this sector. The data collection also included 50 travel agencies, both in Lebanon and in Malta. A quantitative survey of 70 questions was disseminated to a random sample to test stakeholder behavior.

The research aims to gather information on the effects of the coronavirus pandemic on the tourism industry, the precautionary measures, and the expectation of tourism in the future so that professionals can plan what will happen after COVID-19, as per the WHO recommendations.

The main findings presented in this chapter have shown that COVID-19 had a devastating influence on the world economy, mainly on the tourism and travel sector. However, the future of the tourism sector will change by integrating technology, touchless contact, social distancing, and hygiene precautions, focusing on domestic travel, virtual tourism, and luxury travel. It may support the tourism sector through sustainable development goals (SDGs).

1 INTRODUCTION

The tourism and travel sector is a large and thriving industry that recorded in 2019 around 5 billion tourist arrivals worldwide, with an annual growth rate of 4% from the previous year [1,2]. This confirms that the tourism sector is a leading economic sector for several countries worldwide due to increased wages, paid holidays, increased air traffic, the opening of borders, and the development of tourism marketing [3].

Besides, the UNWTO was looking forward to a growth of 3%–4% predicted for 2020, showing a cautious optimism due to major sporting events. These included cultural events such as Expo 2020 Dubai and sport events such as the Tokyo Olympics, which were expected to have a positive impact on the industry [1,2]. Unfortunately, by the end of January 2020, the expectation of tourism growth has changed completely, with the appearance of COVID-19. In the past decades, the world has experienced many significant epidemics/pandemics such as AIDS, H1N1, and NDM1, where traveling had a role in contributing to the transmission of those diseases by affecting public health negatively [3,4]. However, there were never implications on the global economy and tourism sector, such as the ones presented by the COVID-19 pandemic, which caused a loss of 1 trillion USD in the tourism annual export revenues [4]. By April 2020, the majority of worldwide destinations had travel restrictions to limit the spread of COVID-19 [4].

A total of 97 destinations (45%) refused incoming tourists, 65 destinations (30%) have partially or fully canceled international flights, and 39 destinations (18%) are undergoing lockdown measures in a more

COVID-19: Tackling Global Pandemics through Scientific and Social Tools. https://doi.org/10.1016/B978-0-323-85844-1.00003-9

differentiated manner by banning the entry for passengers from specific countries of origin, which threaten about 110 million direct tourism jobs (Fig. 13.1) [1,2].

The main purpose of this chapter is to evaluate the current situation and the impact of the coronavirus pandemic on the tourism industry, the scenario of the tourism sector, the recovery, and finally, the future of the tourism industry and how the coronavirus will change it. The following four points will be raised through the study.

First, all city and village mayors, governors, leaders, statesmen, and all development agencies, the political elite, as well as their representatives, need to stop pursuing their interests and going for quick gains or feeding corrupt mechanisms while we are fighting a pandemic. A previous study determined this.

Second, proofs of the contributions for and against valorizing sites were dug up, if one takes the account of the destroyed harbor (BEY 194), which was reported in Mr. Scandre Hachem's article; the arguments are aligned with the literature review, as many articles were discussing this issue [55].

Third, to date, only a limited number of studies were conducted on heritage management by scholars (this research differs in the sense that the model from those studies were adapted to a different case) [50,52–55]. The same applies for the tourism industry with respect to COVID-19 and the aftermath of this pandemic on the entire industry, both in Lebanon and worldwide [4,5]. Many past research, as well as recent studies, documented this [4,5,40,41,46,51].

Fourth, all stakeholders involved with valorizing any resource for economic growth have been identified objectively, and now, with the pandemic, more than ever, people need to spend wisely and consider ecotourism, sustainable development, or other movements beneficial for the planet and for greater goals [30,32–34,40,41,46,48,49,52].

2 LITERATURE REVIEW

All previous studies that were done on the subject state how important the tourism sector is, irrespective of the nature of the site or locality one visited before or during the lockdown period [4,43].

According to the BBC (and/or CNN) the COVID-19 vaccine should be ready by mid-next year. Yet, world economic growth will start by 2025. If one stays pragmatic when a global pandemic is killing several tens of thousands of people and is severely influencing most sectors pervasively, especially in tourism and hospitality, there should be a dim hope for an immediate economic boom [1,5]. Moreover, once the tourism sector got affected a worldwide deficit of 1 trillion was

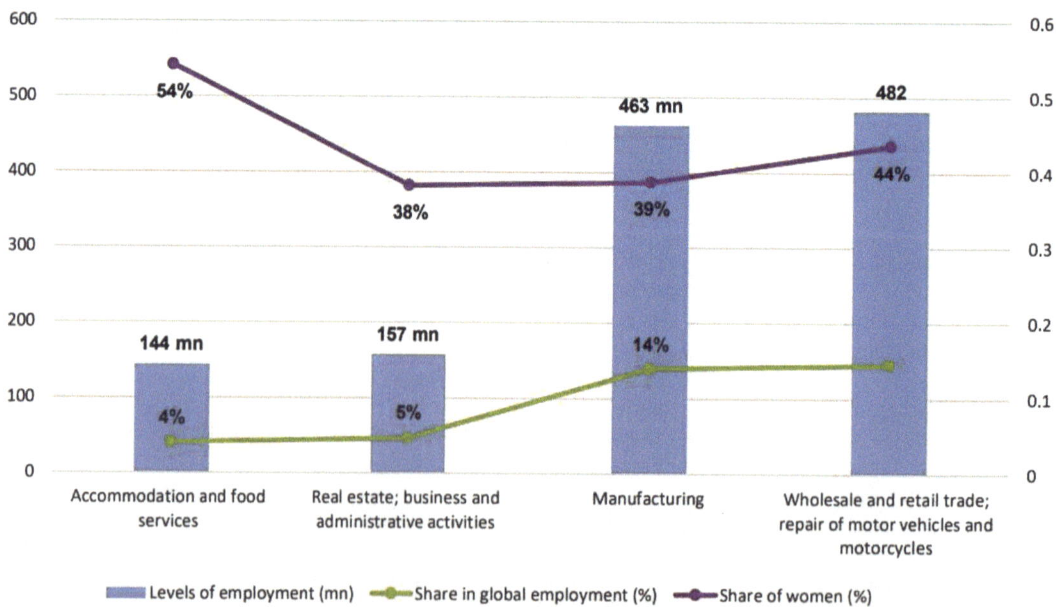

FIG. 13.1 Workers in the sector at most risk. (ILO. *The impact of COVID-19 on the tourism sector*; April 2020. From: https://www.ilo.org/wcmsp5/groups/public/—ed_dialogue/—sector/documents/briefingnote/wcms_741468.pdf. Accessed 4 November 2020.)

reported [1,19,42]. So the tourism potential of any coastal site (such as that of the Levantine coast) that attracted tourists for sightseeing, or for luxury holidays, will definitely be severely affected under the COVID-19-induced recession, as travels to destinations were halted [19].

Many studies were carried out on coastal and other sites for evaluating their tourism potential, in terms of regular tourism potential (coastal or others, like in the Mediterranean) or in terms of reactivity toward COVID-19 as a massive pandemic, as declared as such by the WHO, and discussed in a scientific publication, tracing the origins of the Coronavirus epidemic to Wuhan, China, in late 2019 [46,50,51]. The internationalization of this pandemic took global proportions by early to mid-2020, according to experts, and all tourist sites, worldwide, were affected [40,46,50,51].

As such, traveling had a role in contributing to the transmission of the disease by affecting public health negatively [4]. Yet, previous epidemics/pandemics never had implications on the global economy and the tourism sector as the COVID-19 pandemic did. Due to the COVID-19 pandemic, loss in export revenues of 1 trillion USD from tourism was reported, which helps better understand how COVID-19 will change the tourism sector [4,5].

Previous research in this subject, such as those proposed initially by Dr. Viana Hassan, included to assess the tourism potential in Lebanon and to assess how tourism activities can benefit it economically. In 2020, the entire annual global tourism revenues recorded a global deficit of about 1 trillion due to COVID-19 [4,5,46,50]. The main problem with Lebanon is that it is incapable of properly using its natural and cultural assets in spite of very rich potential.

Many studies on the coastal and rural tourism potential in the Mediterranean and worldwide sites demonstrated that sustainable tourism development in Lebanon is dependent on the availability of all essential resources [32,33,40,41,52]. However, Lebanon lacks an adequate administrative organization and the management of its limited resources. The importance of human capital in the development of any tourist destination was brought forward in several studies, once one considers the importance of human capital [40,52].

2.1 Positive Cases of Valorization: Tyre and Jbeil

First is the importance of economic development and tourism in Tyre, as any coastal city [40]. Jbeil did this before Sour (yet not necessarily in a good way). True, Lebanon is plagued with mismanagement and corruption, leading to a useless management of resources. This has to change either after COVID-19 or once all the country's leaders get held accountable and get tried for their crimes, such as the recent Beirut Harbor blast set Lebanon backward, and their lack of managing the pandemic [58,59]. This research shows in what way archeologists can help and the link between economic development and tourism, or any other player who should consider working with national or world heritage sites for enhancing tourism activities (see Figs. 13.2–13.11). After COVID-19 the industry will definitely change [46,50,52–54]. Why? Because many

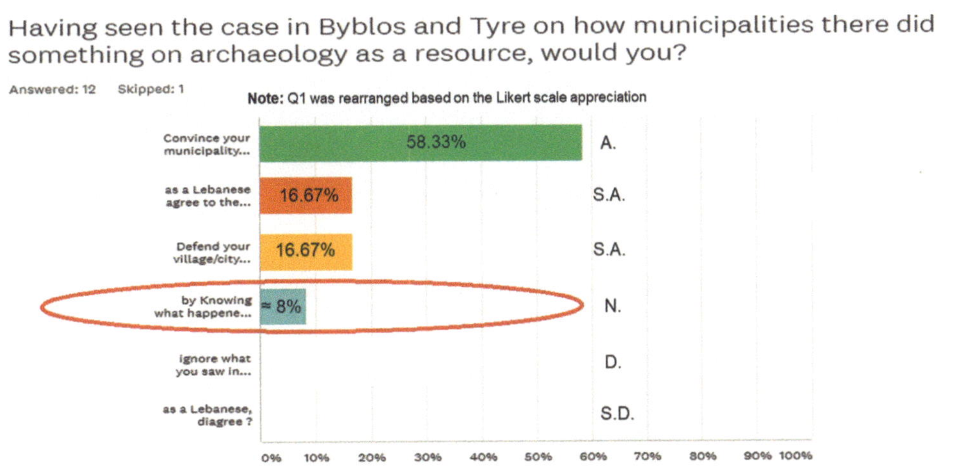

FIG. 13.2 Respondents in a pilot study questionnaire, following a previous literature review research estimating the role of archeology in tourism potential yielded these results. So they mostly agreed with the above statement (92%) and only 8% did not take a stance, or were neutral. The results of that study may concord with the current research, and on potential evaluations in the future of the industry. (see: [40,50,52,61]).

Adopting Bey 194 as an example, if you were involved in any way (as an individual), or showing any concern at all, with all due respect to all parties involved, would you take a firm stance in your village/city if the same thing happened to your heritage?

Answered: 11 Skipped: 2

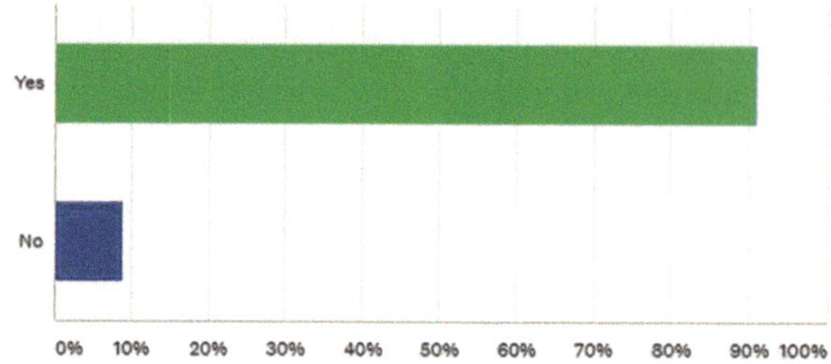

FIG. 13.3 The responses to the above question (two of the initial pilot study) show that 92% agreed and 8% disagreed (see: [55, 61]).

Did you know that a study was conducted in Tyre (and surroundings) on how can the Unions of the Tyrian Municipalities attempted to find ways to use cultural sites (and other existing heritage), as valuable resources (like assets) to improve tourism there?

Answered: 13 Skipped: 0

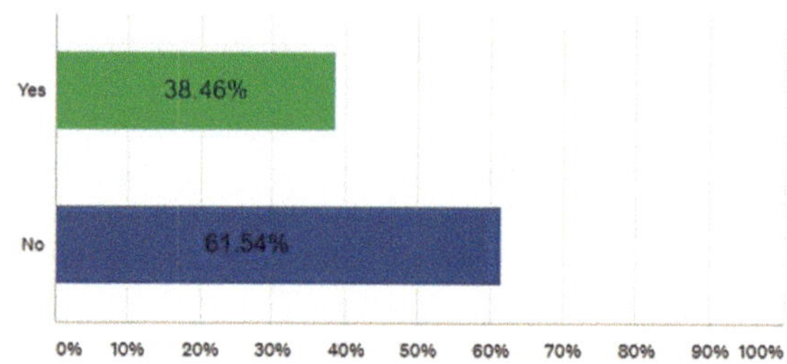

FIG. 13.4 At the time of the initial pilot study, roughly 38% of the respondents were aware of the existence of the studies mentioned in the above statement and 62% were not. This should be interesting for the academics, scholars, or tourism experts to consider the future of tourism in Lebanon after coronavirus disease 2019 (COVID-19) (see: [40,50,52,61]).

In a similar case in Byblos (like with Tyre, with its municipality and the Union of the Jbeil municipalities), it has found ways to present their city (or groups of cities) in the cultural scene, and hence improve tourism. did you actually know that this does improve our economy?

Answered: 13 Skipped: 0

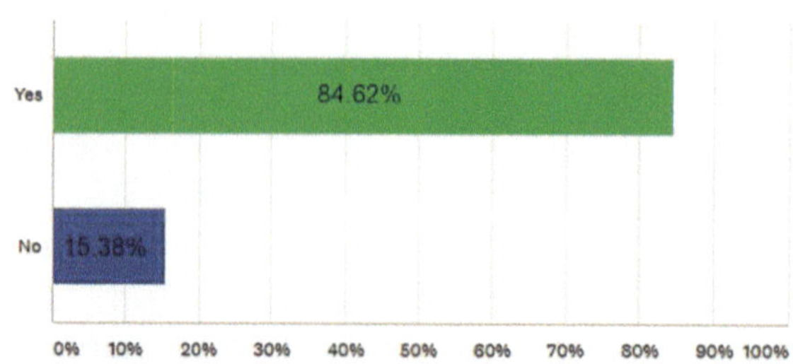

FIG. 13.5 The above statement certifies that the public knows that "preserving and valorizing heritage sites" do enhance tourism and improve a country's GDP; 85% of the respondents agreed and 15% disagreed (see: [4,40,50,52,61]).

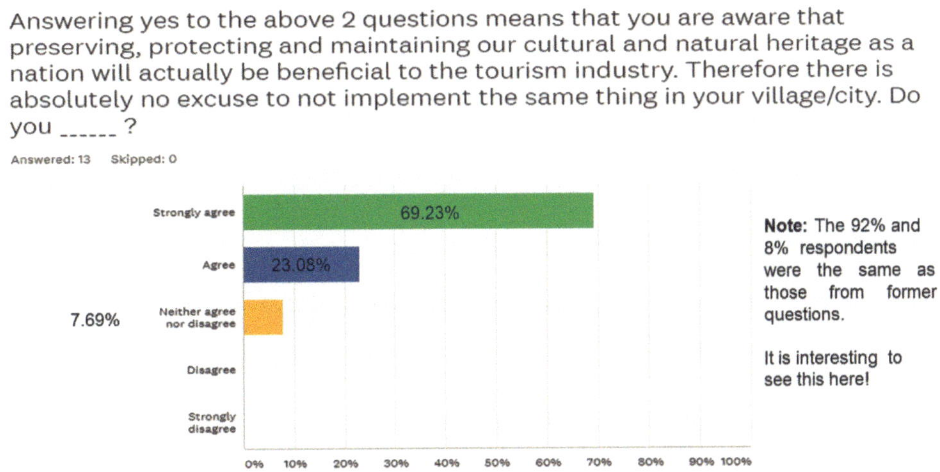

FIG. 13.6 The response reconfirms the previous dataset, with 92% of the respondents agreeing and 8% being neutral, in the former questions; they may have been the ones who disagreed or took no stance (see: [4,40,52,61]).

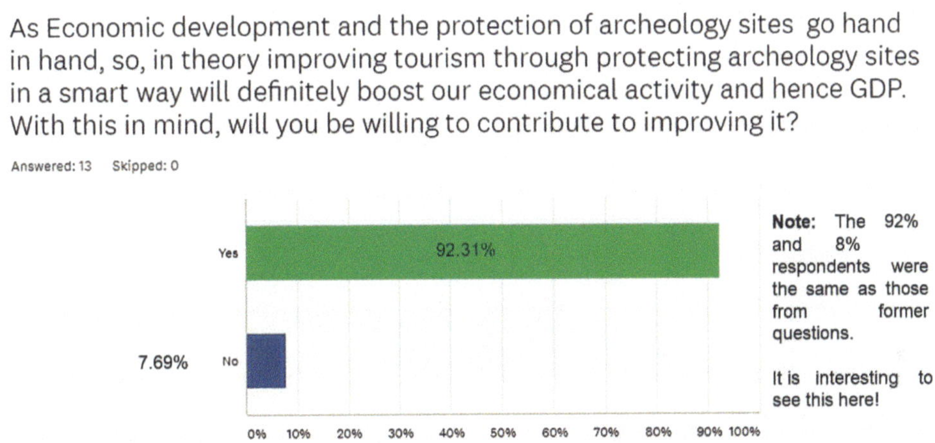

FIG. 13.7 This statement again had 92% of the respondents agree to it and 8% to disagree to it. This confirms that the industry needs assistance, as the respondents notified it would help the industry like in a much needed situation induced by the coronavirus infection worldwide, and if one refers to the islands' economies dependent on tourism, Lebanon could be a candidate if the Rabbit Island in Tripoli becomes again accessible for eco-tourism, the apparent new norm of the industry after coronavirus disease 2019 (COVID-19) (see: [4,61]).

experts noted that municipalities and political elites voluntarily dismiss valorizing resources for long-term goals, favoring short-term profits.

2.2 Negative or Inexistent Cases of Valorization: Beirut and Others

The clear absence of resource handling through archeology, or proper tourism management or economic development led to cases like BEY 194 or the so-called "Phoenician Harbor", the Plot 1398 (Minet El-Hosn), in spite of a panel of 10 specialists sending appeals to the Culture Ministry, before its dismantling on June 27, 2012. This was following the arguments of one of the specialists on the other side who deemed that these ancient slipway structures were not worth valorizing [53,54].

Beirut, in this case, is one big mess. Development destroys heritage on a daily basis for the benefits of soulless construction projects, like the notorious Beirut Souks and other development ventures SOLIDÈRE

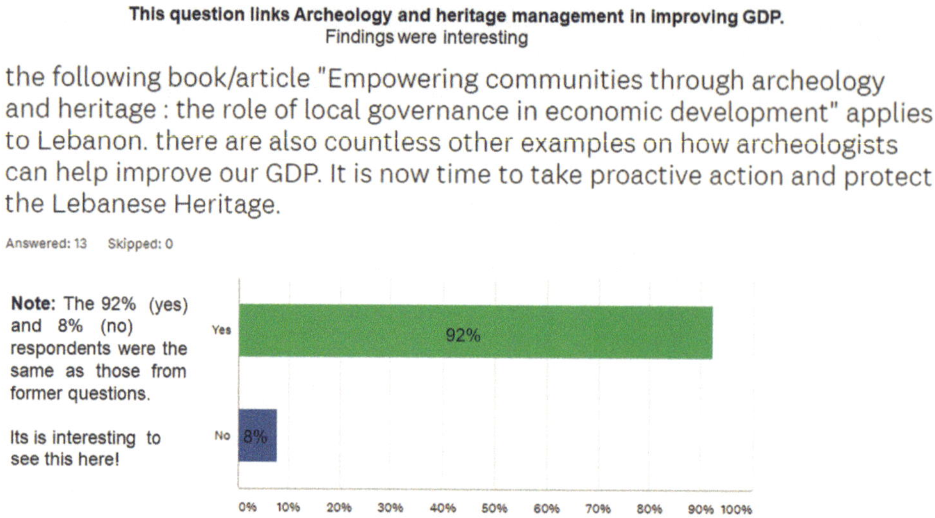

FIG. 13.8 From the initial pilot study, the following question links archeology and heritage management in improving a country's GDP. The responses were consistent with the same answers for the former questions, with 92% agreeing and 8% disagreeing. If one considers the study on island economies and tourism, Lebanon could benefit as well (see: [4,61]).

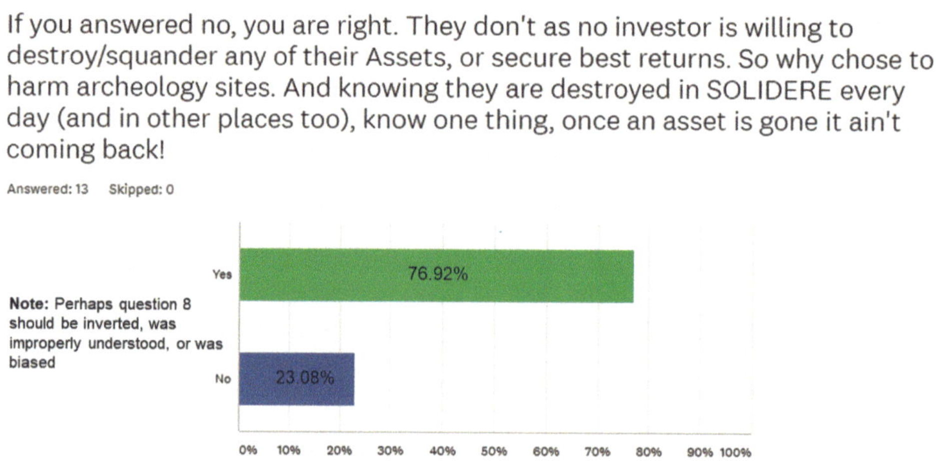

FIG. 13.9 For this question, which addresses that generally "once assets are gone they will not come back," assuming that once an asset (building, valuable item, and so on) gets destroyed it is gone forever, 77% of the respondents agreed and 23% disagreed [61].

handled since 1992. They do not valorize the heritage to preserve it for the benefits of tourism but destroy it [53,54].

Therefore BEY 194 is the best case of what to never allow to happen again in future plans to valorize resources to enhance tourism. The lesson learnt that SOLIDERE and the development sector can apply in any future development projects is to include archeology as valorized resources to enhance tourism if they consider their activity to be effective for improving long-term development and just securing their shot-term profit [55].

So to better understand how tourism sites can have their potential evaluated, one could take a look at a few studies that evidenced the importance of the role of archeology and geosciences in heritage management that ultimately contributes to tourism management, yet in terms of COVID-19 the industry will change

If you answered no, you are right. They don't as no investor is willing to destroy/squander any of their Assets, or secure best returns. So why chose to harm archeology sites. And knowing they are destroyed in SOLIDERE every day (and in other places too), know one thing, once an asset is gone it ain't coming back!

Answered: 13 Skipped: 0

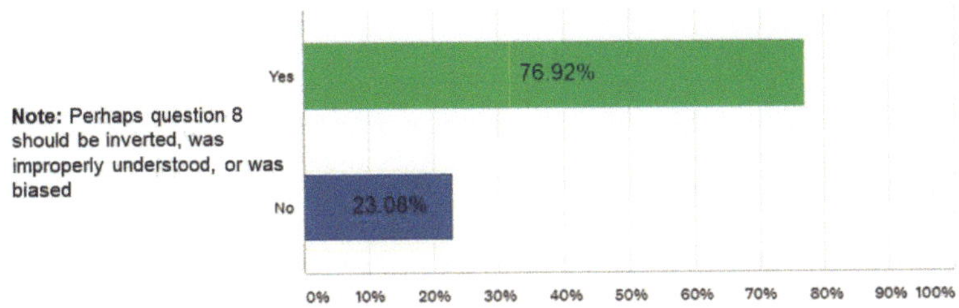

FIG. 13.10 To the statement "once an asset is gone, it is not coming back," 77% of the respondents agreed and 23% disagreed [61].

Should the DGA and the DGU work together as one unit to improve our situation? (If they do, sites will be protected, and our economy will improve). Note that the same will apply for the oil/gas reserves too.

Answered: 13 Skipped: 0

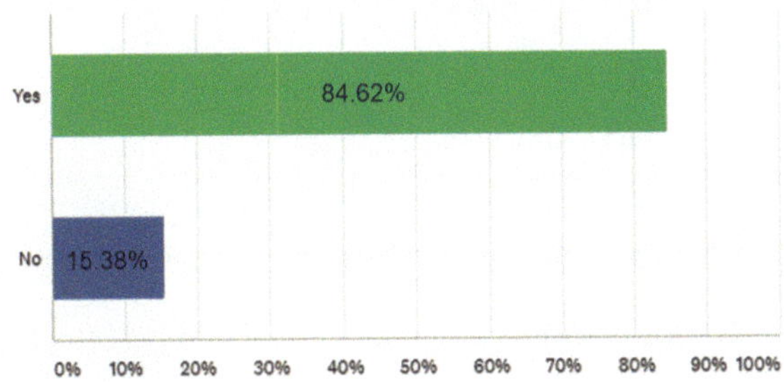

FIG. 13.11 For the last question of the pilot study, showing results similar to those of question 4, 85% of the respondents were in favor of the above statement and 15% were against it [61].

[4,40,50,52]. Figs. 13.2–13.11 show the statistical results of a pilot study following a sound literature review check, to discuss the pilot study results, which have already been tested for Cronbach alpha and were internally correlatable (Table 13.1).

TABLE 13.1
The Cronbach Alpha Test [56,61].

Cronbach alpha	0.964878
Split-half (odd-even) correlation	1
Split-half with Spearman-Brown adjustment	1
Mean for test	500
Standard deviation for test	315
KR21 (use only 0 and 1 to enter data for this)	1.38546
KR20 (use only 0 and 1 to enter data for this)	1.38546

3 METHODOLOGY

In this chapter, the author will carry out data collection based on different secondary sources, such as the gathering of statistical data from the UNWTO scientific framework [1,2]. In addition to the secondary data, primary qualitative data will be gathered, permitting the author to go deeper into the issue at hand. An online unstructured survey, conducted through "Google

Docs", was addressed to 100 employees in the tourism and the aviation industries as well as 50 travel agencies in Lebanon and Malta [5]. These exclude the two interviews that were conducted on two leaders of the tourism industry.

Fig. 13.12 shows the research onion method that explains the abovementioned methodology in more detail. Thus data collection, which focuses on research input and human opinion and provides a better insight into the impact of COVID-19 in the tourism industry, was done through qualitative methodologies. Moreover, how this epidemic will change the future of the tourism sector is what this research wants to find out [1,6].

Thus this explains why we went for mixed-qualitative research methodologies, utilizing both questionnaires and interviews. As a result, one can understand how, by adopting a descriptive and an exploratory nature, this research selected this approach that involves an interpretivist view as well as an inductive approach, whereby some specific arguments are to be generalized through the adoption of "surveying" as the preferred data collection method. This method is best summarized in Fig. 13.12 and by gathering the arguments of most research methodologists would do, once such studies are required. This is because various statistical techniques are recommended for interpreting unstructured and qualitative surveying data, either electronically (SPSS, Excel, NVIVO, and so on) or manually, once abstract arguments and human opinions are considered as opposed to confirmatory data requiring testing hypotheses [6,7,22].

For the purpose of correlation, one needs to refer to the Cronbach alpha test, so we will definitely need to use this test for our dataset following the methodology here, while ensuring the data is both correlatable and internally valid (it is used for reliability, the triangulation is used for validity) [56].

Using the research onion (Fig. 13.12), we decided to conduct a quantitative survey as well, addressing 70 questions, which was used to collect information about the stakeholders' behaviors, through semistructured surveys as our means to gather primary data

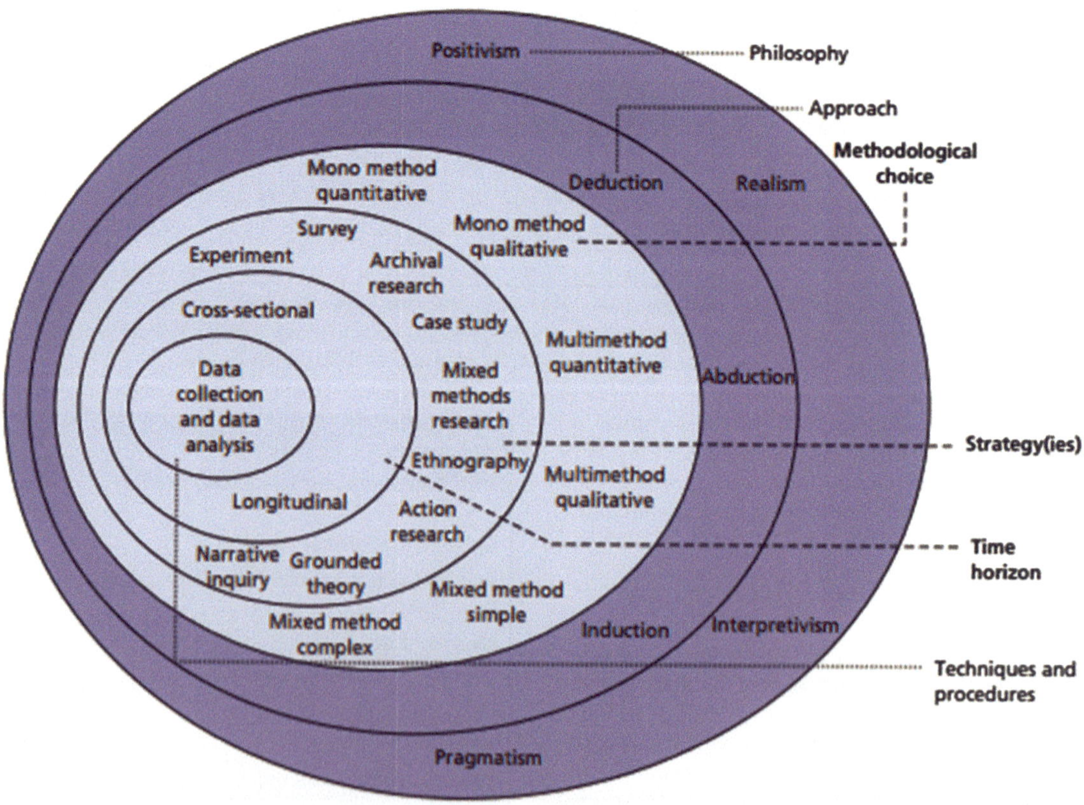

FIG. 13.12 Research onion. (Adapted from Saunders M, et al. *Research methods for business students*. 7th ed. Pearson; 2015.)

[4,10,57]. The initial interview question and the first two surveys are shown in the Appendix (but not the last survey, as it is ongoing). Moreover, we ought to pursue the investigation, utilizing complex mixed methods [57]. This should be followed by a Cronbach alpha statistical research on a software such as SPSS or Excel to verify the validity and accuracy of the data; Table 13.1 shows the internal correlation of the pilot study [56]. The survey confirmed the pilot study [56,57,61], and some secondary data were included in the research as provided by the discussions in the literature review [61].

4 FINDINGS

4.1 Tourism and Coronavirus Disease 2019

COVID-19 has been described by the of the International Monetary Fund's (IMF) Managing Director, Mrs. Kristalina Georgieva, as "a crisis like no other". Dr. Kenneth Rogoff also cited a similar quote on the virus [5,7,23]. It spreads primarily through saliva droplets or nasal mucus ejection when an infected person coughs or sneezes [8,60]. Due to the unavailability and the lack of sound tests in many countries, there was a quick upsurge in the propagation of confirmed coronavirus cases from its epicenter. This has led to quasiparalysis of the tourism sector, having voyagers struggling to return home and devastating economies that are mostly dependent on tourism [9–12,44]. This argument was also brought about by Maria Niestadt.[1] [12].

According to the United Nations World Tourism Organization estimates, global international tourist arrivals could decline 20%–30% in 2020 as compared to 2019 (Fig. 13.13). In contrast, the 2003 severe acute respiratory syndrome (SARS) epidemic triggered a decline of 0.4% (Fig. 13.14).

Moreover, the tourism and travel industry will lose 75 million jobs worldwide, with 6.4 million jobs in the European Union, and will take about 10 months to start returning to normal [13,14]. Additionally, and according to the International Air Transport Association (IATA)[2], the world air travel may incur losses of over 252 billion USD, as 2 million passenger flights were canceled until June 30, 2020 [15,45,47]. Within the hospitality industry, also affected by the major economic crunch, small family businesses have been mainly affected by the COVID-19 pandemic [5,15,45,46]. Worldwide, many hotels had to shut down operations or, like many restaurants, bars, and guesthouses, had to permanently (or temporarily) fire thousands of workers, and all those businesses in the industry sectors will need up to 10 months to start recovering [3,11,13,14,44].

4.2 Tourism and Recovery

By mid-May, some countries, which rely mostly on tourist dollars, started lifting travel restrictions and opened safe travel corridors, such as Seychelles, Maldives, Aruba, the Bahamas, and so on. After the restrictions are lifted, how will the tourism sector recover? As per the WTTC, which represents the worldwide tourism and travel private sector, there must be a range of new global actions to recover and rebuild the trust between the industry and its consumers [13,14].

For the safe reopening of the hospitality sector, including food and beverages, the key measures include social distancing between guests and employees, hygiene measures (masks, gloves, washing hands), integrating technologies (contactless payments), room service using no-contact delivery methods, and revisiting guidelines for cleaning teams, with specific focus on high-frequency touch points, in addition to particular attention placed to seating and queue management in line with social distancing requirements and promoting e-menus [13,14]. According to the American Hotel and Lodging Association (AHLA), new safety and cleaning guidelines should be used when reopening amid the coronavirus pandemic.

First, they encourage contactless check-in, payment processes, and access room without interacting with another person [16].

Second, as per the new AHLA guidelines, hotel cleaning staff should not enter any guest's room during a stay, except when explicitly requested. Room service should be limited, as delivered foods and "grab and go" items ought to be the favored way of delivering food. As the general manager of the Snow King Resort in Jackson Hole suggested, as the new norm, the typical food delivery service could be substituted with a contactless delivery method "in the form of a market bag left at the door" [17].

After lifting some restrictions in the tourism sector, few people are traveling solely for the sake of tourism, so in order to appeal to them, some countries like Thailand are proposing self-quarantine at very competitively priced packages at Pattaya and Bangkok hotels, with rates slashed by 20% [18,19]. This policy seems to be followed by other hotel chains around the world. For example, the Dorsett Hotels chain is selling 2- to 3-week quarantine packages at nine locations, according to the *Wall Street Journal* some of Hong Kong's high-

[1]Niestadt, M. (2020), COVID-19 and the tourism sector, At a Glance, EPRS, European Parliament.

[2]See https://www.iata.org/.

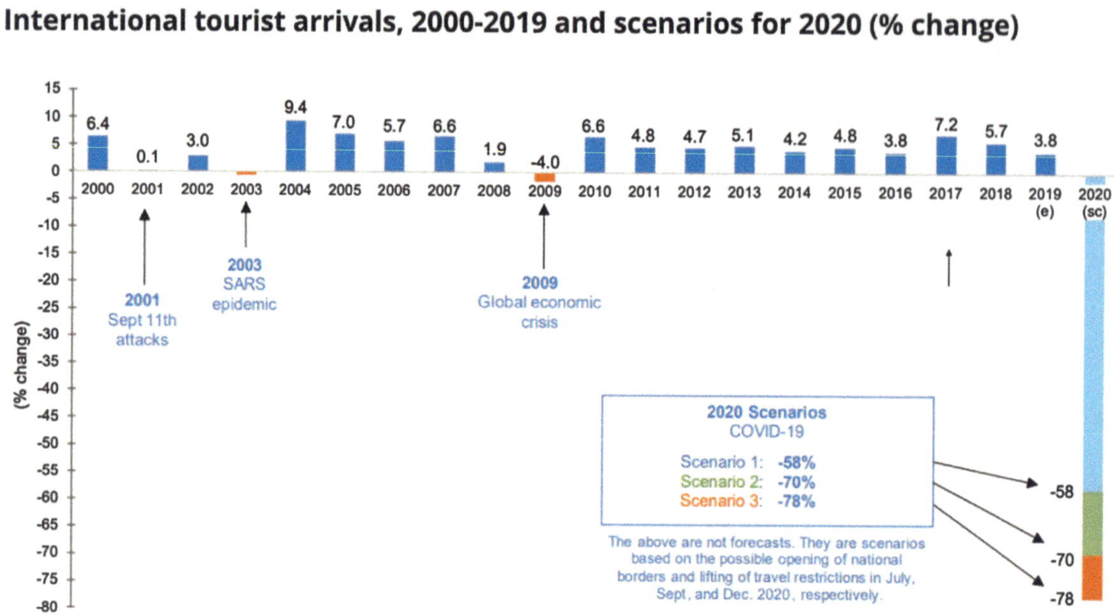

FIG. 13.13 Largest blow to tourism ever could slash 1 billion arrivals. *COVID-19*, coronavirus disease 2019; *SARS*, severe acute respiratory syndrome.

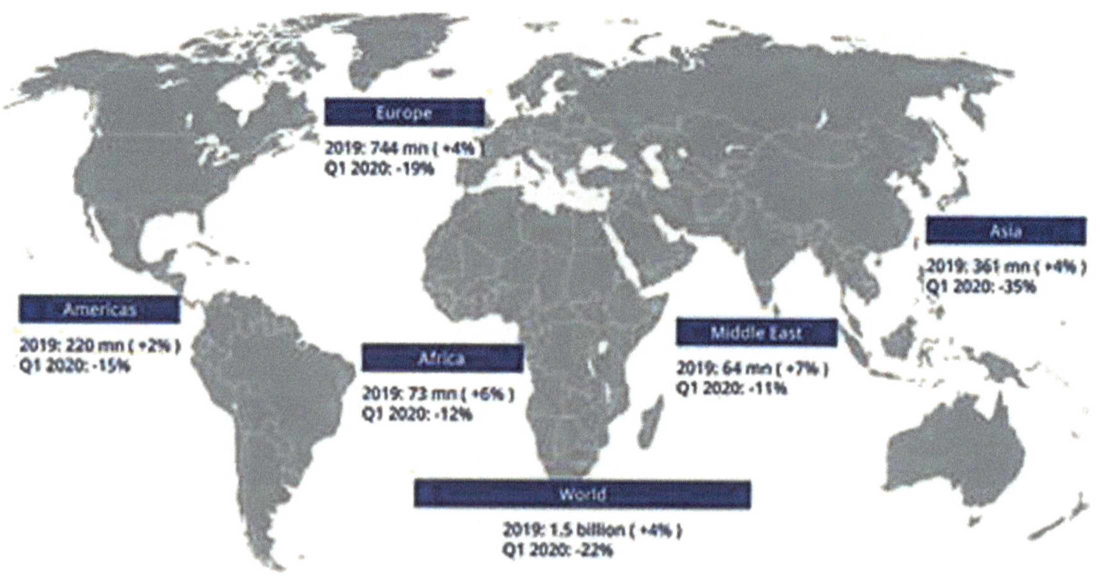

FIG. 13.14 International tourist arrivals by region in Quarter 1 (Q1) 2020. (UNWTO, 2020.)

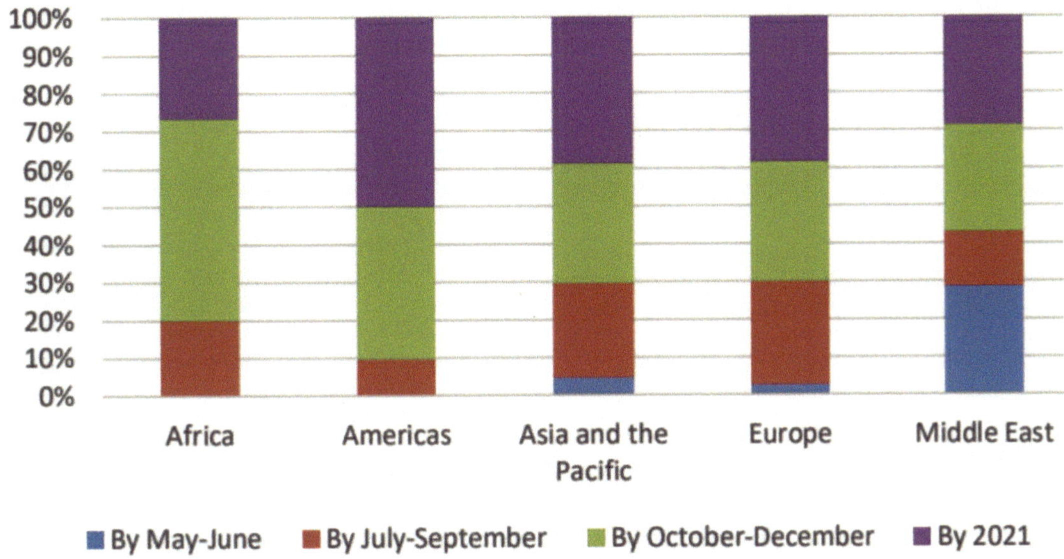

FIG. 13.15 Expectation for tourism recovery. (UNWTO. *Impact assessment of the covid-19 outbreak on international tourism*; 2020a. Retrieved from: https://www.unwto.org/impact-assessment-of-the-covid-19-outbreak-on-international-tourism. Accessed 4 November 2020; UNWTO. *International tourism growth continues to outpace the global economy*, January 2020; 2020b. Retrieved from: https://www.unwto.org/international-tourism-growth-continues-to-outpace-the-economy. Accessed 4 November 2020.)

end hotels.[3] Similar to Singapore, the Park Lane hotel in Hong Kong has a 2-week package, including full boarding, for $2525.[4]

During the UNWTO webinar "The Impact of the Coronavirus Pandemic on Global Tourism" on May 25, 2020, his Excellency, Ali Waheed, the tourism minister of the Maldives declared that the tourism sector will open on July 1st to tourists, with competitive rates [1,20,48]. The precautionary measures taken upon arrival to the airport will be higher. In case if they have infected cases, an isolated room can be prepared for the tourists, such as luxury villas with pool, so that tourists can still enjoy the holiday [21].

For the aviation sector, until now, the WTTC did not announce any key measures [13,14]. Hence, we interviewed the Group CEO of AIRX, Mr. Hazzoury, to get more information about the restrictions they followed [22]. According to him, and based on the IMF, the key measures included following hygiene measures before departure and upon arrival, contactless payment, checking crowds for fever-grade body temperatures, and performing PCR test for their crew before and after the flights [23,49].

It was also pointed out that some negotiations with some luxury tour operators in the United Kingdom who are planning to arrange a luxury holiday to the Maldives, using the private jet, which avoids the tourist's self-quarantine for 14 days [9,24].

Finally, per UNWTO, the travel and tourism sector will recover depending on each continent (Fig. 13.15), with all the continents recovering by 2021, except Africa by October to December 2020 [1,2].

5 FUTURE OF TOURISM

Worldwide, in 2030, there will be 1.8 billion tourists per year, as the UNWTO declared, due to the cheap airfares, affordable hotel rooms, and rideshares [1,2]. According to forecaster Ray Hammond, the number of airline passengers will double in the next two decades, and consumers will want faster and more productive travel experiences [25]. However, after COVID-19, the future of the travel and tourism industry will change dramatically.

In the beginning, governments may more directly subsidize the tourism and travel industries through bailout packages for countries and require a coronavirus-infection-free certificate. They must also facilitate travelers' arrivals by improving online visa application formalities and requiring a COVID-19 test

[3]See https://www.wsj.com.
[4]See www.parklane.com.hk.

certificate [26]. In the longer term, the tourism and travel sector will start to be turning again; there will have been many changes [27,49].

Firstly, at the airport, the entire process of checking in could take up to 4 h, involving social distancing, sanitation, and integrated digital technologies (self-check-in, self-baggage drop, face masks, surgical gloves, UV sanitation, contactless payments, and immunity passports) [28].

Secondly, domestic tourism and family travel will explode in conjunction with the growth of midrange hotels and decline of 5-star hotel occupancies. Activity and experiential vacations will be essential [29].

Lastly, the choices of the samples may be lined culturally, as both Lebanon and Malta (cf. Motya in the Phoenician era) share cultural and common ancestry, which could be a reason why these two locations were selected to start with [50].

According to the questionnaire addressed to 50 travel agencies in both Lebanon and Malta on how they see the future of the tourism industry, the majority found that domestic tourism will be the first sector to recover. Most tour operators will organize domestic packages because the prices will not be as high as the international package. Furthermore, it will be the first type of tourism that will help the economy of a country, in a cyclic way, once that country considers working on SDGs [22,30].

Following some of our findings, 75% of travel agencies declare that this sector will work via the internet, as they are becoming increasingly computer-proficient and tech-savvy surfers, whereas 15% found that the tourism sector will need at least 4 years or more to recover from this effect [8,9,31,60].

Besides, the results of the second questionnaire addressed to 100 employees in the tourism sector (among hotels, airports, airlines, guides, and so on) showed that the tourism industry could align more closely with SDGs, especially the environmental aspect. With the publication of 'the Brundtland Report,' the importance of SDGs was noticed. This report states that through implementing SDGs, the needs of the present generation will not compromise the possibility for future generations to meet their own needs [32–35,49].

On the environmental aspect, we noticed several improvements: reduction in pollution levels and greenhouse gas emissions, closure of many power generation plants and factories, and so on [36].

Finally, for tourist sites, the results showed that the technology would be integrated, such as using touchless contact; at some point, they will replace guides with digital ones and enforce social distancing, which will reduce the impact of mass tourism, such as coastal sites or any other destination having multiple visitors [37].

6 CONCLUSION

This research aimed to investigate the stages of the tourism industry within the scope of COVID-19. After COVID-19 spread to all the parts of the world, it not only had a devastating impact on the global economy but also has exposed some critical measures on the tourism and travel sector, especially such as closing the tourism businesses, closing borders, travel restrictions, and flight cancellations.

According to the Organization for Economic Cooperation and Development (OECD), and the International Labor Organization (ILO),[5] the pandemic and the global efforts to contain it could cause the international tourism economy to decrease by 45%–70%, depending on the duration of the crisis and the speed of the sector's recovery [3,5,38]. However, the tourism sector is known for its resistance to crises and epidemics, which was the case of the 2003 SARS epidemic and the 2008–09 worldwide financial crisis.

Yet, COVID-19 will change the image and future of tourism by requiring to take more precautions during travel and by integrating technology, touchless contact, and social distancing [3,39]. On the other hand, "building back better" in the postpandemic period could align the tourism industry more closely with the SDGs, especially in the environmental aspect [36,38]. Moreover, it has highlighted some of the conflicts and difficulties within the concept of SDGs and has recommended some radical solutions to the challenges of sustainability, in coastal tourism sites, for instance [40,41].

Finally, one of the research's main limitations, aside from the time constraints, was that s the subject is rather recent, most of the references come from online sources and not from scientific articles published in journals. This was partly due to the lockdown and the lack of published peer-reviewed research about COVID-19 and tourism, if one considers the cases of coastal tourism or rural sites [40,41,52].

In the future, there will be a wide range of research opportunities, such as the impact of the COVID-19 crisis on SDGs and how information and communication technologies are used to address safety concerns, in high-profile sites, such as coastal tourism, urban, or rural sites, and/or any other type of sites, especially if they could have yielded large potential [40,41,52–55].

[5]See [36].

APPENDIX 1: QUESTIONNAIRE ADDRESSED TO 50 TRAVEL AGENCIES IN MALTA AND LEBANON

Dear participants, through this online questionnaire, your answers will be helpful to study the impact of COVID-19 on the tourism industry and the future of travel agencies and tourism. Your response will only be used for survey purposes.

Instructions: Please put a tick in the box next to the answer of your choice or write in the space provided.

1. Does COVID-19 affect the travel agencies in any harmful way?
   ○ Yes
   ○ No
   If yes, how?
   ....

2. Do you often get disturbed from groups of different sizes visiting your village?
   ○ Yes
   ○ No
   If yes, please state why?
   ....

3. Do you think the future of travel agencies will change?
   ○ Yes
   ○ No
   If yes, how?
   ....

4. During COVID-19, how do you manage to revive your agency?
   ....

5. Is there any change in your packages?
   ○ Yes
   ○ No
   If yes, explain.

6. In your opinion, do you think the future of the travel agencies will change?
   ○ Yes
   ○ No
   If yes, how?

7. Are their suitable precautions taken by your agency?
   ○ Yes
   ○ No
   If yes, which type of precautions.

8. Are you still organizing packages to danger zones like Paris, London, etc.?.
   If yes, how do you protect your clients and how they avoid the quarantine as requested?

9. Do you have clients requesting a private jet to travel? If so, how do you organize it and is it safer than normal jet?

APPENDIX 2: INTERVIEW QUESTIONS

1. What is the role of your company?
2. What are the restrictions taken by your company? Are they similar to national airline restrictions or different?
3. What has made the travel with a private jet different than in a normal airline during COVID-19?
4. As we know, the aviation sector has been affected, is it the same the private jet sector?
5. How your crew takes precautions as they travel to several affected countries considered as red zone?
6. In your opinion, how will the aviation sector and the travel agencies revive after COVID-19?
7. Based on your experience, what is your expectation for the tourism industry in the future?
8. Any recommendations ...

APPENDIX 3: QUESTIONNAIRE ADDRESSED TO 100 EMPLOYEES IN THE TOURISM AND AVIATION SECTOR

Dear participants, through this online questionnaire, your answers will be helpful to study the impact of COVID-19 on the tourism industry and how the tourism sector will revive after this pandemic. Your response will only be used for survey purposes.

Instructions: Please put a tick in the box next to the answer of your choice or write in the space provided.

Demographic information:

Gender:
○ Male
○ Female

Age group:
○ Less than 18 years
○ 18–25 years
○ 26–40 years
○ Above 40 years

Level of education:
○ University
○ College

○Technical
○Primary school
○Secondary school
○Other

Impact of COVID-19 on the tourism industry:

1. Do you think the future of the tourism sector will change in the future?
   ○Yes
   ○No
   If yes, why?
   ....

2. Does COVID-19 affect the employees of the tourism sector in any harmful way?
   ○Yes
   ○No
   If yes, how and How long?
   ....

3. Are there suitable solutions to revive this sector during COVID-19?
   ○Yes
   ○No

4. Do you think the future of the tourism industry will completely change?
   ○Yes
   ○No
   ○I don't know
   If yes, how?
   ....

5. In your opinion, what should be done and established to revive this sector?
   ....

6. Is there any economic impact on the society?
   ○Yes
   ○No

REFERENCES

[1] UNWTO. Impact assessment of the covid-19 outbreak on international tourism. 2020. Retrieved from: https://www.unwto.org/impact-assessment-of-the-covid-19-outbreak-on-international-tourism. [Accessed 4 November 2020].

[2] UNWTO. "International tourism growth continues to outpace the global economy. 2020. January 2020. Retrieved from: https://www.unwto.org/international-tourism-growth-continues-to-outpace-the-economy. [Accessed 4 November 2020].

[3] OECD. OECD policy responses to coronavirus (COVID-19): tourism policy responses to the coronavirus (COVID-19). 2020. Retrieved from: http://www.oecd.org/coronavirus/policy-responses/tourism-policy-responses-to-the-coronavirus-covid-19-6466aa20/. [Accessed 4 November 2020].

[4] Hassan V. Relation between tourism and health: case study AIDS in Lebanon. 2017. Atiner Conference Paper Series, No: TOU2017-2262. Retrieved from: https://www.atiner.gr/paper-series. [Accessed 4 November 2020].

[5] Rogoff K. Mapping the COVID-19 recession. 2020. Retrieved from: https://www.project-syndicate.org/commentary/mapping-covid19-global-recession-worst-in-150-years-by-kenneth-rogoff-2020-04. [Accessed 13 August 2020].

[6] Saunders M, et al. Research methods for business students. 7th ed. Pearson; 2015.

[7] Sekaran U, Bougie R. Research methods for business: a skill-building approach. 7th ed. NY: John Wiley & Sons; 2016.

[8] World Economic Forum. Latin America and Caribbean travel & tourism competitiveness landscape report: assessing regional opportunities and challenges in the context of COVID-19. 2020. Retrieved from: https://www.weforum.org/reports/latin-america-and-caribbean-travel-tourism-competitiveness-landscape-report-in-the-context-of-covid-19. [Accessed 4 November 2020].

[9] WHO. Updated WHO recommendations for international traffic in relation to COVID-19 outbreak. 2020. Retrieved from: https://www.who.int/news-room/articles-detail/updated-who-recommendations-for-international-traffic-in-relation-to-covid-19-outbreak. [Accessed 4 November 2020].

[10] Li R, Pei S, Chen B, Song Y, Zhang T, Yang W, Shaman J. Substantial undocumented infection facilitates the rapid dissemination of novel coronavirus (SARS-CoV2). Science 2020;368.

[11] European Centre for Disease Prevention and Control (ECDC). COVID-19 Situation update worldwide. 2020. Retrieved from: https://www.ecdc.europa.eu/en/geographical-distribution-2019-ncov-cases. [Accessed 4 April 2020].

[12] Niestadt M. COVID and the tourism sector. European Parliamentary Research Service; 2020. Retrieved from: https://www.europarl.europa.eu/RegData/etudes/ATAG/2020/649368/EPRS_ATA(2020)649368_EN.pdf. [Accessed 4 November 2020].

[13] WTTC. The world travel & tourism Council (WTTC) represents the travel & tourism sector globally. 2020. Retrieved from, https://wttc.org/en-gb/COVID-19/Global-Protocols-for-the-New-Normal. [Accessed 4 November 2020].

[14] WTTC. WTTC unveils "Safe Travels" – new global protocols to restart the travel & tourism sector. 2020.

Retrieved from, https://wttc.org/News-Article/WTTC-unveils-Safe-Travels-new-global-protocols-to-restart-the-Travel-Tourism-sector. [Accessed 4 November 2020].

[15] National Geographic. How hard will the coronavirus hit the travel industry?. 2020. Retrieved from, https://www.nationalgeographic.com/travel/2020/04/how-coronavirus-is-impacting-the-travelindustry/. [Accessed 8 April 2020].

[16] AHLA. Enhanced industry-wide hotel cleaning guidelines in response to COVID-19. 2020. Retrieved from: https://www.ahla.com/sites/default/files/safestayguidelinesv3_081420_0.pdf. [Accessed 24 March 2020].

[17] Warren K. 5 significant changes travelers can expect to see right away in hotels like Marriott, Hilton, and Wyndham as they open back up. Business Insider; 2020. Retrieved from, https://www.businessinsider.com/major-changes-hotels-marriott-hilton-wyndham-pandemic-2020-5?. [Accessed 22 May 2020].

[18] Kucheran K. Thailand reopening Phuket for tourism Oct 1—visitors must stay 30 days. 2020. Retrieved from: https://www.traveloffpath.com/thailand-reopening-phuket-for-tourism-oct-1-visitors-must-stay-30-days/. [Accessed 4 November 2020].

[19] Spinks R. The travel industry needs to talk about ventilation in coronavirus fight. 2020. Retrieved from, https://skift.com/2020/08/17/the-travel-industry-needs-to-talk-about-ventilation-in-coronavirus-fight/. [Accessed 24 March 2020].

[20] Aiham A. 'Maldives will beat COVID-19', says tourism minister Ali Waheed. 2020. Retrieved from: https://edition.mv/news/15620. [Accessed 4 November 2020].

[21] Saiidi U. Global Traveler: the Maldives reopened to tourists — and its hotels have some ideas to lure you back. 2020. Retrieved from: https://www.cnbc.com/2020/07/16/is-it-safe-to-visit-the-maldives-heres-what-hotels-are-doing.html. [Accessed 4 November 2020].

[22] Creswell J-W. Qualitative inquiry& research design: choosing among five approaches. 2nd ed. SAGE Publications; 2007.

[23] IMF. Policy responses to COVID-19. 2020. Retrieved from: https://www.imf.org/en/Topics/imf-and-covid19/Policy-Responses-to-COVID-19. [Accessed 4 November 2020].

[24] Lauren. "How to travel the Maldives on a budget: it's possible!". 2020. Retrieved form: https://www.neverendingfootsteps.com/travel-the-maldives-on-a-budget/. [Accessed 4 November 2020].

[25] Elliot C. Does the travel industry have a future?. 2020. Retrieved form: https://www.forbes.com/sites/christopher elliott/2020/03/15/does-the-travel-industry-have-a-future/#4cc668462367. [Accessed 4 November 2020].

[26] Kravenchko A. The future of tourism post- COVID 19. United Nations ESCAP; 2020. Retrieved from: https://www.unescap.org/blog/future-tourism-post-covid-19. [Accessed 4 November 2020].

[27] Fallows J, Wadhwa V, Iyer P, Potts R, Becker E, Crabtree J, De Juniac A. The future of travel after the coronavirus pandemic. 2020. Retrieved from, https://foreignpolicy.com/2020/06/13/travel-tourism-coronavirus-pandemic-future/. [Accessed October 2020].

[28] Rodriguez C. Future air travel: four-hour process, self-check-in, disinfection, immunity passes. 2020. Retrieved from, https://www.forbes.com/sites/ceciliarodriquez/2020/05/future-air-travel-four-hour-process-self-check-in-disinfection-immunity-passess/#351829a13709. [Accessed 4 November 2020].

[29] Wood J. Travel and tourism — 12 predictions of our future. Travel Daily Media; 2020. Retrieved from, https://www.traveldailymedia.com/travel-and-tourism-12-predictions-of-our-future/. [Accessed 4 November 2020].

[30] Knowles H, Westcott M. Chapter 7: Travel services. In: Westcott M, editor. Introduction to tourism and hospitality in BC; 2020.

[31] WCED. Our common future. The World Commission on Environment and Development; 1987. p. 43.

[32] Hall C. Historical antecedents on sustainable development: new labels on old bottles? In: Hall C-M, Lew A-A, editors. Sustainable tourism: a geographical perspective. Longman; 1998.

[33] Weaver D. Sustainable tourism. Theory and practice. Elsevier Butterworth-Heinemann; 2006.

[34] Törn A. Sustainability of nature-based tourism. In: Acta Universitatis Ouluensis, series A, vol. 298; 2007.

[35] Schraufnagel D-E, et al. Health benefits of air pollution reduction. Focused Rev (Ann ATS) 2019;16:1478—87.

[36] ILO. The impact of COVID-19 on the tourism sector. 2020. Retrieved from: https://www.ilo.org/wcmsp5/groups/public/—ed_dialogue/—sector/documents/briefingnote/wcms_741468.pdf. [Accessed 4 November 2020].

[37] EPA. Sources of greenhouse gas emissions. 2018. Retrieved from: https://www.epa.gov/ghgemissions/sources-greenhouse-gas-emissions. [Accessed 4 November 2020].

[38] World Economic and Social Survey. Sustainable development challenges. Department of Economic and Social Affairs; 2013. United Nations Report, 2013.

[39] Ridgley B. The future of tourism post-COVID19. 2020. Retrieved from: https://blog.mobility.here.com/future-tourism-post-covid19. [Accessed 4 November 2020].

[40] Corluka G, et al. Classification of tourist season in costal tourism. UTMS J Econ 2016;7:71—83.

[41] Gössling S, Scott D, Michael Hall C. Pandemics, tourism and global change: a rapid assessment of COVID-19. J Sustainable Tourism 2020;28:1—20.

[42] IATA Economics. COVID-19 updated impact assessment. March 24, 2020. https://www.iata.org/en/iata-repository/publications/economic-reports/third-impact-assessment/. [Accessed 4 November 2020].

[43] Cooper A. How hard will the coronavirus hit the travel industry. 2020. Retrieved from: https://www.hptourism.org/latest-news/how-hard-will-the-coronavirus-hit-the-travel-industry/. [Accessed 4 November 2020].

[44] Whitley A. How the coronavirus will change airlines and how we fly. 2020. Retrieved from, https://www.bloomberg.com/news/features/2020-04-24/coronavirus-

travel-covid-19-will-change-airlines-and-how-we-fly. [Accessed 4 November 2020].

[45] Ziady H. 2020 is a catastrophe for tourism businesses. Here is what the industry needs to get back on its feet. 2020. Retrieved from: https://edition.cnn.com/2020/05/13/business/travel-and-tourism-recovery-coronavirus/index.html. [Accessed 4 November 2020].

[46] Chaterjee S. COVID-19: tackling global pandemic through scientific and social tools. Elsevier; 2020.

[47] WHO. Public health considerations while resuming international travel. 2020. Retrieved from: https://www.who.int/news-room/articles-detail/public-health-considerations-while-resuming-international-travel. [Accessed 4 November 2020].

[48] Weaver D. Mass tourism and alternative tourism in the Caribbean. In: Wallingford H-D, editor. Tourism and the less developed world: issues and case studies. UK: CABI; 2001.

[49] McChesney I-G. The Brundtland Report and sustainable development in New Zealand. Information Paper, 25. Centre for Resource Management Lincoln University and University of Canterbury; 1991.

[50] Aubet M-E. The phoenicians and the west: politics, colonies and trade. Cambridge: Cambridge University Press; 2001.

[51] Huang C, et al. Clinical features of patients infected with 2019 novel coronavirus in Wuhan, China. Lancet 2020; 395.

[52] Saule M. Rural tourism as a tool for local development, diagnosis of the French area of Haute-Correze. 2004. Master's thesis submitted to the Bournemouth University, Dorset, England.

[53] Blackman D-J. The context: dockyards in antiquity. In: Blackman D-J, Lentini M-C, editors. Ricoveri per navi militari nei porti del mediterraneo antico i medievale. Edipuglia; 2010. p. 13–6.

[54] Blackman D-J. Rock-cut slipways in Beirut: some comments. 2011. Open letter submitted to the Lebanese Ministry of Culture, Beirut, on the 19th of December 2011.

[55] Hachem S. Conserver, restaurer et mettre en valeur les ports phéniciens du Liban. 2020. Retrieved from: https://libnanews.com/conserver-restaurer-et-mettre-en-valeur-les-ports-pheniciens-du-liban-scandre-hachem/. [Accessed 4 November 2020].

[56] Cronbach LJ. Coefficient alpha and the internal structure of tests. Psychometrika 1951;16:297–334.

[57] Creswell J-W. Educational research: "planning, conducting, and evaluating quantitative and qualitative research. 4th ed. Pearson; 2012.

[58] Gerges F-A. "Will Beirut's blast be a catalyst for change"?. 2020. Retrieved from: https://www.devdiscourse.com/article/politics/1159968-will-beiruts-blast-be-a-catalyst-for-change. [Accessed 14 August 2020].

[59] El-Bacha F. Coronavirus: L'erreur libanaise. 2020. Retrieved from: https://libnanews.com/coronavirus-lerreur-libanaise/. [Accessed 14 August 2020].

[60] World Economic Forum. Blogs and opinions. 2020. Retrieved from: https://reports.weforum.org/travel-and-tourism-competitiveness-report-2019/blogs-and-opinions/. [Accessed 4 November 2020].

[61] Bellos GSG. Estimating the impact of the effective valorization of cultural and natural resources for the economic development of Beirut. 2019. MBA Thesis submitted to the School of Business, at the Lebanese International University, Summer, 2019.

Index

Note: Page numbers followed by "f" indicate figures, "t" indicate tables.

FSC
www.fsc.org
MIX
Papier | Fördert gute Waldnutzung
FSC® C083411

Kolibri 360 GmbH
Ferdinand-Jühlke-Straße 7
99095 Erfurt, Deutschland
produktsicherheit@kolibri360.de